Lecture Notes in Biomathematics

Managing Editor: S. Levin

71

Mathematical Topics in Population Biology, Morphogenesis and Neurosciences

Proceedings of an International Symposium held in Kyoto, November 10–15, 1985

Edited by E. Teramoto and M. Yamaguti

Springer-Verlag
Berlin Heidelberg New York London Paris Tokyo

Editors

Ei Teramoto
Department of Biophysics, Faculty of Science,
Kyoto University, Sakyo-ku, Kyoto 606, Japan

Masaya Yamaguti
Department of Mathematics, Faculty of Science,
Kyoto University, Oiwake-cho, Kitashirakawa Sakyo-ku,
Kyoto 606, Japan

Mathematics Subject Classification (1980): 60, 65, 35, 34, 92-06, 92A05, 92A10, 92A15

ISBN-13: 978-3-540-17875-0 e-ISBN-13: 978-3-642-93360-8
DOI: 10.1007/978-3-642-93360-8

Softcover reprint of the hardcover 1st edition 1987

2146/3140-543210

PREFACE

This volume represents the edited proceedings of the International Symposium on Mathematical Biology held in Kyoto, November 10-15, 1985. The symposium was organized by an international committee whose members are: E. Teramoto, M. Yamaguti, S. Amari, S.A. Levin, H. Matsuda, A. Okubo, L.M. Ricciardi, R. Rosen, and L.A. Segel. The symposium included technical sessions with a total of 11 invited papers, 49 contributed papers and a poster session where 40 papers were displayed. These Proceedings consist of selected papers from this symposium.

This symposium was the second Kyoto meeting on mathematical topics in biology. The first was held in conjunction with the Sixth International Biophysics Congress in 1978. Since then this field of science has grown enormously, and the number of scientists in the field has rapidly increased. This is also the case in Japan. About 80 young Japanese scientists and graduate students participated this time. The sessions were divided into 4 categories: 1) Mathematical Ecology and Population Biology, 2) Mathematical Theory of Developmental Biology and Morphogenesis, 3) Theoretical Neurosciences, and 4) Cell Kinetics and Other Topics. In every session, there were stimulating and active discussions among the participants. We are convinced that the symposium was highly successful in transmitting scientific information across disciplines and in establishing fruitful contacts among the participants. We owe this success to the cooperation of all participants.

This symposium was sponsored by the Japan Society for the Promotion of Science, the Research Institute for Mathematical Sciences of Kyoto University, the Commemorative Association for the Japan World Exposition, the Inamori Foundation, and the Inoue Foundation for Science, with support from the Biophysical Society of Japan, the Mathematical Society of Japan and the Japan Ethological Society.

Kyoto, 1986

E. Teramoto

M. Yamaguti

LIST OF AUTHORS

AMARI, Shun-ichi	Faculty of Engineering, Univ. of Tokyo, 7-3-1 Hongo Bunkyo-ku, Tokyo 113, JAPAN
BANKS, H. Thomas	Lefschetz Center for Dynamical Systems, Division of Applied Mathematics, Brown University, Providence, RI 02912, U.S.A.
BARNA, G.	Central Research Inst. for Physics, Hungarian Academy of Sciences, H-1525 Budapest, P.O.B. 49, HUNGARY
CHENG, Louis Y.	Dept. of Civil Engineering, University of California, Berkeley, Berkeley, CA 94720, U.S.A.
CLARK, Colin W.	Institute of Applied Mathematics, University of British Columbia, Vancouver, B.C., V6T 1Y4, CANADA
COHEN, Dan	Department of Botany, The Institute of Life Sciences, The Hebrew University of Jerusalem, Jerusalem 91904, ISRAEL
DIEKMANN, Odo	Centre for Mathematics and Computer Science, Kruislaan 413, 1098 SJ Amsterdam, and Institute of Theoretical Biology, University of Leiden, Groenhovenstraat 5, 2311 BT Leiden, THE NETHERLANDS
ÉRDI, Peter	Central Research Inst. for Physics, Hungarian Academy of Sciences, H-1525 Budapest, P.O.B. 49, HUNGARY
GANNON, B. J.	Dept. of Anatomy and Histology, The Flinders Univ. of South Australia, School of Medicine, Flinders Medical Centre Bedford Park, South Australia 5042, AUSTRALIA
GINZBURG, Lev. R.	Dept. of Ecology and Evolution, State University of New York, Stony Brook, NY 11794, U.S.A.
GOLDBETER, Albert	Faculté des Sciences, Université Libre de Bruxelles, Campus Plaine, C.P. 231, B-1050 Bruxelles, BELGIUM
INUTSUKA, Hiroki	Dept. of Biology, Faculty of Science, Kyushu University, 6-10-1 Hakozaki Higashi-ku, Fukuoka 812, JAPAN
ISHII, Kazushige	College of General Education, Nagoya University, Furo-cho Chikusa-ku, Nagoya 464, JAPAN
IWASA, Yoh	Dept. of Biology, Faculty of Science, Kyushu University, 6-10-1 Hakozaki Higashi-ku, Fukuoka 812, JAPAN
JONES, Michael E.	Dept. of Anatomy and Histology, The Flinders Univ. of South Australia, and Dept. of Anaesthetics, Repatriation General Hospital, Daw Park, South Australia, AUSTRALIA
KAN-ON, Yukio	Information Processing Center, Hiroshima University, 1-1-89 Higahi-Senda-cho Naka-ku, Hiroshima 730, JAPAN
KAWASAKI, Kohkichi	Science and Engineering Research Inst., Doshisha University, Higashi-iru Imadegawa-dori Kamigyo-ku, Kyoto 602, JAPAN
KLINGLER, Martin	Max-Planck-Institut Max-Planck-Istitut fur Entwicklungsbiologie, Spemannstr. 35, 7400 Tubingen, W. GERMANY

KOBUCHI, Youichi	Dept. of Biophysics, Faculty of Science, Kyoto University, Oiwake-cho Kitashirakawa Sakyo-ku, Kyoto 606, JAPAN
KURAMOTO, Yoshiki	Dept. of Physics, Faculty of Science, Kyoto University, Oiwake-cho Kitashirakawa Sakyo-ku, Kyoto 606, JAPAN
LEVIN, Simon A.	Section of Ecology and Systematics, and Ecosystem Research Center, Cornell University, Ithaca, N.Y. 14853-0239, U.S.A.
LI, Wei	Shanghai Inst. of Biochemistry, Academia Sinica, Yo-Yang Road 320 Shanghai, 200031, CHINA
LÜCK, Hermann B.	Lab. de Botanique Analytique, et Structuralisme végétal, Faculté des Sciences et Techniques de St-Jérôme, C.N.R.S.-ER 161, Rue Henri Poincaré 13397 MARSEILLE cedex 13, FRANCE
LÜCK, Jacqueline	Lab. de Botanique Analytique, et Structuralisme végétal, Faculté des Sciences et Techniques de St-Jérôme, C.N.R.S.-ER 161, Rue Henri Poincaré 13397 MARSEILLE cedex 13, FRANCE
MANGEL, Marc	Depts. of Agricultural Economics, Entomology, and Mathematics, University of California, Davis, CA 95616, U.S.A.
MARTIEL, J. L.	Faculté des Sciences, Université Libre de Bruxelles, Campus Plaine, C.P. 231, B-1050 Bruxelles, BELGIUM
MARUYAMA, Minoru	Central Research Laboratory, Mitsubishi Electric Coop. JAPAN
MATSUDA, Hirotsugu	Dept. of Biology, Faculty of Science, Kyushu University, 6-10-1 Hakozaki Higashi-ku, Fukuoka 812, JAPAN
MATSUDA, Hiroyuki	Information Processing Center of Medical Sciences, Nippon Medical School, 1-1-5 Sendagi Bunkyo-ku, Tokyo 113, JAPAN
MEINHARDT, Hans	Max-Planck-Institut Max-Planck-Istitut fur Entwicklungsbiologie, Spemannstr. 35, 7400 Tubingen, W. GERMANY
MIMURA, Masayasu	Dept. of Mathematics, Faculty of Science, Hiroshima University, 1-1-89 Higashi Senda-machi Naka-ku, Hiroshima 730, JAPAN
MIURA, Robert M.	Depts. of Mathematics and Pharmacology and Therapeutics, Institute of Applied Mathematics, Lefschetz Center for Dynamical Systems, University of British Columbia, Vancouver, B.C. V6T 1Y4, CANADA
MURPHY, K. A.	Division of Applied Mathematics, Brown University, Providence, RI 02912, U.S.A.
MURRAY, James D.	Centre for Mathematical Biology, University of Oxford, Oxford, England, OX1 3LB, U.K.
ODELL, Garret M.	Dept. of Mathematical Sciences, Rensselaer Polytechnic Institute, Troy, N.Y. 12181, U.S.A.
OGITA, Naofumi	The Inst. of Physical & Chemical Research, 2-1 Hirosawa Wako-shi, Saitama 351-01, JAPAN
OKUBO, Akira	Marine Sciences Research Center, State University of New York, Stony Brook, N.Y. 11794, and Ecosystems Research Center, Cornell University, Ithaca, N.Y. 14853-2701, U.S.A.

OSTER, George F.	Depts. of Biophysics and Entomology, University of California, Berkeley, Berkeley, CA 94720, U.S.A.
OTHMER, Hans G.	Department of Mathematics, University of Utah, 233 Widtsoe Building, Salt Lake City, Utah 84112, U.S.A.
PATE, E. F.	Dept. of Mathematics, Washington State University, Pullman, Washington, 99163, U.S.A.
RICCIARDI, Luigi M.	Dipartimento di Matematica e Applicazioni, Università di Napoli, Via Mezzocannone 8 - Cap. 80134, Napoli, ITALY
RINZEL, John	Mathematical Research Branch NIDDK, National Institutes of Health, Bldg.31 Rm 4B-54 Bethesda, Md 20892, U.S.A.
SAKAGUCHI, Hidetsugu	Dept. of Physics, Faculty of Science, Kyoto University, Oiwake-cho Kitashirakawa Sakyo-ku, Kyoto 606, JAPAN
SASAKI, Akira	Dept. of Biology, Faculty of Science, Kyushu University, 6-10-1 Hakozaki Higashi-ku, Fukuoka 812, JAPAN
SEGEL, Lee A.	Dept. of Applied Math., The Weizmann Inst. of Science, P.O.B. 26, Rehovot 76100, ISRAEL
SHIGESADA, Nanako	Dept. of Biophysics, Faculty of Science, Kyoto University, Oiwake-cho Kitashirakawa Sakyo-ku, Kyoto 606, JAPAN
SHINOMOTO, Shigeru	Research Institute for Fundamental Physics, Kyoto University, Oiwake-cho Kitashirakawa Sakyo-ku, Kyoto 606, JAPAN
SUGIHARA, George	Scripps Institution of Oceanography, University of California, San Diego, La Jolla, California 92093, U.S.A.
TAKIGAWA, Shinya	Dept. of Mathematics, Faculty of Science, Hiroshima University, 1-1-89 Higashi Senda-machi Naka-ku, Hiroshima 730, JAPAN
TAMACHI, Nobuo	Dept. of Biology, Faculty of Science, Kyushu University, 6-10-1 Hakozaki Higashi-ku, Fukuoka 812, JAPAN
TERAMOTO, Ei	Dept. of Biophysics, Faculty of Science, Kyoto University, Oiwake-cho Kitashirakawa Sakyo-ku, Kyoto 606, JAPAN
XU, Jinghua	Shanghai Inst. of Biochemistry, Academia Sinica, Yo-Yang Road 320 Shanghai, 200031, CHINA
YAMAGUTI, Masaya	Dept. of Mathematics, Faculty of Science, Kyoto University, Oiwake-cho Kitashirakawa Sakyo-ku, Kyoto 606, JAPAN

TABLE OF CONTENTS

Evolution

II. MATHEMATICAL THEORIES OF PATTERN AND MORPHOGENESIS

Morphogenesis and Pattern Formation

Pattern Formation in *Dictyostelium discoideum*

III. THEORETICAL NEUROSCIENCES AND RELATED PROBLEMS IN PHYSIOLOGY

Neurosciences

Physiology and Related Problems

Chaos and Fractals

Masaya Yamaguti

Department of Mathematics, Faculty of Science
Kyoto University, Kyoto 606, JAPAN

My talk is a survey of the mathematical work done in our laboratory in the past 10 years. We are not mathematical biologists; rather, we work in mathematics, while maintaining a close relationship with Teramoto's group, which is a true mathematical biology group. Mathematical facts may not seem very exciting, but I feel they are important for the study of mathematical modelling in science.

One example of our collaboration with Teramoto's group is the study of chaos arising in some discretizations of ordinary differential equations (O.D.E's.) that I shall discuss later. This study originates from the work of Robert May.

1. Predator-prey systems.

Oshime considered the system of differential equations of Lotka-Volterra prey-predator type:

$$\dot{x}_i = x_i(e_i + \sum_{j=1}^{n} P_{i,j}\, x_j) \qquad (i = 1,2, \cdots, n. \quad n \geq 3)\,, \qquad (1)$$

under the conditions

$$P_{i,i+1} > 0 \quad (\forall i), \qquad P_{i,j} \leq 0 \quad (j \neq i+1)\,.$$

He classified completely the asymptotic behavior of the solution of the initial value problem for (1) as follows:

<u>THEOREM</u> (Oshime) Put

$$Q = \begin{pmatrix} P_{12} & \cdots & P_{1n} & P_{11} \\ P_{22} & \cdots & P_{2n} & P_{21} \\ \vdots & & \vdots & \vdots \\ P_{n2} & \cdots & P_{nn} & P_{n1} \end{pmatrix} \quad \text{and} \quad R = \begin{pmatrix} e_1 & P_{12} & \cdots & P_{1n} \\ e_2 & P_{22} & \cdots & P_{2n} \\ \vdots & \vdots & & \vdots \\ e_n & P_{n2} & \cdots & P_{nn} \end{pmatrix}$$

Let λ_0 be the smallest real eigenvalue of the matrix Q. Then the following holds:

(1) If $\lambda_0 > 0$, then some solutions to (1) blow up in finite time. If we denote by t_0 the time for blowing up, then

$$\varepsilon < x_i(t)/x_j(t) < \varepsilon^{-1} \quad \forall\, i,j, \quad \forall\, t \in (t_0 - \eta,\, t_0)$$

for some $\eta > 0$ and $\varepsilon > 0$.

(2) If $\lambda_0 < 0$, then all solutions to (1) remain bounded. Moreover, there exists a certain constant $M > 0$ independent of the initial data such that

$$\overline{\lim_{t\to\infty}}\, x_i(t) \le M \quad \text{for any } i.$$

(3a) If $\lambda_0 = 0$ and $(-1)^{n+1}\det(R) > 0$, then all solutions blow up. Moreover,

$$\varepsilon < x_i(t)/x_j(t) < \varepsilon^{-1} \quad \forall\, i,j, \quad \forall\, t > t_0,$$

for some $\varepsilon > 0$ and $t_0 > 0$. Here t_0 depends on the initial data.

(3b) If $\lambda_0 = 0$ and $(-1)^{n+1}\det(R) < 0$, then all solutions to (1) remain bounded. Moreover, there exists a certain constant $M > 0$ independent of the initial data such that

$$\overline{\lim_{t\to\infty}}\, x_i(t) \le M \quad \text{for any } i.$$

(3c) If $\lambda_0 = 0$ and $(-1)^{n+1}\det(R) = 0$, then all solutions to (1) remain bounded. Now the upper bound depends on the initial data. Actually the equilibrium points whose components are all positive form a set

$$\{c + s\,\gamma \;;\; s \in \mathbb{R}^1\} \cap \mathring{\mathbb{R}}^n_+,$$

where $\gamma \in \mathring{\mathbb{R}}^n_+$ and $c \in \mathbb{R}^n$.

2. Chaos arising from the discretization of ordinary differential equation (Yamaguti, Matano, Ushiki, Hata).

The generalized logistic model

$$x_{n+1} = ax_n(1 - x_n) , \tag{2}$$

which comes from the logistic differential equation:

$$\frac{dx}{dt} = (\varepsilon - hx)x$$

by Euler's discretization has been studied in ecology by R. May (1975). By the theorem of Li-Yorke, if Δt is fairly large $(3.8284\cdots < 1 + \varepsilon\Delta t \leq 4)$, then the discrete dynamical system:

$$x_{n+1} = ((1 + \varepsilon\Delta t) - \Delta t\, h\, x_n)x_n \tag{3}$$

has chaotic orbits. This fact was generalized by Yamaguti-Matano (1979) for scalar autonomous equations as follows:

<u>THEOREM</u> (Yamaguti-Matano) We consider scalar differential equations of the form

$$\frac{du}{dt} = f(u),$$

where $f(u)$ is continuous in $\mathbb{R}$. We assume that this equation has at least two equilibrium points, one of which is asymptotically stable; that is, (possibly after a linear transformation)

(1) $f(0) = f(\bar{u}) = 0$ for some $\bar{u} > 0$,

(2) $f(u) > 0 \quad (0 < u < \bar{u})$,

(3) $f(u) < 0 \quad (\bar{u} < u < \infty)$.

Then there exist two constants $0 < c_1 < c_2$ such that for any $c_1 < \Delta t \leq c_2$, the difference equation

$$x_{n+1} = x_n + \Delta t\cdot f(x_n)$$

is chaotic in the sense of Li-Yorke in some invariant finite interval.

For the autonomous system, M. Hata (1982) obtained the following. Both results are based on the assumption that the autonomous differential system is genuinely non-linear.

<u>THEOREM</u> (Hata) We consider n-dimensional autonomous differential equations of the form

$$\frac{du}{dt} = F(u),$$

where F is continuously differentiable in $\mathbb{R}^n$. Suppose that there exist $\bar{u} \neq \bar{v}$ such that $F(\bar{u}) = F(\bar{v}) = 0$, $\det F'(\bar{u}) \neq 0$, and $\det F'(\bar{v}) \neq 0$. Then there exists a positive constant c such that for any $\Delta t > c$, the difference equation

$$u_{n+1} = u_n + \Delta t \cdot F(u_n)$$

is chaotic in the sense of Li-Yorke.

This direction of our research has been extended (Yamaguti 1981) to yield the result by S. Ushiki and M. Yamaguti on the study of the central difference scheme,

$$x_{n+1} = x_{n-1} + 2\Delta t\, x_n(1 - x_n) , \qquad (4)$$

which is used to solve the logistic equation. We observed and analysed a very singular solution which is completely different from the solution of (2). Such a singular solution is illustrated in Fig. 1 with $\Delta t = 0.1$ and initial condition $x_0 = 0.5$, $x_1 = 0.525$. Finally,

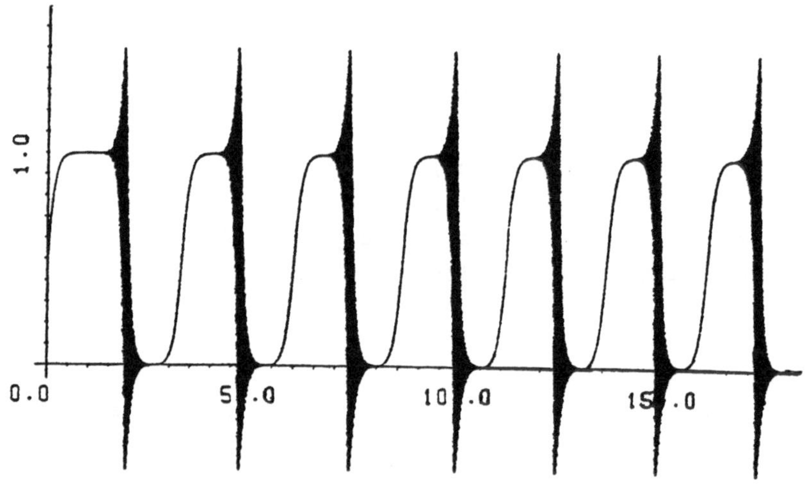

Fig. 1

S. Ushiki succeeded in proving that the orbit of the initial value problem for (4) is always chaotic for any mesh size Δt, by using the notion of "pseudo-horseshoe", which is a generalization of Smale's approach.

3. Predictability of one-dimensional chaos.

J. Kigami (1985) provided a mathematical framework for the work of J.H. Vandermeer (1982). Vandermeer's work treated population dynamics of insects that predate seeds from trees. The model is expressed by the equation:

$$x_{n+1} = f(x_n) = a\,b\exp(-bx_n/m), \tag{5}$$

where x_n is the population size in the n-th year and a, b and m are positive biological constants. We can see easily from the graph of (5) on (x_n, x_{n+1}) that an extremely large population x_n (which is called population flush) leads to a very small population x_{n+1}. This feature of model (5) reflects the following qualitative discussion: the supply of resource (total number of seeds from trees) is almost constant every year; therefore population flush at a given year leads to a serious struggle for existence, and consequently the next year's population is very small. If we continue this iteration, the rare population caused by population flush grows slowly each year and finally, after some years of rarity, population flush takes place again. Hence the population will repeat population flush and years of rarity will again follow. Vandermeer concluded in his paper that "The number of years of rarity that can be predicted from the size of the preceding population flush of the dynamics of the model (5) is sufficiently chaotic". Kigami provided a precise quantitative interpretation for Vandermeer's discussion for the general one-dimensional iteration map:

$$x_{n+1} = f_\lambda(x_n), \tag{6}$$

where $f_\lambda(x)$ is a family of unimodal maps that represent a population model with some fluctuation.

Finally I will discuss recent research by M. Hata and M. Yamaguti on the fractal set. Our study began from the very simple mathematical fact (Yamaguti and Hata 1983) that the famous Weierstrass function

W(x), which is continuous and nowhere differentiable, is just a generating function of a discrete chaotic dynamical system:

$$x_{n+1} = 4x_n(1 - x_n) . \qquad (7)$$

More precisely, we have

$$W(x) \equiv \sum_{n=0}^{\infty} a^n \cos(2^n x)$$
$$= (1 - a)^{-1} - 2 \sum_{n=0}^{\infty} a^n \psi^n (\sin^2(\tfrac{\pi x}{2})),$$

where ψ^n is the n-hold iteration by $\psi(x) = 4x(1 - x)$. We extended this fact to other dynamical systems and obtained some functional equations to describe these kind of continuous functions.

At this stage, we discovered some old works done by G. de Rham which studied extensively similar functional equations (Hata and Yamaguti 1984). M. Hata obtained last year a nice definition of a self-similar set which includes all deterministic (self-similar) fractals in Mandelbrot's book. I will explain this more precisely. Let E be a complete metric space, f_λ ($\lambda \in \Lambda$ finite or countable) a family of contractions that map E into itself. Then we call a set $K \subset E$ a self-similar set if K is non-empty, compact and satisfies the following set-theoretical equation:

$$K = \bigcup_{\lambda \in \Lambda} f_\lambda(K) . \qquad (8)$$

This definition was based on the idea of G. de Rham and R.F. Williams. Later we remarked that this definition is already found in J.E. Hutchinson (1981). Hutchinson showed that if Λ is finite, then there exists a unique self-similar (compact and non-empty) solution to the equation (8). Hata (1985) generalized this result to the case in which Λ is countable under some additional conditions. The equation (8) can be reduced to the fixed-point problem of a certain operator that maps the set of all (non-empty) compact subsets into itself.

Hata also studied the topological structures (e.g. connectedness, the number of end points, parameterization, the topological dimension and the Hausdorff dimension, etc.) of such self-similar sets in general. Indeed he proved the following

<u>THEOREM</u> (Hata). Let Λ be a finite set. Then the unique compact solution K of (8) is a locally connected continuum if and only if the set K satisfies the following chain condition: for any $i \neq j \in \Lambda$, there exists a set of indices $r_1, r_2, \cdots, r_n \in \Lambda$ such that

$$K_i \cap K_{r_1} \neq \phi, \quad K_{r_1} \cap K_{r_2} \neq \phi, \quad \cdots, \quad K_{r_n} \cap K_j \neq \phi, \tag{9}$$

where $K_\lambda = f_\lambda(K)$.

<u>THEOREM</u> (Hata). Suppose that Λ is finite, each f_λ is one to one, and the unique compact solution K satisfies the chain condition (9). Then the set K is either acyclic, a simple arc, or it has an infinite number of end points.

Finally he also obtained several very interesting examples of self-similar sets in $\mathbb{R}^2$, as illustrated in Fig. 2.

Fig. 2a

Fig. 2b

References

Hata, M. 1982. Euler's finite difference scheme and chaos in R^n. Proceedings of the Japan Academy, **58**, Ser. A, 178-181.

Hata, M. 1985. On the structure of self-similar sets. Japan Journal of Applied Mathematics, **2**, 381-411.

Hata, M. and Yamaguti, M. 1984. The Takagi function and its generalization. Japan Journal of Applied Mathematics, **1**, 183-199.

Hutchinson, J.E. 1981. Fractals and self-similarity. Indiana University Mathematical Journal, **30**, 713-747.

Kigami, J. 1985. Some predictability in one-dimensional chaotic dynamical systems arising in population models. Adv. Applied Mathematics **6**, 188-208.

May, R.M. 1975. Biological populations obeying difference equations: stable points, stable cycles, and chaos. J. Theor. Biol. **49**, 511-524

Vandermeer, J.H. 1982. On the resolution of chaos in population models. Theor. Pop. Biol. **22**, 17-27.

Yamaguti, M. 1981. Chaos in numerical analysis of ordinary differential equations. Physica D, **3D**, 618-626.

Yamaguti, M. and Hata, M. 1983. Weierstrass's function and chaos. Hokkaido Mathematical Journal, **XII**, 333-342.

Yamaguti, M. and Matano, H. 1979. Euler's finite difference scheme and chaos. Proceedings of the Japan Academy, **55**, Ser. A, 78-80.

RECURRENT THEMES IN MATHEMATICAL BIOLOGY

by
Simon A. Levin
Cornell University
Ecosystems Research Center
and
Section of Ecology and Systematics
Ithaca, New York 14853

1. Introduction:

Organisms have elaborate control systems for the regulation of function, and even whole ecosystems exhibit feedback mechanisms that regulate the flow of materials and energy. The fact that similar control mechanisms and similar patterns exist in a wide variety of biological systems has made it natural to seek to abstract those features that are common and to develop mathematical descriptions that help to catalogue them. This search has drawn many mathematicians, physicists, and cyberneticists to the study of biology. Fittingly, many such individuals are represented in this volume, because Prof. Teramoto has been a central figure in the development of biophysics and mathematical biology in Japan, and in introducing physicists and mathematicians to the complexities and challenges of biology.

[1] This paper is a slight revision of "The role of mathematics in biology," originally presented to the Danish Mathematical Society in Vingsted in 1981 and recorded in the proceedings of that meeting (Levin 1981c). It is dedicated to Alexander Lerner, the distinguished Soviet cyberneticist.

Research in mathematical biology has increased dramatically in the past twenty years. This is evident not only in the appearance of an increasing number of biologically-inspired papers in mathematics journals, but also in a heightened influence of mathematics upon biological investigation. This dual nature of mathematical biology--as part of mathematics as well as part of biology--has long been a characteristic of the most highly developed branch of the subject: theoretical population genetics. Its more general applicability reflects the maturation of the field.

But in what sense is there such a field? Although there obviously is a place for "biological mathematics"--mathematics inspired by biology--mathematical biology sensu strictu first must be good biology, and that requires a fundamental commitment to the problem area. It follows that mathematical molecular biologists and mathematical ecologists, for example, have no more in common with each other than do other molecular biologists and other ecologists except a similar philosophical orientation and a shared set of mathematical methods and formulations. It is the parallel techniques and perspective which sustain the interactions among mathematical biologists, and support the existence of the field.

The solutions to many biological problems require the integration of understanding at several different scales. For example, neural tissue is composed of nerve cells. To what extent can our knowledge of how those cells respond to stimuli assist our understanding of the behavior of aggregates (see for example Arbib 1978, Cowan and Ermentrout 1978)? Control of the development of an organism is governed by "decisions" localized in both space and time. However, the result of the process will be reliably similar enough to some archetypical pattern that systematists can separate even closely related species. Ecosystems involve complex trophic linkages and flows among large numbers of species, but exhibit regularities in and regulation of biogeochemical fluxes, population dynamics, and successional patterns. How can we understand the evolution of such systems, given that natural selection is mediated at much lower levels of organization? Can knowledge of the physiological responses to stress of organisms or species or small microcosms--the type of information most readily available in evaluating toxic effects--help in the assessment of the responses of whole ecosystems similarly stressed?

In each of these, one seeks to understand the organization of large-scale systems of smaller units, and how pattern emerges at various scales. In each, one wishes to relate the behaviors of single units, and of large-scale ensembles of such units (Levin 1978). In each, and throughout mathematical biology, we are faced with the need to understand complex interactions and processes involving mobile species; on multiple temporal, spatial, and structural scales; and possibly complicated by stochastic effects and delays. Adrift in a common sea, mathematical biologists cling together, benefitting from each other's progress.

Of course, these scientific issues are not unique to problems in biology, but bear strong analogy to similar problems in the physical sciences. Questions associated with pattern formation, regular oscillations, stochastic fluctuations, and turbulence in ensembles of interacting units relate to collective or synergistic phenomena which transcend disciplinary boundaries. These have brought theoretical biologists into common cause with scientists from other disciplines.

This explains, for example, why consideration of a purely chemical system such as the Belusov-Zhabotinskii reaction (Zaikin and Zhabotinskii 1970, Winfree 1974, 1980) has been of such interest to theoretical biologists, and why investigations of it are reported in journals and monographs on mathematical biology (e.g. Tyson 1976, Levin and Segel 1985). This reaction scheme shows regular and strikingly visible and homogeneous oscillations when well-stirred but exhibits complex spiral, target, scroll, and other patterns when the chemicals involved are not stirred but are allowed to diffuse in a petri dish. The development of such complex patterns simply from the interaction of diffusion and oscillation makes the Belusov-Zhabotinskii system a convenient prototype, which lends itself to mathematical analysis. Oscillation and diffusion represent recurrent themes in mathematical biology and elsewhere, and similar methods of analysis are likely to bear fruit independently of the source of the application. The role of mathematical models is to abstract in such a way as to facilitate that analysis by relating a specific problem to the broader and more general literature. In the next two sections, some examples will be given.

2. Oscillation

Since the consideration by Vito Volterra (1926) of the fluctuations of the Adriatic fisheries by examination of coupled differential equations of the form

$$\frac{dx}{dt} = rx(1 - cy),$$
$$\frac{dy}{dt} = sy(bx - 1), \tag{1}$$

problems associated with explaining and analyzing fluctuations have received considerable attention in the mathematical biological literature. The system of equations (1), which represents the coupled oscillations of a prey species (x) and its predator (y), may easily be shown to possess the invariant

$$H = s(bx - \ln x) + r(cy - \ln y). \tag{2}$$

That is, if (1) is satisfied, then $\frac{dH}{dt}=0$ and hence H must remain constant for any solution. The curves H= constant are closed, and further consideration of (1) shows that solutions must cycle, following those curves in the counterclockwise direction. This property of (1) makes larger-scale versions, involving many predators and prey, ready candidates for statistical mechanical treatment using the method of Gibbs ensembles (Kerner 1959, Leigh 1966, Goel et al. 1971).

However, the solutions of (1) are metastable: analogously to harmonic oscillations, they oscillate with fixed "amplitudes" (measured by H) determined by initial conditions and subject to permanent modification by infinitesimal perturbations. Moreover, the system (1) may be shown to be "structurally unstable." That is, introducing infinitesimal changes in the model's structure will lead to qualitatively different behavior (e.g. spiralling convergence to a steady state). Of course, Volterra was aware of this; but his goal was to show that the predator-prey interaction could lead to oscillation. Furthermore, modifications of his model to introduce more complicated nonlinearities (e.g. Kolmogorov 1936, Rosenzweig and MacArthur 1963) produced a class of more realistic models which could be shown by arguments based on the Poincare-Bendixson theory (see for example Hirsch and Smale 1974) to lead to stable oscillations of the limit cycle type (Kolmogorov 1936; May 1972, 1973; Albrecht et al. 1973).

Problems involving periodic solutions arise in a variety of other biological applications: for example, biological clocks (Winfree 1980), delay population models, epidemic models (Hethcote 1973), or the repetitive firing of a "space-clamped" neuron under sustained stimulation (Rinzel 1978; see Hodgson, 1983). For a variety of reasons, the most productive approach is often to try to separate regions of parameter space which yield periodic oscillations from those in which a stable equilibrium point results, and to do so by the technique of Hopf bifurcation theory (see for example Marsden and McCracken 1976) in which periodic solutions may be shown to bifurcate from point solutions. In general, for a nonlinear vector differential equation

$$\frac{d\underset{\sim}{x}}{dt} = \underset{\sim}{F}(\underset{\sim}{x}), \tag{3}$$

the stability of equilibrium solutions (solutions of $\underset{\approx}{F}(\underset{\sim}{x}) = 0$) may be studied by examining the eigenvalues of the linearization (Jacobian) matrix

$$A = \frac{\partial(F)}{\partial(\underset{\sim}{x})} = \frac{\partial(F_1, \dots F_n)}{\partial(x_1, \dots x_n)} \tag{4}$$

at the equilibrium. If all eigenvalues of that matrix have negative real part, the solution is stable; if any has positive real part, the solution is unstable. Thus instability arises when parameters are changed in such a way that the eigenvalue of $\underline{A}$ with largest real part crosses the real axis. If that crossing is by means of a pair of complex eigenvalues ($\underline{A}$ is real, so complex eigenvalues occur in pairs), and if certain other technical conditions are satisfied, then the linearized system corresponding to A describes an oscillation, and a periodic solution will exist for the nonlinear problem in a neighborhood of that critical parameter. Whether or not that periodic solution emerges as a stable one depends on other conditions; often these can be checked only with considerable labor.

Discrete time analogues also exist. Indeed, perhaps the most innocent looking dynamical model is the stock-recruitment model which is at the basis of much theory in fisheries science. Ricker (1954) in his classic paper on stock and recruitment set out the general framework, and carried out insightful analysis. Let N_t represent the stock size of a fish population in generation t. Assume that individuals live a single year; then the stock size in generation (year) $t + 1$ is identical to the number of new recruits to the population. If immigration is ignored, N_{t+1} may be related to N_t by a function F, assumed to be continuous:

$$N_{t+1}=F(N_t). \tag{5}$$

Equation (5) is presented as an autonomous relationship, although stochastic and other temporal influences may modify this. There are a variety of candidate functions F which have been advanced, and the choice among them often is inconclusive. Further, there is almost always considerable scatter in the estimates of both stock and recruitment. Because of these problems, the stock-recruitment fit in general is not a very dependable device, and not an adequate basis for management. Nonetheless, analyses of (5) and of modifications of it have constituted a pervasive, and some would argue pernicious, influence on management theory (Ludwig and Walters 1985). Similar models have been used to describe the dynamics of insect populations with non-overlapping generations.

The simple iteration scheme (5) can admit surprisingly complex behaviors, especially when, as is often assumed, F attains a maximum at an intermediate value of N. For example, the model

$$N_{t+1}=N_t\exp[r(1 - N_t/K)] , \tag{6}$$

which utilizes the stock-recruitment relationship attributed to Ricker, has for all values of r a steady state $\bar{N} = K$. However, as a linearized analysis will show, this steady state is stable to small perturbations only for $0 < r < 2$. As r is increased beyond the value 2, a stable 2-point oscillation will bifurcate from the equilibrium point (Ricker 1954); this bifurcation is a Hopf bifurcation (Guckenheimer et al. 1977, Wan 1978) and occurs because the single eigenvalue of the linearization about K exceeds unity in magnitude. As May (1974, 1975, 1978) has discussed, as r is increased further, a sequence of period-doubling bifurcations occurs; the threshold values of r become closer together, being bounded above by r =

2.692. Beyond $\underline{r}$ = 2.692, aperiodic solutions appear, and solutions become highly sensitive to initial conditions. Moreover, beyond $\underline{r}$ = 3.102, a cycle of length 3 appears; and as Li and Yorke (1975) have shown for the more general scheme (5), if there exists a cycle of period 3, then there exists one of period $\underline{n}$ for any choice of $\underline{n}$. Further, there is an uncountable set of initial conditions which are not even asymptotically periodic under (5). The region beyond r = 2.692 in (6) is called the chaotic region. Similar results apply for the quadratic relationship

$$N_{t+1}=N_t\,[\,1 + r(1 - N_t/K)\,]\,; \tag{7}$$

in this case the critical "chaotic" value of $\underline{r}$ is 2.570 (May 1974, 1978).

Li and Yorke (1975) were inspired by the work of Lorenz (1963a, b, 1964), who showed that analysis of a particular set of ordinary differential equations motivated by a problem in meteorology could be reduced in part to a map of the form (5), and further that apparently chaotic behavior could result. For such maps, Sarkovskii (1964) provided a very complete analysis, including not only those results already discussed, but also a dissection of the structure of the system regarding periodic solutions within the chaotic region. However, Sarkovskii's paper, in Russian, was not well-known in the West until fairly recently (Stefan 1976). An earlier treatment of (7) with $\underline{r}$ = 4, within the context of ergodic theory, may be found in Ulam and von Neumann (1947).

Equation (5) is only the simplest system within which such complicated behaviors can emerge. Internal structure of various kinds can lead to such dynamics (see, e.g., Oster et al. 1976, Guckenheimer et al. 1977, Oster 1978, Levin and Goodyear 1980, Levin 1981a), and similar phenomena occur in continuous dynamical systems. Indeed, one of the classic examples is Lorenz's (1963a, b) already-mentioned analysis of the system

$$\begin{aligned} \dot{x} &= \sigma y - \sigma x \\ \dot{y} &= -xz + rx - y \\ \dot{z} &= xy - bz, \end{aligned} \tag{8}$$

which was proposed as a model of turbulence. (The equations arise in an approximation to the appropriate partial differential equations in the classical Benard problem.) This system of equations has a complicated, non-periodic "strange" attractor. Considerable recent work on the theory of turbulence has built on this and other models; noteworthy is the work of Hopf (1948), Ruelle and Takens (1971), and Ruelle (1975). Moreover, the application of such ideas to population models has stimulated new mathematical research which has in turn provided fresh insights to these more classical investigations.

The analysis of strange attractors and chaotic behavior is currently one of the most active research areas in applied mathematics, and its relevance to biology and to the topics of this volume has been established long ago. Prof. Yamaguti's excellent paper in this volume discusses some of the most beautiful aspects of the theory.

3. Diffusion and reaction

Models of diffusion and reaction are ubiquitous in mathematical biology (see Levin and Segel 1985, or simply the other papers in this volume). For example, there is a voluminous literature in population biology, dating back to the early part of the century, involving diffusion models for dispersing species. However, it was Turing (1952), in considering problems of pattern formation in developmental biology, who first emphasized the extra richness which emerged when one considered systems of interacting and diffusing species. These are governed by equations of the form

$$\frac{\partial u}{\partial t} = F(u, v) + D_u\left(\frac{\partial^2 u}{\partial x^2} + \frac{\partial^2 u}{\partial y^2}\right)$$

$$\frac{\partial v}{\partial t} = G(u, v) + D_v\left(\frac{\partial^2 v}{\partial x^2} + \frac{\partial^2 v}{\partial y^2}\right) \qquad (9)$$

where, in Turing's usage, _u_ and _v_ denote the concentrations of _morphogens_, substances which diffuse and interact and provide the information guiding morphogenesis. Turing showed how diffusion, generally thought of as an homogenizing influence, could destabilize an activator-inhibitor interaction if the diffusion coefficients of the two species were sufficiently different. In theoretical developmental biology this work has influenced thinking to this day regarding the fundamental problem of the breaking of symmetry (Keller and Segel 1970, Gierer and Meinhardt 1974, Kaufmann et al. 1978). Turing (1952) suggested that destabilization could lead to nonuniform dissipative structures, and this was verified by the nonlinear calculations of Segel and Levin (1976).

Segel and Jackson (1972), Levin (1974), Okubo (1974), and Levin and Segel (1976) applied Turing's ideas to ecological problems. More generally, and as already mentioned, diffusion models have long played an important role as models of population dispersal (see reviews in Levin 1976a, b, 1981b; Okubo 1980, Kareiva 1982a, b, Levin and Segel 1985). Their use has increased since the fundamental papers of Skellam (1951) and Kierstead and Slobodkin (1953); but some of the earliest ecological investigations were actually carried out by geneticists (e.g. Dobzhansky and Wright 1943), and attention to diffusion models in the population genetics literature first focused attention on a number of important mathematical questions. In particular, the equation

$$\frac{\partial p}{\partial t} = \frac{\sigma^2}{2} \frac{\partial^2 p}{\partial x^2} + sg(x)f(p) \tag{10}$$

was introduced as a model for the change of gene frequency p along a spatial gradient defined by x (Fisher 1937, Haldane 1948). Here $s\,f(p)$ defines the basic selection regime, with s the force of selection; $g(x)$ modifies selection according to spatial influences. Biologically-inspired questions led to a number of fundamental mathematical investigations of this equation (e.g. Kolmogorov et al. 1937; Malecot 1948; Aronson and Weinberger 1975, 1977; Conley 1975; Nagylaki 1975, 1978; Fife 1979). In particular:

(i) Do there exist stable nonuniform stationary solutions ("clines") under appropriate boundary conditions? (For reviews see Nagylaki 1978, Levin 1981.)

(ii) Under what conditions (including initial conditions) do there exist travelling wave solutions (waves of advance of advantageous alleles)? Are they stable? Can the asymptotic speed(s) of propagation be computed? (See Fife 1979).

Consideration of the maintenance of genetic variation in the absence of selection--the neutral gene theory--has also led to diffusion models, but with a different interpretation (see Ewens 1979). The Kolmogorov or Fokker-Planck equation

$$\frac{\partial}{\partial t}f(x;\ t) = -\frac{\partial}{\partial x}[m(x)f(x;\ t)] + \frac{1}{2}\frac{\partial^2}{\partial x^2}[v(x)f(x;\ t)] \tag{11}$$

emerges from consideration of the stochastic process describing changes in gene frequency with time. In (15), $f(x;\ t)$ is the probability density function for allele frequency. Stochastic models have been important also in ecology, epidemiology, and neurobiology.

In the neurobiological literature, a classic deterministic application of diffusion models for transport involves the description of the propagation of the nerve impulse down an axon. In work which later won for them a Nobel prize, Hodgkin and Huxley (1952) empirically justified a set of four equations: one describing the voltage as a function of position, the other three describing auxiliary variables. These equations were consistent with the fundamental aspects of nerve behavior: in response to a stimulus, there will be an all-or-none response (firing) of the neuron, followed by a refractory period in which the neuron cannot be stimulated to fire.

Fitzhugh (see 1969, 1973) and Nagumo et al. (1962) presented simplifications of these equations which still retained the essential behavior. In non-dimensional form, and after a change of variables, these became

$$\begin{aligned} \frac{\partial v}{\partial t} &= \frac{\partial^2 v}{\partial x^2} - f(v) - w \\ \frac{\partial w}{\partial t} &= b(v - \gamma w) \end{aligned} \tag{12}$$

(Rinzel 1978). Here $\underline{v}$ is voltage and $\underline{w}$ is the "recovery" variable of the axon, generally assumed to change on a slower time scale.

Within the context of the equations (12), a set of well-defined mathematical questions can be posed (for particular forms of the function), for example regarding the existence and stability of solitary pulses and periodic wave solutions under constant stimulus. Some of these Fitzhugh (see 1973) and Nagumo et al. (1962) addressed; others have provided meat for numerous mathematical investigations (e.g. McKean 1970; Evans 1972; Rinzel and Keller 1973; Aronson and Weinberger 1975, 1979; Carpenter 1977a, b; Rinzel 1978; Fife 1979). Such studies have benefitted both mathematics and biology.

Mathematics provided critical input at various stages of the investigations by Hodgkin and Huxley and their followers: model formulation; simplification, including reduction in dimension by time scales; non-dimensionalization; and analysis of simplified cases. These are among the major contributions the mathematician has to make. The distinguished probabilist Mark Kac, wrote, "The main role of models is not so much to explain and to predict--though ultimately these are the main functions of science--as to polarize thinking and to pose sharp questions." (Kac 1969).

4. The role of mathematics in biological investigation

In its ability to simplify, to make precise, to explain, and sometimes to provide the basis for prediction, mathematics has left its stamp upon almost every area of biology, and will be at the forefront in new investigations. Mathematics will be indispensable in the examination of problems ranging from molecular biology and biochemistry, in which mathematical descriptions of enzyme interactions, drug binding, and molecular conformational changes have been essential aspects of conceptual development, to ecosystems, in which problems of the management of multi-species fisheries and the stability of complex networks of interacting species have a fundamentally mathematical character. The list of biological areas of application is legion, including neurobiology (e.g. drug receptor interactions, nerve transport, and models of higher function and pattern recognition); immunology; developmental biology; physiology; biomechanics; and epidemiology and host-parasite interactions.

One of the most fundamental areas of mathematical application is evolutionary theory, especially regarding its relationship to population genetics. Evolution is the theme which distinguishes the biologist from the engineer. It fundamentally modifies one's biological thinking to recognize that the systems of interest, rather than being the designs of a master engineer, have evolved under the influence of a number of factors (Jacob 1977). This recognition should also shape our view of adaptation, although optimization arguments too easily seep into evolutionary thought (Levin 1980).

Among the forces guiding evolution are Darwinian natural selection among individuals, stochastic processes, and selection at higher levels of organization (including trait groups). The synthesis of population genetics theory and evolutionary theory that took place during the first part of this century related evolutionary change to changes in gene frequencies through consideration of these basic processes. It made fundamental use of probability, statistics, discrete dynamical systems, and stochastic processes; and it played a major role in stimulating the development of these mathematical subjects. That synthesis repeatedly has been challenged on a variety of issues (including for example the importance of higher level selection), and constantly must be updated to accomodate them. Sorting out the complexities is impossible without a mathematical framework.

In particular, the classical theory--especially as represented by Fisher's fundamental theorem of natural selection (Fisher 1958) and Wright's adaptive surface (Wright 1948)--demonstrated that in the simplest constant fitness models, the mean fitness of a population would increase monotonically to an equilibrium. These fundamental insights are, however, severely limited in their applicability (Levin 1978, 1980). The mathematical theory is a special one, and in realistic situations it is not generally clear how fitness should be measured or what maximization means, especially when one individual's fortune depends on those of others. Indeed, in the latter case any measure of fitness must be "frequency-dependent," varying with gene frequencies, and the mathematical proofs of fitness maximization are invalid. This is the rule rather than the exception in ecological theory, with examples ranging from the obvious cases of aggressive and social behavior and caste formation, to characters affecting seed dispersal in plants (Levin 1980, Levin, Cohen, and Hastings 1984) and parasitic virulence (Levin and Pimentel 1981, Levin 1983). Such considerations have brought new attention to game theoretical approaches to evolutionary biology (Maynard Smith 1976, 1977; Oster and Wilson 1978) as well as the methods of optimization and control theory.

As with other biological problems, evolutionary theory involves complex processes at many levels of integration, and on numerous temporal, spatial, and structural scales. It is not surprising that mathematics is well integrated into population genetics theory, and that theoretical consideration of fundamental questions invariably involves a mathematical formalism.

5. Summary

In this paper I have tried to highlight some mathematical problems in theoretical biology, and to provide examples of the diversity of mathematical techniques and the pervasive role mathematics plays in biological thought. The appropriate mathematical methods in biology nearly span the mathematical spectrum, and the examples discussed already--primarily from dynamical systems theory, probability and statistics, stochastic processes, and the theory of games--are only illustrative. Optimization and control play a role in understanding of physiological regulatory processes, and in the management of renewable resources (e.g. Clark 1976). Network analysis and graph theory are of value in consideration of ecosystem structure and function, and in the reconstruction of evolutionary phylogenies. Pattern analytic techniques are important in applications from ecology to medical data analysis, including three-dimensional reconstructions from x-rays (Gordon, Herman, and Johnson 1975, Katz 1978). There are no limitations to the types of mathematics which may be useful.

The primary functions of mathematics are to abstract, to simplify, to organize information, and to provide a framework in which ideas may be manipulated and their consequences discovered. In simplifying, one encounters similar problems and needs similar techniques in a wide variety of applications; those techniques often include such basic tools as dimensional analysis and recognition of multiple time scales (see for example Ludwig et al. 1978, Steele and Henderson 1981). More difficult issues, such as questions of pattern formation, have arisen in a variety of contexts, and have contributed new directions to mathematical research as well as influencing biological thought.

Without question, biology has led to some elegant mathematical formulations and stimulated work upon many mathematical problems. Biology brings a freshness to familiar models, a motivation for assumptions, and an insight into appropriate questions and results. It is not surprising that increasing numbers of mathematicians are being drawn into this field.

Conversely, mathematics is having broad impact upon biological areas, from molecular biology to ecosystems, from respiratory physiology to behavior, from comparative morphology to medical data analysis. Whether as predictive devices or as metaphors, mathematical models have proven essential in the consideration of the complexities of biological systems. This distinguished symposium is testimony to that progress, and to the diversity of influence of mathematical thinking.

Acknowledgment: I am pleased to acknowledge the support of the National Science Foundation under Grants MCS-8001618 and DMS-8406472.

REFERENCES

Albrecht, F., H. Gatzke, and N. Wax. 1973. Stable limit cycles in prey-predator populations. Science 181:1073-1074.

Arbib, M. 1978. Segmentation, schemas, and cooperative computation. Pages 118-155 in S.A. Levin (ed.), Studies in Mathematical Biology I: Cellular Behavior and the Development of Pattern. Mathematical Association of America, Washington, D.C.

Aronson, D.G. and H.F. Weinberger. 1975. Nonlinear diffusion in population genetics, combustion, and nerve propagation. Pages 5-49 in J. Goldstein (ed.), Partial Differential Equations and Related Topics. Lecture Notes in Mathematics 446. Springer-Verlag, Heidelberg.

Aronson, D.G. and H.F. Weinberger. 1978. Multidimensional nonlinear diffusion arising in population genetics. Adv. Math. 30:33-76.

Carpenter, G.A. 1977a. A geometric approach to singular perturbation problems with applications to nerve impulse equations. J. Diff. Eq. 23:335-367.

Carpenter, G.A. 1977b. Periodic solutions of nerve impulse equations. J. Math. Anal. Appl. 58:152-173.

Clark, C.W. 1976. Mathematical Bioeconomics: The Optimal Management of Renewable Resources. Wiley, New York.

Conley, C.C. 1975. Traveling wave solutions of nonlinear diffusion equations. In P.F. Hilton (ed.), Proc. Conf. Structural Stability, Catastrophe Theory, and Their Applications in the Sciences. Springer-Verlag, New York.

Cowan, J. and G.B. Ermentrout. 1978. Some aspects of the eigenbehavior of neural nets. Pages 67-117 in S.A. Levin (ed.), Studies in Mathematical Biology I: Cellular Behavior and the Development of Pattern. Mathematical Association of America, Washington, D.C.

Dobzhansky, T., J.R. Powell, C.E. Taylor, and M. Andregg. 1979. Ecological variables affecting the dispersal behavior of Drosophila pseudoobscura and its relatives. Am. Nat. 114-325-334.

Evans, J.W. 1972. Nerve axon equations III. Stability of the nerve impulse. Indiana Math. J. 22:577-593.

Ewens, W.J. 1979. Mathematical Population Genetics. Springer-Verlag, Heidelberg.

Fife, P.C. 1979. Mathematical Aspects of Reacting and Diffusion Systems. Lecture Notes in Biomathematics 28. Springer-Verlag, Heidelberg.

Fitzhugh, R. 1969. Mathematical models of excitation and propagation in nerve. Pages 1-85 in H.P. Schwan (ed.), Biological Engineering. McGraw-Hill, New York.

Fitzhugh, R. 1973. Dimensional analysis of nerve models. J. Theor. Biol. 40:517-541.

Fisher, R.A. 1937. The wave of advance of advantageous genes. Ann. Eugen. London 7:355-369.

Fisher, R.A. 1958. The Genetical Theory of Natural Selection. Second rev. ed. Dover, New York.

Gierer, A. and H. Meinhardt. 1974. Biological pattern formation involving lateral inhibition. Pages 163-184 in S.A. Levin (ed.), Some Mathematical Questions in Biology, VI. Lectures on Mathematics in the Life Sciences, Vol. 7. American Mathematical Society, Providence, Rhode Island.

Goel, N.S., S.C. Maitra, and E.W. Montroll. 1971. On the Volterra and Other Nonlinear Models of Interacting Populations. Academic Press, New York.

Gordon, R., G.T. Herman, and S. Johnson. 1975. Image reconstruction from projections. Sci. Am. 233:56-61, 64-68.

Guckenheimer, J., G. Oster, and A. Ipaktchi. 1977. Dynamics of density dependent population models. J. Math. Biol. 4:101-147.

Haldane, J.B.S. 1948. The theory of a cline. J. Genet. 48:277-284.

Hethcote, H.W. 1973. Asymptotic behavior in a deterministic epidemic model. Bull. Math. Biol. 35:607-614.

Hirsch, M.W. and S. Smale. 1974. Differential Equations, Dynamical Systems and Linear Algebra. Academic Press, Inc., New York.

Hodgkin, A.L. and A.F. Huxley. 1952. A quantitative description of membrane current and its application to conduction and excitation in nerve. J. Physiol. (London) 117:500-544.

Hodgson, J.P.E. 1983. Oscillations in mathematical biology. Lecture Notes in Biomathematics 51. Springer-Verlag, Heidelberg.

Hopf, E. 1948. A mathematical example displaying features of turbulence. Commun. Pure Apple. Math. 1:303-322.

Jacob, F. 1977. Evolution and tinkering. Science 196:1161-1166.

Kac, M. 1969. Some mathematical models in science. Science 166:695-699.

Kareiva, P. 1982a. Experimental and mathematical analyses of herbivore movement: quantifying the influence of plant spacing and quality on foraging discrimination. Ecol. Monogr. 52

Kaufmann, S.A., R.M. Shymko, and K. Trabert. 1978. Control of sequential compartment formation in Drosophila. Science 199:259-270.

Keller, E.F. and L.A. Segel. 1970. Initiation of slime mold aggregation viewed as an instability. J. Theor. Biol. 26:399-415.

Kerner, E.H. 1959. Further considerations on the statistical mechanics of biological associations. Bull. Math. Biophys. 21:217-255.

Kierstead, H. and L.B. Slobodkin. 1953. The size of water masses containing plankton blooms. J. Mar. Res. 12:141-147.

Kolmogorov, A.N. 1936. Sulla teoria di Volterra della lotta per l'esisttenza. Giorn. Istituto Ital. Attuari. 7:74-80.

Kolmogorov, A., I. Petrovsky, and N. Piscounov. 1937. Etude de l'equation de la diffusion avec croissance de la quantite de matiere et son application a un probleme biologique. Moscow Univ. Bull.

Leigh, E. 1966. The ecological role of Volterra's equation. Pages 1-61 in M. Gerstenhaber (ed.), Some Mathematical Problems in Biology: Lectures on Mathematics in the Life Sciences 1. American Mathematical Society, Providence, Rhode Island.

Levin, S.A. 1974. Dispersion and population interactions. Am. Nat. 108:207-228.

Levin, S.A. 1976a. Population dynamic models in heterogeneous environments. Annu. Rev. Ecol. Syst. 7:287-310.

Levin, S.A. 1976b. Spatial patterning and the structure of ecological communities. Pages 1-35 in S.A. Levin (ed.), Some Mathematical Questions in Biology 7: Lectures on Mathematics in the Life Sciences 8. American Mathematical Society, Providence, Rhode Island.

Levin, S.A. 1978. Introduction. Pages ix-xii in S.A. Levin (ed.), Studies in Mathematical Biology I: Cellular Behavior and the Development of Pattern. Mathematical Association of America, Washington, D.C.

Levin, S.A. 1980. Some models for the evolution of adaptive traits. Pages 56-72 in C. Barigozzi (ed.), Vito Volterra Symposium on Mathematical Models in Biology. Lecture Notes in Biomathematics 39. Springer-Verlag, Heidelberg.

Levin, S.A. 1981a. Age-structure and stability in multiple-age spawning populations. In T.L. Vincent and J. Skowronski (eds.), Renewable Resource Management. Springer-Verlag, Heidelberg.

Levin, S.A. 1981b. Models of population dispersal. In S. Busenburg and K.L. Cooke (eds.), Differential Equations and Applications to Ecology, Epidemics, and Population Problems. Academic Press, New York. To appear.

Levin, S.A. 1981c. The role of mathematics in biology. Proc. Landsmoedet on Mathematikken I Denmark, pp. 455-478. Danish Mathematical Society, Copenhagen.

Levin, S.A. 1983. Some approaches to the modelling of coevolutionary interactions. pp. 21-65 in M. Nitecki, (ed.), Coevolution U. Chi. Press.

Levin, S.A. and C.P. Goodyear. 1980. Analysis of an age-structured fishery model. J. Math. Biol. 9:245-274.

Levin, S.A. and D.P. Pimentel. 1981. Selection of intermediate rates of increase in parasite-host systems. Am. Nat. 117-308-315.

Levin, S.A., D. Cohen, and A. Hastings. 1984. Dispersal strategies in patchy environments. Theoretical Population Biology 26(2):165-191.

Levin, S.A. and L.A. Segel. 1976. An hypothesis for the origin of planktonic patchiness. Nature 259:659.

Levin, S.A. and L.A. Segel. 1985. Pattern generation in space and aspect. SIAM Review, 27:45-67.

Li, T.-Y. and J. Yorke. 1975. Period three implies chaos. Am. Math. Monthly 82:985-992.

Lorenz, E.N. 1963a. Deterministic nonperiodic flows. J. Atmos. Sci. 20:130-141.

Lorenz, E.N. 1963b. The mechanics of vacillation. J. Atmos. Sci. 20:448-464.

Lorenz, E.N. 1964. The problem of deducing the climate from the governing equations. Tellus 16:1-11.

Ludwig, D., D.D. Jones, and C.S. Holling. 1978. Qualitiative analysis of insect outbreak systems: the spruce budworm and forest. J. Anim. Ecol. 47:315:332.

Ludwig, D. and C.J. Walters. 1985. Are age-structured models appropriate for catch-effort data? In press.

Malecot, G. 1948. Les Mathematiques de l'Heredite. Masson et Cie, Paris. (English translation published in 1969 by W.H. Freeman, San Francisco.)

Marsden, J. and M. McCracken. 1976. The Hopf Bifurcation. Springer-Verlag, Heidelberg.

May, R.M. 1972. Limit cycles in predator-prey communities. Science 177:900-902.

May, R.M. 1973. Stability and Complexity in Model Ecosystems. Princeton Univ. Press, Princeton, New Jersey.

May, R.M. 1974. Biological populations with non-overlapping generations: stable points, stable cycles, and chaos. Science 186:645-647.

May, R.M. 1975. Biological populations obeying difference equations: stable points, stable cycles, and chaos. J. Theor. Biol. 49:511-524.

May, R.M. 1978. Mathematical aspects of the dynamics of animal populations. Pages 317-366 in S.A. Levin (ed.), Studies in Mathematical Biology II: Populations and Communities. Mathematical Association of America, Washington, D.C.

Maynard Smith, J. 1976. Evolution and the theory of games. Am. Sci. 64:41-45.

Maynard Smith, J. 1977. Evolution and the theory of games. In W. Matthews (ed.), Mathematics in the Life Sciences. Lecture Notes in Biomathematics. Springer-Verlag, Berlin-New York.

McKean, H.P.M. 1970. Nagumo's equation. Adv. Math. 4:209-223.

Nagumo, J.S., S. Arimoto, and S. Yoshizawa. 1962. An active pulse transmission line simulating nerve axon. Proc. IRE 50:2061-2070.

Nagylaki, T. 1975. Conditions for the existence of clines. Genetics 80:595-615.

Nagylaki, T. 1978. The geographical structure of populations. Pages 588-623 in S.A. Levin (ed.), Studies in Mathematical Biology II: Populations and Communities. Mathematical Association of America, Washington, D.C.

Okubo, A. 1974. Diffusion-induced instability in model ecosystems. Chesapeake Bay Institute, The John Hopkins Univ. Tech. Rep. 86.

Okubo, A. 1980. Diffusion and Ecological Problems: Mathematical Models. Biomathematics 10. Springer-Verlag, New York.

Oster, G. 1978. The dynamics of nonlinear models with age structure. Pages 411-438 in S.A. Levin (ed.), Studies in Mathematical Biology II: Populations and Communities. Mathematical Association of America, Washington, D.C.

Oster, G., D. Auslander, and T. Allen. 1976. Deterministic and stochastic effects in population dynamics. Trans. ASMG. J. Dyn. Sys., Meas. and Control 98:44-48.

Oster, G.F. and E.O. Wilson. 1978. Caste and Ecology in the Social Insects. Princeton Univ. Press, Princeton, New Jersey.

Rinzel, J. 1978. Integration and propagation of neuroelectric signals. Pages 1-66 in S.A. Levin (ed.), Studies in Mathematical Biology I: Cellular Behavior and the Development of Pattern. Mathematical Association of America, Washington, D.C.

Rinzel, J. and J.B. Keller. 1973. Traveling wave solutions of a nerve conduction equation. Biophys. J. 13:1313-1337.

Rosenzweig, J.L. and R.H. MacArthur. 1963. Graphical representation and stability conditions of predator-prey interactions. Am. Nat. 97:209-223.

Ruelle, D. 1975. The Lorenz attractor and the problem of turbulence. In Quantum Dynamics Models and Mathematics. Bielefeld, September 1975.

Ruelle, D. and F. Takens. 1971. On the nature of turbulence. Commun. Math. Phys. 20:167-192, 23:343-344.

Sarkovskii, A.N. 1964. Coexistence of cycles of a continuous map of a line into itself. Ukr. mat. Z. 16:61-71. In Russian.

Segel, L.A. and J.L. Jackson. 1972. Dissipative structure: an explanation and an ecological example. J. Theor. Biol. 37:545-559.

Segel, L.A. and S.A. Levin. 1976. Application of nonlinear stability theory to the study of the effects of diffusion on predator-prey interactions. Pages 123-152 in R.A. Piccirelli (ed.), Topics in Statistical Mechanics and Biophysics. AIP Conf. Proc. 27. American Institute of Physics, New York.

Skellam, J.G. 1951. Random dispersal in theoretical populations. Biometrika 38:196-218.

Steele, J.H. and E.W. Henderson. 1981. A simple plankton model. Am. Nat. 117-676-691.

Stefan, P. 1976. A theorem of Sarkovskii on the existence of periodic orbits of continuous endomorphising of the real line. Ms. Bures sur Yvette IHES.

Turing, A. 1952. The chemical basis of morphogenesis. Philos. Trans. R. Soc. London Ser. B 237:37-72.

Tyson, J.J. 1976. The Belousov-Zhabotinskii Reaction. Lecture Notes in Biomathematics 10. Springer-Verlag, Heidelberg.

Ulam, S.M. and J. von Neumann. 1947. Abstract. Bull. Am. Math. Soc. 53:1120.

Volterra, V. 1926. Variazioni e fluttuazioni del numero d'individui in specie animale conviventi. Mem. R. Accad. Nazionale del Lincei (Ser. 6) 2:31-113.

Wan, Y.H. 1978. Computation of the stability condition for the Hopf bifurcation of diffeomorphism on R^2. SIAM J. Appl. Math. 34:167-175.

Winfree, A.T. 1974. Rotating chemical reactions. Sci. Am. 230:82-95.

Winfree, A.T. 1980. The Geometry of Biological Time. Biomathematics 8. Springer-Verlag, Heidelberg.

Wright, S. 1949. Adaptation and selection. Pages 365-389 in G.L. Jepson, G.G. Simpson, and E. Mayr (eds.), Genetics, Paleontology, and Evolution. Princeton Univ. Press, Princeton, New Jersey.

Zaiken, A.N. and A.M. Zhabotinskii. 1970. Concentration wave propagation in two-dimensional liquid-phase self-oscillating systems. Nature 225:535-537.

I. MATHEMATICAL ECOLOGY AND POPULATION BIOLOGY

The Structure of Populations and Communities

Dispersal

Evolution

Fantastic Voyage into the Deep: Marine Biofluid Mechanics

Akira Okubo

Marine Sciences Research Center
State University of New York
Stony Brook, NY 11794-5000 USA
and
Ecosystems Research Center
Cornell University
Ithaca, NY 14853-2701 USA

1. Introduction

Since life originated in the ancient sea, the interplay between the environmental fluid and organisms has become an inseparable part of physiology, ecology and evolution of organisms. Yet our understanding of the mechanisms and processes in the interaction remains irritatingly poor. An often hostile attitude between biologists and physicists/mathematicians is in part at fault.

In recent years, however, we have begun to overcome the barrier, and biofluid mechanics has become a fast growing field in marine sciences. In this paper we discuss the basic equations of fluid mechanics governing the motion of marine organisms and examine some characteristic parameters pertinent to the problem. Furthermore we present some examples in marine biology, where the interaction between environmental flow and organisms plays a major role in understanding the problem.

2. Basic equations

In almost all motions in marine ecology we may assume that the environmental flow is incompressible, Newtonian, and Boussinesqian (variations in the fluid density are neglected in so far as they influence the inertia). Thus the equations of motion for the environmental water read

$$\frac{D\mathbf{u}}{Dt} \equiv \frac{\partial \mathbf{u}}{\partial t} + \mathbf{u}\cdot\nabla\mathbf{u} = -2\boldsymbol{\Omega} \times \mathbf{u} - \frac{1}{\rho_o}\nabla p + \frac{\rho-\rho_o}{\rho_o}\,\mathbf{g} + \nu\,\nabla^2\mathbf{u} \qquad (1)$$

$$\nabla \cdot \mathbf{u} = 0 \qquad (2)$$

where $\mathbf{u}$: fluid velocity, $\boldsymbol{\Omega}$: earth's rotation vector, p: pressure, ρ: fluid density, ρ_o: reference density, $\mathbf{g}$: acceleration of gravity vector, ν: kinematic viscosity of sea water. The force terms on the right-hand side of (1) represent respectively the Coriolis force, pressure-gradient force, buoyancy, and viscous forces per unit mass. The above set of equations is subject to appropriate initial and boundary conditions.

An organism is usually considered to be an entity separated from the environmental water and can, for the most part, be taken to be an inpenetrable region on whose boundary ∂B the fluid has no slip. In other words if $\underset{\sim}{r}\,(t,\underset{\sim}{r}_o)$ is the position of a boundary point initially at $\underset{\sim}{r}_o$, then the condition on the fluid at the organism surface is

$$\underset{\sim}{u}(\underset{\sim}{r}\,,t) = \dot{\underset{\sim}{r}}\,(t,\underset{\sim}{r}_o), \qquad \underset{\sim}{r}_o \in \partial B \tag{3}$$

i.e., no relative velocity at the boundary.

In general the boundary exhibits a complicated form varying in time, which, in turn, depends upon the response of the fluid to its changes of shape. We usually can avoid this "active" boundary condition by prescribing the organism boundary.

If the number of organisms is very great and organism dimensions are very small compared with the dimensions of the system of interest, then a seawater-organism suspension may conveniently be regarded as a continuum. In this case, however, the organism concentration should not be too great to violate the assumption of the Newtonian nature of the flow of the mixture. In essence we deal with a dilute system with an effective viscosity that depends upon the organism concentration. The behavioral motion of the organism in suspension may produce an additional advection velocity in the Navier-Stokes equation.

2.1 Dimensionless form of the equation

Consider a typical problem of environmental flow around an organism. Let f, ℓ, U be some characteristic frequency of flow, characteristic length of organism, and characteristic speed of organism relative to the ambient flow. In terms of these scales the basic equations can be expressed in terms of the dimensionless variables

$$\begin{aligned} t' = ft, \quad \underset{\sim}{x}' = \ell^{-1}\underset{\sim}{x}, \quad \underset{\sim}{u}' = U^{-1}\underset{\sim}{u}, \\ p' = (\rho_o U^2)^{-1}p, \quad \underset{\sim}{\nabla}' = \ell^{-1}\underset{\sim}{\nabla} \end{aligned} \tag{4}$$

These equations then become

$$\sigma\,\partial \underset{\sim}{u}'/\partial t' + (\underset{\sim}{u}'\cdot\underset{\sim}{\nabla}')\underset{\sim}{u}' = -R_o^{-1}\underset{\sim}{\omega}'\mathrm{x}\ \underset{\sim}{u}' - \underset{\sim}{\nabla}'p' + R_e^{-1}\nabla'^2\underset{\sim}{u}' + F^{-1}\underset{\sim}{k} \quad (5)$$

$$\underset{\sim}{\nabla}' \cdot \underset{\sim}{u}' \quad (6)$$

where $\underset{\sim}{\omega}'$ is earth's unit rotation vector and $\underset{\sim}{k}$ is a unit vertical vector. In (5) we define four dimensionless parameters:

$$\sigma \equiv f\ell/U \qquad \text{(frequency parameter)} \quad (7)$$

$$R_o \equiv U/2\,\Omega\,\ell \qquad \text{(Rossby number)} \quad (8)$$

$$R_e \equiv U\ell/\nu \qquad \text{(Reynolds number)} \quad (9)$$

$$F \equiv (\rho-\rho_o/\rho_o)\ g\ell/U^2 \qquad \text{(Froude number)} \quad (10)$$

Generally speaking we may regard σ as a parameter of order 1 or less and the Froude number plays no role in our problem since the buoyancy term can be absorbed into the pressure gradient (Childress, 1981; Batchelor, 1967). As will be seen later, the Rossby number is almost always much larger than unity for the movement of marine organisms. These considerations leave the Reynolds number being the centerpiece of marine biofluid mechanics. In conclusion the basic dynamic equation may be written as

$$\sigma\,\partial \underset{\sim}{u}'/\partial t' + (\underset{\sim}{u}'\cdot\underset{\sim}{\nabla}')\underset{\sim}{u}' = -\underset{\sim}{\nabla}'p + R_e^{-1}\ \nabla'^2\underset{\sim}{u}' \quad (11)$$

2.2 Reynolds number associated with movements of marine organisms

The Reynolds number measures the relative importance of the forces of inertia acting on a unit volume of fluid and the force developed on the volume by the viscous stresses, or putting it another way, the Reynolds number provides an estimate of the relative importance of the non-viscous and viscous forces acting on unit volume of fluid.

Fig. 1 shows the relation between the Reynolds number and the body length of marine organisms ranging from bacteria to whales (Okubo and Mitchell, 1986). The calculation of the Reynolds number takes U as a typical swimming speed for animals and bacteria and as settling velocity for phytoplankton. For the sake of curiosity the figure includes human data of height 2 m and swimming speed of 1 m/sec.

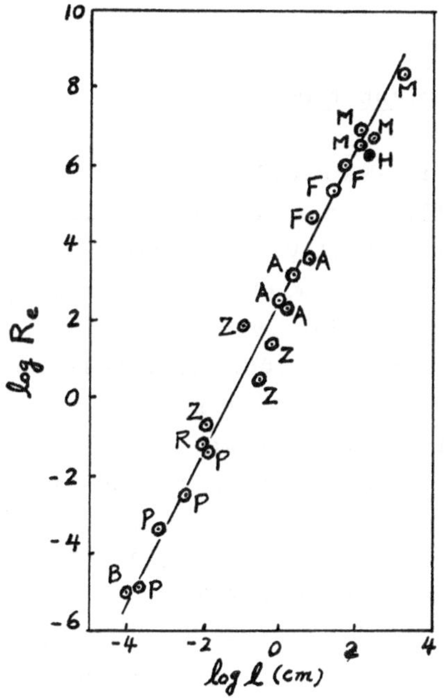

Fig. 1. Reynolds number (R_e) versus organism size (ℓ)
M: mammal, F: fish, A: amphypod, z: zooplankton,
R: protozoa, P: phytoplankton, B: bacteria, H: man.

The overall trend in Fig. 1 is clear. The Reynolds number increases systematically with the size of organism and the number varies over 13 orders of magnitude in the range of marine organisms. An empirical formula is given by

$$R_e = 269\ \ell^{1.86} \tag{12}$$

where ℓ is measured in cm.

Since $R_e = U\ell/\nu$ and $\nu = 0.01\ \text{cm}^2/\text{sec}$, the empirical relation implies that

$$U(\text{cm/sec}) = 2.69\ \ell^{0.86}\ (\text{cm}) \tag{13}$$

There may be some energetical/physiological basis in the relationship between the mean speed of locomotion and organism size. In any case the substitution of (13) into (8) results

$$R_o = O(10^4 \sim 10^5) \tag{14}$$

where the lower limit is applied to $\ell = O(10\ m)$ and the upper limit to $\ell = O(1\mu)$.

2.3 Laminar versus turbulent flow

When a fluid flows in an orderly fashion, the flow is called laminar; when the flow flows in a stochastic fashion, the flow is called turbulent. For a neutrally stratified fluid a Reynolds number determines whether a given flow will be laminar or turbulent. The critical Reynolds number for transition from a laminar to turbulent flow is usually of the order of 1,000. Thus swimming fish can leave behind a turbulent wake, while the movement of marine bacteria is characteristic of a laminar flow regime.

The system of equations mentioned previously can be directly applied to a low Reynolds number laminar flow. In turbulence, on the other hand, a description of the flow at all points and times is not feasible simply because turbulence consists of random velocity fluctuations. The basic equations must be treated with statistical methods. Thus we make an attempt to develop equations governing mean velocities and pressure. This leads to the closure problem of turbulence, a notoriously difficult problem to solve, characteristic of all nonlinear stochastic systems.

In this paper we will not treat a turbulent flow in biofluid mechanics, with the only exception of the use of a quadratic drag law that is characteristic of resistance in a high Reynolds number flow. Instead our main concern is the problem of low Reynolds number flows.

3. Bioconvection and biologically-induced circulation

Dense populations of free-swimming organisms such as ciliates and flagellates often exhibit curious pattern-like variations in spatial concentration. An example is seen in dense cultures of *Tetrahymena pyriformis* where the organisms form polygonal cellular patterns with the associated motion of the suspending fluid. Platt (1961) has called this phenomenon "bioconvection".

Childress et al. (1975) and Levandowsky et al. (1975) developed a fluid dynamical model for pattern formation by negatively geotactic microorganisms and determined the conditions for the occurrence of bioconvection. The basic equations are the same as (1) and (2) with the exception that the Coriolis force is neglected. The buoyancy term is now expressed in terms of the organism concentration c, i.e.,

$$(\rho - \rho_o)/\rho_o \underset{\sim}{g} = (1 + \alpha c)\, g\, \underset{\sim}{k} \tag{15}$$

The equation for the organism concentration is written as

$$\frac{DC}{Dt} = -\underset{\sim}{\nabla}\{c\, u_s(c,z)\underset{\sim}{k}\} + D_m \nabla^2 c \tag{16}$$

where $u_s(c,z)\underset{\sim}{k}$ is the directed swimming velocity in the vertical direction and its speed depends on the concentration and the vertical coordinate, z, and D_m is the diffusivity of the suspension. The governing equations are subject to appropriate conditions at both surface and bottom boundaries.

The basic pattern-free equilibrium of concentration is obtained, which depends only on the vertical coordinate, z. A linear stability analysis of the equilibrium leads to a critical value of a Rayleigh number R_a:

$$R_a = g\alpha c_o h^3/\nu D_m = g\alpha c_o D_m^2/\nu U_o^3 \tag{17}$$

where c_o and U_o are the values of c and u_s at $z = 0$.

Fujita and Watanabe (1986) conducted a numerical study of the basic equations proposed by Childress et al. (1975) and Levandowsky et al. (1975) and found that the bioconvection could be led into chaotic behavior through a single-frequency oscillatory mode to a sequence of period-doubling bifurcation by increasing the Rayleigh number.

The bioconvection due to negative geotaxis of microorganisms can be reinforced by <u>gyrotaxis</u>, i.e. directed locomotion resulting from the orientation of cell's axis by compensating gravitational and viscous torques (Kessler, 1984, 1985a,b). The convection pattern due to this effect has been named "gyrotactic buoyant convection". At a sufficient average cell concentration, this bioconvection can cause localized intermittent concentration pulses.

Although these studies of bioconvection are limited to laboratory cultures, a similar phenomenon may occur in natural waters. In particular, when an extremely dense population of algal cells exists in the sea under calm conditions, the occurrence of bioconvection seems likely. The gyrotactic buoyant convection of microorganisms may occur even under moderate wind conditions associated with the Langmuir circulation in the surface layer (Kessler, private communication).

It has been known for sometime that the Antarctic krill, <u>Euphausia superba</u> form swarms, whose horizontal extent may exceed ten kilometers ("superswarms"). The krill's concentration in such large swarms presents an interesting fluid dynamical problem since krill's density is about 2% heavier than the surrounding seawater. Hofmann et al. (1986) speculate that the density difference between the krill swarm and the ambient water should be sufficient to induce a water circulation within the swarm.

The higher density associated with the krill swarm produces a downward bend of constant pressure surfaces, and thus gives rise to an inward-directed horizontal pressure gradient field in the swarm. This pressure force can generate a water circulation in such a manner that the centrifugal acceleration balances the pressure gradient. The force balance, for a radially symmetric steady swarm, is expressed by

$$- v^2/r = -\rho_1^{-1} \, \partial p/\partial r \tag{18}$$

where v is the circular velocity, r the radial distance from the center of swarm, p the pressure and ρ_1 is the density of the volume containing water and krill. All other forces in the equation of motion are ignored.

Using the hydrostatic approximation we can express the radial pressure gradient in terms of the density gradient, and the circular velocity is given by

$$v^2 = - gr \, \rho_1^{-1} \int_{-H}^{z} \partial \rho_1/\partial r \, dz \tag{19}$$

where z is taken positive upwards and z = -H is the bottom of the swarm. The knowledge of the radial distribution of krill concentration will provide us with the evaluation of the velocity. Thus, a swarm thickness of 50 m and densities of 0.1 kg/m^3 can produce velocities of 3 to 4 cm/sec. Thicker and higher densities swarms could produce velocities approaching 30 cm/sec.

The biologically-induced circulation would have some interesting implications. At swarm densities of 2 kg/m^3, oxygen depletion in the water due to krill respiration would limit the duration of swarming to a few hours. The biologically-induced swirling flow together with a vertical plane circulation would provide a mechanism to transport oxygenated water from the environment into the swarm.

4. Settling of organisms and byssus thread drift

For a flow of which Reynolds number is much less than unity

inertial terms are neglected, leading to a linear system of fluid dynamical equations. Thus in a steady flow (1) - (2) reduce to

$$0 = - \nabla p + \eta \nabla^2 u \tag{20}$$

$$\nabla \cdot u = 0 \tag{21}$$

For a given body shape of organism we can solve (20) and (21) subject to a no-slip condition at the body and, say, uniform pressure and velocity at infinity.

From the solution we can calculate the total drag on the organism by integrating pressure and shear stress over the surface. For a spherical body of radius a moving with a uniform velocity of U or at rest in a free stream of uniform speed U, the total drag is found to be

$$D_1 = 6 \pi a \eta U \tag{22}$$

This is the famous formula by Stokes (1851). It is strictly valid only for $R_e << 1$, but in reality it is farely accurate up to about $R_e = 1$.

The law of drag is conveniently expressed by

$$D_2 = 1/2 \; C_D \rho A U^2 \tag{23}$$

where C_D: drag coefficient, A: projection area. For a given body form the drag coefficient depends only upon the Reynolds number. At low values of R_e, $C_D = 24/R_e$, which reduces (23) to (22) as $R_e = 2a \rho U/\eta$ and $A = \pi a^2$.

Fig. 1 indicates that the marine organism whose size is less than 1 mm or so may be regarded as a Stokes particle characterized by the drag law (22). When this force of resistance balances the immersed weight of an organism, the spherical organism falls with a constant terminal velocity, i.e. the settling velocity, w_s, and it is given by

$$w_s = 2(\rho_1 - \rho) a^2 g/9\eta \tag{24}$$

The calculated settling velocity for marine organisms of radius $a = 1 \sim 250 \ \mu$ varies from 10^{-6} to 0.1 cm/sec. For an excellent review on the settling velocity of phytoplankton in the sea, consult Smayda (1970).

Except for microzooplankton, most zooplankton and post-larval bivalves possess moderate Reynolds numbers for which the quadratic drag law (23) is appropriate with C_D ranging from 1 to 10, and the settling velocity of these organisms is of the order of 1 cm/sec or 1,000 m/day. This implies that floatation of post-larval bivalves is a serious problem in their dispersal.

Recent observations (Sigurdsson et al., 1976; Prezant and Chalermwat, 1984; Lane et al., 1985) indicate that the byssus or long mucous threads excreted by young post-larval bivalves act as draglines to sustain the animal into a water column. The method of transportation is analogous to the gossamer flight of young spiders.

If we assume that the drifting thread is cylindrical of radius r and the drag force on the thread is entirely viscous ($R_e << 1$), the drag on a thread of length ℓ is given by

$$D_3 = 4\,\pi\eta\ell\, u/\ell_n(4\ell\eta/r^2\,\rho\, u) \qquad (25)$$

where u is an imposed axial velocity. The force balance is obtained by

$$4/3\ \pi a^3(\rho_1-\rho)g = C_D/2\ \pi\ a^2\ \rho\ w_s^2 + 4\ \pi\ \eta\ \ell\ w_s/\ell_n(4\ell\eta/r^2\ \rho\ w_s) \qquad (26)$$

where the post-larval shell is assumed to be a sphere of radius a and a quadratic law is used for the drag of the shell.

Calculations show that the settling velocities of post-larval bivalves without drifting threads range from 1 to 3 cm/sec for the size of 0.5 to 2 mm, while settling velocities with drifting threads of the order of 10 cm in length are about 0.03 to 0.3 cm/sec for the size range. The calculated contribution of the thread to the total drag is approximately one order of magnitude greater than that of the larval body. Thus the drifting threads are highly effective in enhancing the dispersal of young mussels.

5. Zooplankton feeding currents and prey detection

Our understanding of the mechanisms involved in the foraging process of zooplankton has recently increased dramatically. Thus microcinematographic techniques of recent development allow us, for the first time, direct observations of small copepod feeding on the even smaller algae (Alcaraz et al., 1980; Rosenberg, 1980; Paffenhöfer et al., 1982; Koehl and Strickler, 1981; Koehl, 1983; Strickler, 1982;

Price et al., 1983; Price and Paffenhöfer, 1984; Vanderploeg and Paffenhöfer, 1985). These observations reveal that the copepod is not feeding passively but rather capable of perceiving the presence of an algal particle at a distance (100μ ∿ 1 mm) away from the animal, reorients itself, and adjusts the current with respect to the particle so as to draw the prey into the feeding current more efficiently. However, the sensory mechanisms controlling particle detection and size selection are still mostly unknown. Chemoreception and mechanoreception have been suggested. Both mechanisms involve fluid mechanical consideration characteristic of the low Reynolds number flow.

Chemoreception depends primarily upon a chemical leaked from an algal cell. To be specific algal cell is surrounded by a zone of associated water ("active space", Andrews, 1983) within which the concentration of simple sugars, amino acids and other excudates exceeds background levels. The feeding current of copepods is characterized by a shear flow of the low Reynolds number, $R_e = 0(1)$. As a result, the active space is deformed and an initially spherical isoconcentration surface is elongated in the direction of the flow to such a degree that the feeding animal may be able to detect the presence of the prey by chemoreception at a distance.

The problem is mathematically treated by the advection and diffusion equation of the chemical concentration c

$$\partial c/\partial t = -\underset{\sim}{u}\underset{\sim}{\nabla}c + D\nabla^2 c \qquad (27)$$

which can be solved under given feeding current field (Andrews, 1983). This model predicts that the distance of detection is a function of the radius of prey, detection threshold, the position of the prey in the feeding current, and the molecular diffusivity of the exudate. An example of calculation shows a detection distance of the order of 1 mm, which agrees well to the distance of early detection. One of the drawbacks in this model is the assumed value of the threshold concentration of chemicals. Despite the fact that an individual primary chemoreception may show an electrophysical response to very low concentration, behavioral responses have not been observed until the animal is exposed to concentrations of amino acids as high as 10^{-2} to 10^{-4} mols, which are far greater than the assumed threshold concentration (Legier-Visser et al., 1986).

A mechanoreception model for prey detection and selection has been proposed by Legier-Visser et al. (1986). The model is based on the pressure field induced by a prey particle. If we assume that the particle is a sphere embedded in a low Reynolds number flow, we can use the Stokes solution of the pressure field

$$\Delta p = 3\ \eta\ a\ U \cos \theta / r^2 \qquad (28)$$

where Δp: pressure difference between uninterrupted flow (in the absence of particle) and flow deformed by the particle of radius a, η: viscosity of fluid, θ: angle between the particle and the mechanoreceptor of copepod, r: distance from the center of the particle to the antenna of copepod measured perpendicular to the antenna, and U is the flow velocity far away from the particle.

Because of the pressure distribution along the antenna, the position of the particle with respect to the animal can be determined by the animal's array of mechanoreceptors as the prey approaches. The copepod can then reorient itself in such a way that the largest pressure difference the antennal mechanoreceptors perceive is above those mechanoreceptors located in the area of the feeding current corresponding to the highest probability of prey capture.

Letting ξ be the distance from the forward edge of the wall of the approaching particle to the antenna, we rewrite (28) and solve for ξ,

$$\xi = (\lambda/\Delta p)^{1/2}\ a^{1/2} - a \qquad (29)$$

with $$\lambda \equiv 3\ \eta\ U \cos \theta$$

For a given minimum pressure sensitivity of copepod mechanoreceptor, Δp, and λ, the distance ξ increases with the prey size a, reaches a maximum value of $\lambda/4\Delta p$ at $a = \lambda/4\Delta p$, and decreases toward zero as a increases. In other words both very small and very large particles would not be detected by the copepod until they come in the very vicinity of the mechanoreceptors.

From this model we expect that copepods would have the greatest detection ability in a narrow range of particle size. This maximum detection distance and optimal size of prey are determined in part by the intensity of the feeding current which may, in fact, be altered by the animal (Rosenberg, 1980; Price et al., 1983; Price and Paffenhöfer, 1984).

6. Microzones (microscale nutrient patches)

Many studies show that peaks of phytoplankton cell numbers in the depth profile are coincident with those of bacteria. Most marine bacteria occur free in the environmental water rather than attached to particles (plankton cells) while obtaining nutrients from the local source, presumably micro-scale nutrient patches arisen from the excretion and autolysis of phytoplankton. On the other hand, glycollate uptake by diatoms is found to be enhanced by the presence of bacteria. Mutual associations of phytoplankton and bacteria are suggested.

It has been proposed that bacteria cluster around particulate nutrient sources or phytoplankton cells and suggested that there exist oceanic microzones having utilizable dissolved organic carbon orders of magnitude above background concentrations. Mitchell et al. (1985) calculated the size of the microzones on the bases of the diffusion equation (27). For a steady-state distribution of nutrient around a phytoplankton cell, the radius of the microzone, R, is estimated by

$$R = Q/4\pi Dc^* \tag{30}$$

where Q: the total nutrient flux per cell, D: molecular diffusivity, c^*: the threshold concentration (10% above the background concentration). Given $Q = 10^{-17}$ mol/s, $D = 10^{-5}$ cm^2/s and $c^* = 10^{-12}$ mol/cm^3, Mitchell et al., calculated $R \sim 1$ mm.

Interestingly enough this value of the microzone is nearly the same as the length scale of Kolmogorov eddies, i.e. the smallest turbulence found in the upper mixed layer of the ocean (Mitchell et al., 1985). It implies that nutrient plumes around phytoplankters would be dispersed by the oceanic turbulence long before reaching a radius of 1 mm. However, in the deep waters of the oceans and in the thermocline below the upper mixed layer, lower turbulence conditions prevail and the characteristic length of the Kolmogorov eddies is of the order of 1 cm. Mitchell et al. suggest that the deep water and particular sheets and layers composing the thermocline are the most likely location for microzones.

Another concept of microzones has been developed in the zooplankton-phytoplankton association. Recent studies indicate that central oceanic regions traditionally thought of biological deserts are, in fact, highly productive. These studies nonetheless reveal that nitrogeneous nutrients in the near-surface waters are below the limit of detectability even though considerable photosynthetic activity is noted (Carpenter and McCarthy, 1975; Sheldon and Sutcliffe, 1978). To resolve this dilemma of "missing mass of nutrients" McCarthy and Goldman (1979) hypothesized that small-scale nutrient plumes or patches either excreted locally by zooplankton or remineralized by bacteria should be continuously utilized by adjecent phytoplankton cells at rates that could keep the concentration averaged over large water volumes at the level below detection limits.

Mathematical modelling of the nutrient plume depends upon the advection-diffusion-reaction equation for the nutrient concentration excreted from zooplankton and uptaken by phytoplankton.

$$\partial c/\partial t + u \; \partial c/\partial x = D\nabla^2 c - (V_m \; c/c+K)N \tag{31}$$

where c: nutrient concentration, D: molecular diffusivity, V_m: maximum uptake rate, K: half-saturation concentration for uptake, N: concentration of phytoplankton cells. The assumption of molecular diffusion is appropriate for the low-Reynolds number regime.

The calculation based on the model has not been conclusive as to whether or not the micropatches play an important role in the coupling. Jackson (1980) pointed out that such patches might dissipate by diffusion too quickly to be used by phytoplankton, but Lehman and Scavia (1982) dismissed this skepticism, showing both theoretically and experimentally that algae can exploit micropatches of nutrient created by zooplankton. Currie (1984) argued that a non-homogeneous nutrient supply regime should actually decrease phytoplankton growth rates, regardless of the physical characteristics of the patches; patchiness cannot account for elevated phytoplankton growth rates in the ocean. The controversy is by no means settled, and further investigations both theoretical and experimental will be necessary.

7. Conclusion

The interface between fluids and organisms is remarkably extensive. This paper has made a brief contact with some examples of the interaction. In fact, there is a long list of questions and problems in physiological ecology, population and community dynamics, and behavioral ecology to which only the interdisciplinary understanding of fluid mechanics and biology will provide the answers.

The unknown abyss still awaits us for exploration. Until then the present paper may serve as a brochure for this "fantastic voyage."

LITERATURE CITED

Alcaraz, M., G.A. Paffenhöfer and J.R. Strickler (1980). Catching the algae: a first account of visual observations on filter feeding calanoids. In: Evolution and Ecology of Zooplankton Communities, 241-248, W.C. Kerfoot (ed.). University Press of New England, Hanover.

Andrews, J.C. (1983). Deformation of the active space in the low Reynolds number feeding current of calanoid copepods. Can. J. Fish. Aquat. Sci. 40: 1293-1302.

Batchelor, G.K. (1967). An Introduction to Fluid Dynamics. Cambridge Univ. Press, London and New York, 615 pp.

Carpenter, E.J. and J.J. McCarthy (1975). Nitrogen fixation and uptake of combined nitrogeneous nutrients by Oscillatoria (Trichodesmium) thiebautii in the western Sargasso Sea. Limnol. Oceanogr. 20: 389-401.

Childress, S., M. Levandowsky, and E.A. Spiegel (1975). Pattern formation in a suspension of swimming microorganisms: equations and stability theory. J. Fluid Mech. 63: 591-613.

Childress, S. (1981). Mechanics of Swimming and Flying. Cambridge Univ. Press, London and New York, 155 pp.

Currie, D.J. (1984). Phytoplankton growth and the microscale nutrient patch hypothesis. J. Plankton Res. 6: 591-599.

Fujita, S. and M. Watanabe (1986). Transition from periodic to non-periodic oscillation observed in a mathematical model of bioconvection by motile micro-organisms. Physica D (in press).

Hofmann, E.E., J.M. Klink, J. Ishizaka and N.D. Nowlin (1986). Biologically-produced currents within a krill swarm. Nature (submitted).

Jackson, G.A. (1980). Phytoplankton growth and zooplankton grazing in oligotrophic oceans. Nature 284:439-441.

Kessler, J.O. (1984). Gyrotactic buoyant convection and spontaneous pattern formation in algal cell cultures. In: Nonequilibrium Cooperative Phenomena in Physics and Related Fields, 241-248, M.G. Velarde (ed.), Plenum Press, New York.

Kessler, J.O. (1985a). Co-operative and concentrative phenomena of swimming micro-organisms. Contemporary Phys. 26: 147-166.

Kessler, J.O. (1985b). Hydrodynamic focusing of motile algal cells. Nature 313: 218-220.

Koehl, M.A.R. and J.R. Strickler (1981). Copepod feeding currents: food capture at low Reynolds number. Limnol. Oceanogr. 26: 1062-1073.

Koehl, M.A.R. (1983). The morphology and performance of suspension - feeding appendages. J. Theor. Biol. 105: 1-11.

Lane, D.J.W., A.R. Beaumont and J.R. Hunter (1985). Byssus drifting and the drifting threads of the young post-larval mussel Mytilus edulis. Mar. Biol. 84: 301-308.

Legier-Visser, M., J.G. Mitchell, A. Okubo and J.A. Fuhrman (1986). Mechanoreception in calanoid copepods: a mechanism for prey detection. Mar. Biol. (in press).

Lehman, J.T. and D. Scavia (1982). Microscale nutrient patches produced by zooplankton. Proc. National Acad. Sci. USA 79: 5001-5005.

Levandowsky, M., S. Childress, E.A. Spiegel and S.H. Hutner (1975). A mathematical model of pattern formation by swimming microorganisms. J. Protozool 22: 296-306.

McCarthy, J.J. and J.C. Goldman (1979). Nitrogeneous nutrition of marine phytoplankton in nutrient-depleted waters. Science 203: 670-672.

Mitchell, J.G., A. Okubo and J.A. Fuhrman (1985). Microzones surrounding phytoplankton form the basis for a stratified marine microbial ecosystem. Nature 316: 58-59.

Okubo, A. and J.G. Mitchell (1986). Relationship between the Reynolds number and body length in marine organisms (in preparation).

Paffenhöfer, G.A., J.R. Strickler and M. Alcaraz (1982). Suspension-feeding by herbivorous calanoid copepods: a cinematographic study. Mar. Biol. 67: 193-199.

Platt, J.R. (1961). "Bioconvection patterns" in cultures of free-swimming organisms. Science 133: 1766-1767.

Prezant, R.S. and K. Chalermwat (1984). Flotation of the bivalve Corbicula fluminea as a means of dispersal. Science 225: 1491-1493.

Price, H.J., G.A. Paffenhöfer and J.R. Strickler (1983). Modes of cell capture in calanoid copepods. Limnol. Oceanogr. 28: 116-123.

Price, H.J. and G.A. Paffenhöfer (1984). Effects of feeding experience in the copepod Eucalanus pileatus: a cinomatographic study. Mar. Biol. 84: 35-40.

Rosenberg, G.G. (1980). Filmed observations of filter feeding in the marine planktonic copepod Acartia clausii. Limnol. Oceanogr. 25: 738-742.

Sheldon, R.W. and W.H. Sutcliffe (1978). Generation times of 3 h for Sargasso Sea microplankton determined by ATP analysis. Limnol. Oceanogr. 23: 1051-1055.

Sigurdsson, J.B., C.W. Titman and P.A. Davies (1976). The dispersal of young post-larval bivalve molluscs by byssus threads. Nature 262: 386-387.

Smayda, T.J. (1970). The suspension and sinking of phytoplankton in the sea. Oceanogr. Mar. Biol. Ann. Rev. 8: 353-414.

Stokes, G.G. (1851). On the effect of the internal friction of fluid on the motion of pendulums. Trans. Cambridge Phil. Soc. 9: 8-106. (Also see H. Lamb, 1945, Hydrodynamics, Dover Publ., New York, p. 594-603).

Strickler, J.R. (1982). Calanoid copepods, feeding currents and the role of gravity. Science 218: 158-160.

Vanderploeg, H.A. and G.A. Paffenhofer (1985). Models of algal capture by the freshwater copepod Diaptomus sicilis and their relation to feed-size selection. Limnol. Oceanogr. 30: 871-885.

On the mathematical synthesis of physiological and behavioural mechanisms and population dynamics

Odo Diekmann

Centre for Mathematics and Computer Science
Kruislaan 413
1098 SJ Amsterdam, The Netherlands
&
Institute of Theoretical Biology
University of Leiden
Groenhovenstraat 5
2311 BT Leiden, the Netherlands

1. Introduction

When investigating the biological world we can concentrate on different levels of organization. For instance, we can look at *individual organisms* and try to understand their *life cycle* and the *physiological processes* that are essential for their functioning. Or we can look at *populations* of such organisms and try to understand or even predict how *numbers* change in the course of time.

In reality these levels are strongly coupled but theoretical as well as experimental studies in ecology and cell kinetics tend to neglect the interplay more or less, with the work of Streifer (1974), Bell & Anderson (1967), Auslander, Oster & Huffaker (1974), Fredrickson, Ramkrishna & Tsuchiya (1967) as early exceptions worth mentioning. A recent attempt to revive the spirit of these older papers and to bridge the gap between the individual and the population level by means of a special class of mathematical models goes under the heading "Dynamics of Structured Populations", where sometimes the adverb "physiologically" is used to emphasize the difference with models incorporating spatial structure (Metz & Diekmann, to appear). In a collective effort we try to study interesting examples in detail, to unravel the general mathematical structure of the class of models and to fit the corresponding class of partial functional differential equations into the framework of dynamical systems theory. This paper is a kind of progress report on work done by Hans Metz, Henk Heijmans, Mats Gyllenberg, Horst Thieme and myself, strongly stimulated by and partly in collaboration with Tom Aldenberg, Frank van den Bosch, Bas Kooijman, Roger Nisbet, André de Roos, Mous Sabelis and John Tyson. In addition, I sketch some lines to the future. This paper, with its concise and, at times, not very precise formulation, is meant as an invitation to read Metz & Diekmann (to appear), Heijmans (1985, to appear) and Clément et al. (in preparation).

2. Modelling

The following is a recipe for the formulation of a structured population model:

(i) Distinguish individuals from one another according to relevant variables. That is, choose finitely many *i-state variables* (*i*- for individual) which, by assumption, contain sufficient information about the past of an individual to fix its future, as far as it concerns population dynamically relevant events. **Examples** of *i*-state variables: age, size, energy reserves, amount of foliage.

(ii) Genuine modelling: specify and describe (mathematically) the *processes* which change the *i*-state and those which lead to the creation or destruction of individuals. The smooth "change" processes describe the trajectories of individuals in the *i*-state space and we usually assume that these are *deterministic* (i.e. identical for all individuals in a given *i*-state) and described by an ordinary differential equation. **Examples:** aging, growth. The "creation-destruction" (including jump) processes are usually *stochastic* (some die some don't) but we will always assume that numbers are so large that we can describe these *chance* processes in terms of *rates* (like most of the time in chemistry and deterministic population dynamics). **Examples:** fission, removal from a chemostat by overflow. (Of course steps (i) and (ii) are not independent and one may actually go repeatedly back and forth between them before ending up with a satisfactory description of the *i*-dynamics).

(iii) Introduce a density function $n = n(t,x)$, t = time, x = i-state, to describe the *p-state* (*p*- for population) and do the necessary *bookkeeping* to derive how the *i*-processes can be expressed at the *p*-level. So $\int_\omega n(t,x)dx$ = number of individuals at time t with i-state in ω, where ω is any (measurable) subset of the i-state space. The ideal is, of course, to be able to compute $n(t,.)$ for $t > t_0$ when $n(t_0,.)$ is given. In finite time intervals all the different processes are interweaved in an intricate manner and this complicates the bookkeeping to such an extent that it becomes impossible to define the mappings $n(t_0,.) \rightarrow n(t,.)$ directly. The old solution to this problem is to take limits, i.e. compare $n(t_0+\delta,.)$ and $n(t_0,.)$ and let $\delta \downarrow 0$ to derive a partial *differential equation*. In small time intervals there will be hardly any compound events and in the limit these can be neglected all together. We end up with a partial differential equation which compactly summarizes the influence of the processes at the individual level on the population as a whole and which is based on the requirement that our bookkeeping should be correct.

Example: Let cells be characterized by their size x. Assume that reproduction occurs through asymmetric division into a part of fixed size x_0 (the daughter) and a part of size $x - x_0$ (the mother). The budding yeast *Saccharomyces cerevisiae* may serve as an example. Let $g(x)$ denote the growth rate of individual cells and $b(x)$ the probability per unit of time that a cell of size x divides. The balance law describing the time-evolution of the size- distribution is

$$\frac{\partial}{\partial t}n(t,x)+\frac{\partial}{\partial x}(g(x)n(t,x)) = -b(x)n(t,x)+b(x+x_0)n(t,x+x_0)$$

$$g(x_0)n(t,x_0) = \int_0^\infty b(\xi)n(t,\xi)d\xi$$

General structure:

$\frac{\partial n}{\partial t}$ + divergence (velocity n) = sources - sinks

ν . velocity $n|_{\partial\Omega_+}$ = source

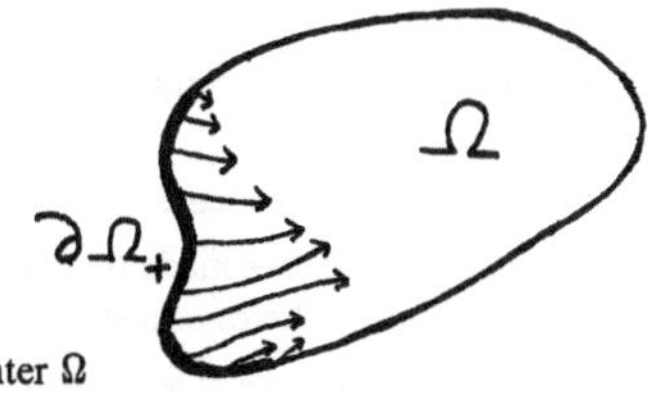

Ω = i-state space

ν = inward normal to $\partial\Omega$

$\partial\Omega_+$ = part of $\partial\Omega$ at which ν. velocity >0, i.e. characteristics enter Ω

Explanation: If we denote the i-variable by x then the trajectories of individuals are governed by $\frac{dx}{dt}$ = velocity. In mathematical terminology these trajectories are (the projections on Ω of) the characteristics. The source terms correspond to the "creation" processes and the sink terms to the "destruction" processes.

If we apply the above scheme to e.g. unicellular organisms (living, say, in a chemostat) we encounter a serious problem which is somewhat hidden in the general formulation: if we would try to incorporate all available knowledge about the individual level we end up, with absolute certainty, with an intractable problem (that is, one for which we cannot distil any interesting information out of the differential equation). So we have to compromise and already in the modelling phase we have to keep in mind that the resulting equation should be amenable to analysis. An obvious idea presents itself: use just one variable to describe the state of individual cells and preferably one which can, at least in principle, be measured. This variable should be a reliable indicator of a much more complicated "true" state which takes account of the detailed physico-chemical composition of the cell. As a mathematical idealization we *declare* it to be the true state, that is, we pretend that all relevant information is contained in this one and only variable. The example above concerning the cell size distribution is based on such considerations.

There exists yet another way out of the problem that realistic models tend to be so complex that they become resistant to mathematical analysis (and, in addition, so parameter rich that an experimentalist has to provide a prohibitive number of estimates). When the initial modelling stage is over we may simplify the model through limiting procedures based on i.a. time scale arguments. It is, in general, safer to first formulate a complicated model and only thereafter take limits then to construct the "limiting" models directly (see, for an example, Diekmann, Nisbet, Gurney & van den Bosch, to appear).

The aim of structured population models is to derive information about the dynamics of populations from information about the dynamics of individuals or vice versa. For instance, in human demography one wants to predict the population growth on the basis of age specific fertility and mortality statistics obtained from civil registration data. In cell kinetics, on the other hand, the experimental inaccessibility of the individual level creates *inverse* problems (e.g. to infer properties of the cell growth rate g and the fission rate b from data about the population growth rate and the stable size distribution, see Painter & Marr (1968), Bell & Anderson (1967), Tyson & Diekmann (to appear), Voorn (1983)).

To further illustrate the applicability of the "structured" approach, I briefly describe two classes of models by listing the *i*-state variables and the "change" and "creation/destruction" processes, while referring to papers in Metz & Diekmann (to appear) for the detailed modelling of these processes.

Kooijman (to appear) describes a *Daphnia* population on the basis of an energy budget. The *i*-state variables are size, age and storage, the "change" processes ingestion, maintenance, growth and storage and the "creation/destruction" processes reproduction (parthenogenetic) and death (due to aging, predation and possibly starvation). In other words, he derives a detailed model for the relation between food input and neonate output which may be used as a basis for population dynamics. The motivation stems from toxicity research. Toxicity tests are usually done under "abundant food" conditions and it is a priori unclear how the results can be extrapolated to "natural" conditions, depending on the particular physiological process being affected by the toxic chemical. In the context of the model one can actually perform the extrapolation (see Kooijman & Metz (1984) for such a study with a simpler model without storage; also see Diekmann, Metz, Kooijman and Heijmans, 1984; recently H.R. Thieme has made a profound mathematical study of the Kooijman- Metz model).

Sabelis & Laane (to appear) describe plant-herbivore-predator systems by characterizing individual patches (one plant or a collection of nearby plants) in terms of the amount of foliage, the number of herbivores in the patch and the number of predatory insects in the patch. The "change" processes are foliage growth, the eating of foliage and the subsequent conversion into new herbivores, the eating of herbivores by predatory insects and the subsequent conversion into new predators. The "creation/destruction" processes are the formation of herbivore colonies and the invasion of a predator into a herbivore colony. The basic question is: what is the relationship between *local* dynamics (in one patch) and *global* dynamics (in a large collection of plants). The motivation stems from biological control. Edelstein (to appear) discusses similar models for plant-herbivore interaction.

3. What is the appropriate p-state space ?

The introduction of Sobolev spaces has made the theory of (elliptic and) parabolic partial differential equations much more elegant and powerful. So the choice of a state space is not without importance.

The convential choice for the *p*-state space is $L_1(\Omega)$. The standard motivation is that integrals yield numbers. The following two observations suggest that this space is too small:

(i) In age-dependent models all individuals are born with age zero. In the L_1-context we can account for the neonate source term by a boundary condition, but when dealing with the variation-of-constants formula (to verify the principle of linearized stability or to study Hopf bifurcation) this creates all kinds of technical problems (see Webb (1985a), Prüss (1983); Schappacher (pers. comm.)). If, on the other hand, we consider the neonate source term as a measure concentrated at age zero (and identify L_1 with absolutely continuous measures) these technical problems disappear as snow in a hot sun (Clément, Diekmann, Gyllenberg, Heijmans & Thieme, in preparation).

(ii) Let the *i*-state be given by age and size and assume that individuals are born with a fixed size (e.g. the asymmetric cell fission model with age as another *i*-state variable).

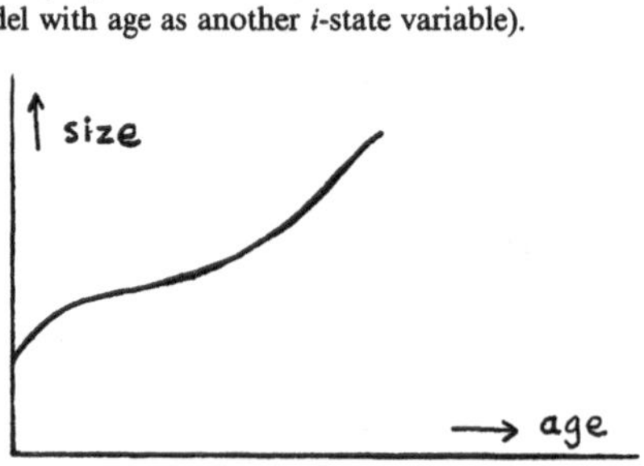

Then the distribution of individuals will concentrate on a line in the age-size plane. Whenever this line is fixed (i.e., whenever the food supply is constant) we can eliminate one variable (say size), but in general this is impossible. So we have to face the situation that with multi-dimensional *i*-state spaces the *i*-state distribution may be concentrated on lower dimensional manifolds.

Thus we are led to choose as our *p*-state space $M(\Omega)$, the space of regular Borel measures, in accordance with the probabilistic interpretation (Feller, 1966) and the standard motivation that integrals yield numbers. Being the dual space of the space of continuous functions on Ω, M has two natural topologies (the norm or total variation topology and the weak * topology) which both seem to play a role. In the case of a one-dimensional i-state space Ω, the absolutely continuous measures will usually constitute an *invariant* subspace, which brings us back to the L_1-setting. However, observation (i) above shows that even then it may be advantageous to consider explicitly the embedding into the larger space M, while observation (ii) shows that in general this subspace need not be invariant.

4. Linear analysis 1: integration-along-characteristics, dual semigroups and the variation-of-constants formula.

Structured population models lead to first order partial functional differential equations. The adjective "functional" is used to express that, as a rule, the source terms contain non-local arguments (e.g. $x+x_0$ in the cell fission equation). The occurrence of these terms complicates the construction of a solution , even in the linear case. A successful strategy to overcome this difficulty is the following. First we simply neglect the source terms (i.e., we put them equal to zero). The resulting easy problem can be solved explicitly by integration along characteristics. When working with measures this is done indirectly: one first solves the pre-dual or backward problem (Feller, 1966) (which is obtained by replacing the divergence (velocity n) term by -velocity . gradient n and changing the boundary condition appropriately) on the space of continuous functions and subsequently defines the forward solution operator by duality; see Heijmans (1984) for an example . Thus we obtain dual semigroups.

Next we add the source terms back in as a perturbation of the generator, replace the full differential equation by a variation-of-constants equation and solve the latter by successive approximations (Clément et al, in preparation). Biologically this amounts to the generation expansion! This program is not yet carried out in detail for cases in which the i-state space has dimension greater than one.

Linear semigroup theory is very useful since there is a known relationship between computable spectral properties of the generator and the large time behaviour of the semigroup (e.g. Webb, 1985a, Greiner, 1984).

5. Linear analysis 2: positivity and stable distributions.

Semigroups generated by population equations leave the cone of positive measures invariant. The structure of the spectrum of the generator of a *positive* semigroup is very special (Greiner, 1981, 1984) and consequently so is its large time behaviour. It appears that for linear problems the combination of positivity and irreducibility yields, as a rule, *stable distributions.* The following references contain some more information on this topic: Greiner (1981, 1984), Heijmans (1985, to appear) and Webb (1985 a,b). Sometimes one can actually compute the stable distribution explicitly and use this computation to analyse the inverse problem (Tyson & Diekmann, to appear).

6. Nonlinear and numerical analysis: interaction through the environment.

Frequently the linear structured models contain parameters, such as the concentration of an essential nutrient S, which will only in carefully controlled laboratory experiments be constant. Under more general conditions one has to supplement the equation for n with a dynamical equation for S such as

$$\frac{dS}{dt} = F(S, L[n])$$

where L is the consumption functional. In other words: we close an input -output loop and obtain a nonlinear feedback system. Our aim is to develop a local stability and bifurcation theory for such systems, by using dual evolutionary systems to describe the open input -output circuit, in combination with fixed point arguments.

In addition we want to develop efficient numerical methods for the study of such systems.

7. References

Auslander, D.M., Oster, G.F. and Huffaker, C.B. (1974). Dynamics of interacting populations. J. Franklin Inst. **297:** 345-376.

Bell, G.I. and Anderson, E.C. (1967). Cell growth and division. I. A mathematical model with applications to cell volume distributions in mammalian suspension cultures. Biophys. J. **7:** 329-351

Clément, Ph., Diekmann, O. Gyllenberg, M., Heijmans, H.J.A.M. and Thieme, H.R. (in preparation). Perturbation theory for dual semigroups.

Diekmann, O., Metz, J.A.J., Kooijman, S.A.L.M. and Heijmans, H.J.A.M. (1984). Continuum population dynamics with an application to *Daphnia magna.* Nieuw Archief voor Wiskunde (4) **2:** 82-109.

Diekmann, O., Nisbet, R.M., Gurney, W.S.C. and Bosch, F. van den (to appear). Simple mathematical models for cannibalism: a critique and a new approach. Math. Biosc.

Edelstein, L. (to appear). Models for plant-herbivore systems. J. Math. Biol.

Feller, W. (1966). An Introduction to Probability Theory and Its Applications. Vol. II, Wiley, New York.

Fredrickson, A.G., Ramkrishna, D. and Tsuchiya, H.M. (1967). Statistics and dynamics of procaryotic cell populations. Math. Biosc. **1:** 327-374.

Greiner, G. (1981). Zur Perron-Frobenius-Theorie stark stetiger Hallbgruppen. Math. Z. **177:** 401-423.

Greiner, G. (1984). A typical Perron-Frobenius theorem with applications to an age-dependent population equation. In: Infinite-dimensional Systems (F. Kappel and W. Schappacher, eds.) Springer Lect. Notes in Math. **1076:** 86-100.

Heijmans, H.J.A.M. (1984). Holling's 'hungry mantid' model for the invertebrate functional response considered as a Markov process. Part III. Stable satiation distribution. J. Math. Biol. **21:** 115-143.

Heijmans, H.J.A.M. (1985). Dynamics of Structured Populations. Ph. D. Thesis, University of Amsterdam.

Heijmans, H.J.A.M. (to appear). Structured populations, linear semigroups and positivity. Math. Z.

Heijmans, H.J.A.M. (to appear). Markov semigroups and structured population dynamics. Proceedings Symposium "Aspects of Positivity in Functional Analysis", Tübingen, 1985.

Kooijman, S.A.L.M. (to appear). Population dynamics on basis of budgets. In: Dynamics of Physiologically Structured Populations (J.A.J. Metz and O. Diekmann, eds.) Springer Lect. Notes in Biomath.

Kooijman, S.A.L.M. and Metz, J.A.J. (1984). On the dynamics of chemically stressed populations: the deduction of population consequences from effects on individuals. Ecotox. Env. Saf. **8:** 254-274.

Metz, J.A.J. and Diekmann, O.,eds. (to appear). Dynamics of Physiologically Structured Populations. Springer Lect. Notes in Biomath.

Painter, P.G. and Marr, A.G. (1968). Mathematics of microbial populations. Ann. Rev. Microbiol. **22:** 519-548.

Prüss, J. (1983). Stability analysis for equilibria in age-specific population dynamics. Nonlinear Analysis, TMA **7:** 1291-1313.

Sabelis, M.V. and Laane, W.E.M. (to appear). Regional dynamics of spider-mite populations that go locally extinct due to food source depletion and predation by Phytoseiid Mites *(Acarina: Tetranychidae, Phytoseiidae).* In: Dynamics of Physiologically Structured Populations (J.A.J. Metz and O. Diekmann, eds.) Springer Lect. Notes in Biomath.

Streifer, W. (1974) Realistic models in population ecology. In: Advances in Ecological Research (A. MacFadyen, ed.) **8:** 199-266.

Tyson, J.J. and Diekmann, O. (to appear). Sloppy size control of the cell division cycle. J. Theor. Biol.

Voorn, W.J. (1983). Statistics of cell size in the steady-state with applications to *Escherichia Coli.* Thesis. University of Amsterdam.

Webb, G.F. (1985 a). Theory of Nonlinear Age-Dependent Population Dynamics. Marcel Dekker, New York.

Webb, G.F. (1985 b). Dynamics of populations structured by internal variables. Math. Z. **189:** 319-336.

A Necessary And Sufficient Assembly Rule for Real Ecosystems

George Sugihara
Visiting Professor of Computer Science
Tokyo Institute of Technology,
O-okayama, 2-12-1, Meguro-ku Tokyo Japan

1. Introduction

I will demonstrate how a simple empirical regularity observed in the topology of real food webs may be used to mathematically deduce a necessary and sufficient constraint operating during system assembly. This differs from the classical bottom-up approach for studying ecosystem structure in that "plausible" assembly rules are not used to induce large scale patterns, but rather, a macroscopic regularity is used to deduce an assembly rule. This is an extension of results reported earlier on patterns in the graphical structure of food webs (Sugihara 1983,1984).

The specific regularity used to motivate this arguement is the observation that all intersection graphs G(C) formed from overlaps in consumer diets are triangulated (contain no circuits of length greater than 3). Such intersection graphs G(C) are nondirected graphs whose vertex set, $V=\{C_1,\ldots,C_p\}$, corresponds to consumer species, and whose nondirected edges E represent overlap in consumer resource use. Thus, Whenever two consumer C_i and C_j (subsets of resources) overlap with respect to at least one resource, an edge is drawn between them, i.e., $E=\{(C_i,C_j)$ whenever $C_i \cap C_j \neq \emptyset\}$. A connected consumer overlap graph G(C) is said to be triangulated, rigid circuit or chordal if every circuitous path $P_n \in G(C)$ of length $n \geq 4$ is shorted by a chord (Dirac 1961). That is, if all generated subgraphs of G(C) contain no more than triangular circuits (Rose et al. 1967).

In what follows, I will show how triangulation in G(C) can be used to induce a partial ordering on the consumer vertex set, which in turn leads to the following necessary and sufficient constraint operating during the assembly of ecological systems.

ASSEMBLY RULE. Species enter G(C) conservatively by joining single guilds (guilds = cliques in G(C)), rather than by bridging multiple guilds.

2. Derivation of Assembly Rule

The following properties of rigid circuit (triangulated) graphs lead to a partial ordering of the vertex set. This particular ordering may be used in gaussian elimination.

PROPERTY 1. (Dirac 1961) Every rigid circuit graph possesses at least one extreme vertex. That is, a vertex whose neighbors form a complete graph or clique.

PROPERTY 2. (Sugihara 1984) If G is a rigid circuit graph, then G-V is also rigid.

A perfect elimination ordering (p.e.o.) is an ordering on the vertex set V

$$\alpha : V \rightarrow \{1,2,\ldots,n\}$$

such that for every pair of vertices (v_i, v_j) not joined by an edge there exists no minimal path from v_i to v_j containing only v_i and v_j and vertices numbered less than $\min[\alpha(v_i), \alpha(v_j)]$ (Rose et al. 1976). Therefore collasping a graph by following a p.e.o. involves eliminating extreme vertices to generate the sequence of graphs

$$G_n, G_{n-1}, \ldots, G_1$$

having n,...,1 vertices.

LEMMA 1. (Rose et al. 1976) G is a rigid circuit graph if and only if it possesses a perfect elimination ordering.

The reverse of a perfect elimination ordering is a perfect addition ordering (p.a.o.),

$$\gamma(v_i)=n-\alpha(v_i)+1.$$

This generates the sequence of graphs

$$G_1,G_2,\ldots,G_n$$

by adding rather than deleting extreme vertices. Clearly, the following must be true.

LEMMA 2. G is a rigid circuit graph if and only if it possesses a p.a.o.

Note that the biological interpretation of a p.a.o. in G(C) is the conservative addition of consumer species to single cliques or guilds in G(C), i.e., the assembly rule stated earlier. Here, cliques in G(C) are taken to correspond to the biological notion of trophic guilds. In practice, they have been found to identify with the intersection or common use of one or more resources (Sugihara 1983).

The observation that real niche overlap graphs G(C) are triangulated guarantees the existence of a p.a.o., hence proves sufficiency. However, to say that such an ordering exists is not to say that real niche overlap graphs were constructed in this way. The following arguement leads to the necessity of such an ordering.

It is conventional wisdom that species tend to enter communities in order of increasing (non-decreasing) specialization. Suppose specialization at the ith stage is associated with fewer niche overlaps or competitors in $G_i(C)$. A specialist, therefore, would have lower point degree (fewer radiating edges) than a generalist. Suppose further that each species subsequent to the ith one is a greater (or equal) specialist with respect ot the species already present at the ith stage $(v_1,\ldots,v_{i-1})$. That is, no subsequent species v_j, $j>i$, will overlap more of the vertices $(v_1,\ldots,v_{i-1})$ than v_i. This rule for adding vertices to a rigid circuit graph has the following property.

THEOREM 1: If $G_1,G_2,\ldots,G_i,\ldots,G_n$ are connected rigid circuit graphs, and each vertex v_i at the ith stage has lower or equal point degree than any other vertex in G_i and is adjacent to more or as many vertices in G_{i-1}, $(v_1,...v_{i-1})$ as any subsequent vertex, then v_i is extreme and $i=1,...,n$ is a p.a.o.

Proof: It should be clear that such a rule for constructing connected triangulated graphs induces a partial ordering P on the vertex set of each G_i corresponding to the inverse of the point degree of vertices in G_i (highest label = lowest point degree). This labeling has the property that for every v_i, $i \neq 1$, there is a strictly non-decreasing path from v_1 to v_i. Furthermore, if at each step G_i is rigid and v_i is adjacent to more vertices in G_i than any subsequent vertex then all vertices in G_i adjacent to v_i, $\{v_{adj}\}_i$, must be adjacent to each other, $\{v_{adj}\}_{adj}$. If not, there must be a vertex v_j, $j>i$, having higher degree relative to the vertices in G_{i-1} than v_i (cf. Sugihara 1984 for details).

Thus this argument for increasing specialization plus the triangulation property of real niche overlap graphs, necessarily leads to perfect addition as the mechanism for sequential construction of G(C). Once again, the biological interpretation of this assembly rule is that incoming species must enter conservatively by attaching as extreme vertices to a single guild (clique in G(C)) rather than by bridging multiple guilds.

3.Validation

It is important to note that necessity here depends on increasing specialization as an assumption. It is desirable, therefore, to seek independent ways of testing perfect addition of specialists as the true mechanism for assembling natural ecosystems. Apart from direct observations on colonization sequences for which data are not generally available, one can obtain indirect evidence for the assembly rule by investigating its predicted topological properties for which data are readily available.

3.1 Intervality and The Absence of Holes

Two other topological regularities that have been discussed in the literature in addition to the rigid property of G(C) are the high frequency of intervality in G(C) (Cohen 1977), and the complete absence of holes observed in the simplical complex representation of the niche K(R) (Sugihara 1983). Both of these properties have been described in depth elsewhere (Cohen 1977, Sugihara 1984) and their technical details will not be repeated here. Of relevance to this discussion however, is the fact these properties are extremely diffucult to reproduce in random systems and therefore place real ecological systems in an exceedingly narrow subset of mathematical possibilities.

One can test the validity of the above assembly rule by seeing if it can succeed in duplicating each of these properties. Numerial implementation of random p.a.o.'s has yielded virtually identical frequencies of intervality in G(C) (63 observed interval graphs, versus 62.36 predicted, out of a pool of 72 food webs). Cohen(1977), who originally discovered this property was unable to duplicate the observed frequencies in over eight different plausible assembly routines. Furthermore, it has been possible to show analytically that a p.a.o. precludes the existence of holes in K(R) (Sugihara 1984). Therefore, this simple assembly rule appears to be capable of reproducing these difficult patterns. This may be taken as strong evidence for its general validity.

3.2 Monotone Root Transitivity

Further evidence can be obtained through the observation of another predicted property not previously reported. This property, defined below, follows as a corrolary to theorem 1.

DEFINITION: A labeled graph G is monotone root transitive (m.r.t.) iff there is a spanning subtree T of G such that there exists a path $P \varepsilon T$ from the root to every terminal branch which is monotone decreasing (increasing) in vertex labels (Figure 1).

If G is decreasing m.r.t. the root is identified with any vertex having the largest label. If G is increasing m.r.t. the root can be any vertex having the smallest label.

Corrolary 1. If G is triangulated and constructed as in theorem 1, and it's vertices are labeled by the inverse of their degree (highest degree = lowest label), then G will be increasing m.r.t.

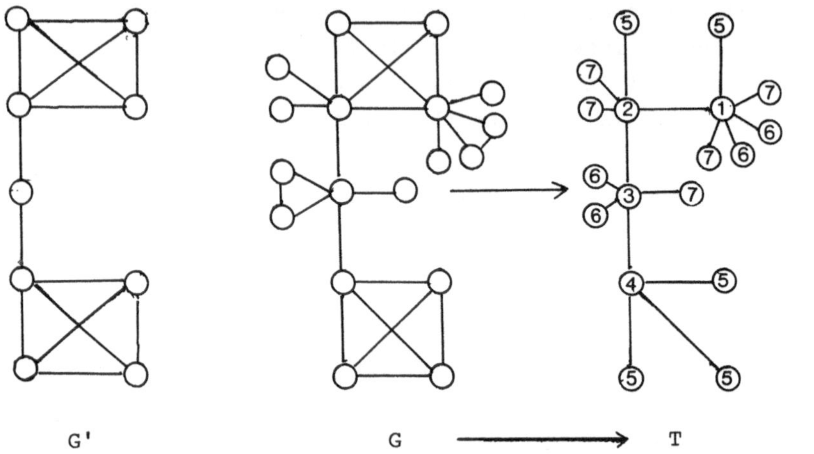

Figure 1. G' is not monotone root transitive. G is m.r.t. The generated spanning tree T of G has vertices labeled inversely to the point degree of G ie., highest point degree recieves the lowest label. G is increasing m.r.t. because there exists a T such that every path from the root (vertex label 1) to any terminal branch is non-decreasing in vertex labels.

Clearly G so labeled is not increasing m.r.t. iff there is a vertex of lower degree on every path between two vertices of higher degree (see Figure 1). The collection of 72 niche overlap graphs constructed from data gathered by Briand(1983) from the open literature, have all been found to be decreasing monotone root transitive, as predicted by corrolary 1.

Although this result appears to further strengthen the assembly rule, at least two specific examples within this food web collection appear contradictory. Both of these cases (Bird's 1930 Aspen Parkland web and the Cochin estuary web) have man as the root of G(C) (i.e. as the first successful colonist) which may not be reasonable. Nonetheless the fact that this deduced assembly rule is capable of reproducing difficult features of real systems, and indeed of predicting new ones, suggests that this simple rule for the conservative addition of species may be a good generalization. Furthermore, the assumption of increasing specialization, required to guarantee necessity, may be a reasonable rule of thumb.

4. Summary

A rule for the conservative addition of species (perfect addition) is derived from the triangulation property of real niche overlap graphs, G(C). This rule is shown to be both necessary and sufficient when the assumption is made that species enter the system sequentially in order of increasing specialization. Three consequences of this rule are discussed and a new predicted property, monotone root transitivity, is suggested and confirmed in the data for real systems.

5. REFERENCES

1) Bird,R.D.,Biotic communities of the aspen parkland of Central Canada. Ecology 11(1930),356-442.
2) Briand,F.,Environmental control of food web structure, Ecology(1983).
3) Cohen,J.E.,Food Webs and Niche Space. Princeton Univ.,Press, Princeton,1978.
4) Dirac,G.A.,On rigid circuit graphs,Abh,Math,Sem.Univ. Hamburg 25 (1961),71-76,
5) Rose,D.J.,R.E.Tarjan,G.S.Leuker,Algorithmic aspects of vertex elimination on graphs,SIAM J.Comput. 5(1976),266-283.
6) Sugihara,G.,Niche Hierarchy: Structure,Organization, and Assembly in Natural Communities. Ph.D.Dissertation, Princeton Univ.(1982).
7) Sugihara,G.,Graph Theory, homology and food webs. In Proceedings of Symposia in Applied Mathematics. S.Levin ed. Springer 30 (1984),83-101.

MARKOVIAN FORAGING MODELS

Colin W. Clark
Institute of Applied Mathematics
University of British Columbia
Vancouver, B.C. V6T 1Y4
Canada

Marc Mangel
Departments of Agricultural Economics, Entomology, and Mathematics
University of California
Davis, CA 95616

The problems facing a foraging animal have been studied on the basis of a variety of mathematical models. MacArthur and Pianka (1966) used marginality arguments familiar in economics to study optimal patch selection and optimal prey selection within patches. Fretwell (1972) discussed the equilibrium distribution of foragers among patches. Charnov (1976) characterized the problem of moving from one patch to another in terms of a Marginal Value Theorem, also familiar in economics. Schoener (1971) modelled territorial behavior. The predictions of these models have been tested in numerous experiments, with varying degrees of agreement between theory and observation (Krebs et al 1983).

While recognizing that many of the problems facing foragers are a consequence of the patchiness of food distributions, these early models were either purely deterministic, or else simply considered long-term averages of stochastic processes, relying in the latter case upon the renewal theorem (Feller 1971). By the late 1970s, however, the limitations of such models were becoming apparent. Oaten (1977) developed a stopping-rule model to study the role of learning in the problem of patch switching, and showed that the predictions of his model could be markedly different from those of Charnov's deterministic model (see also Green 1980, Iwasa et al 1981, Clark and Mangel 1984).

The role of risk preference in foraging behavior was studied by Caraco (1981a), who showed by a simple argument that a forager should be risk averse whenever expected food discoveries exceed requirements, and vice versa (see also Stephens and Charnov 1982). This obvious but important observation stresses the need to consider the influence of the *state* of the forager on foraging and other behavioral decisions.

Markovian Models. Dynamic stochastic optimization models of foraging behavior have been discussed by McFarland and Houston (1981) and McFarland (1982), and in a series of papers by Houston and McNamara (e.g., 1985). Here we shall describe a simple Markov decision process framework for the modelling of foraging and alternative behavior; further details are given in Mangel and Clark (1986).

Let $X(t)$ represent the internal state of the forager at time t. Specifically, we treat $X(t)$ as a single variable--representing, for example, the forager's energy reserves--although a multidimensional state variable would often be more realistic. A continous-time model of state dynamics can be constructed in terms of the jump process

$$dX = \alpha_s(X)dt + Z_s d\pi_s \tag{1}$$

where $\alpha_s(X) < 0$ denotes the rate of decrease in enery reserves due to metabolic processes, Z_s denotes the energy content of food discovered (a random variable with density f_s), and $d\pi_s$ is the jump process specified by

$$\begin{aligned} Pr(d\pi_s = 1) &= \lambda_s dt \\ Pr(d\pi_s = 0) &= 1 - \lambda_s dt \end{aligned} \tag{2}$$

The subscript s refers to the particular foraging strategy $s \in \Omega$ selected at time t. In a model of group foraging strategy, for example (Clark and Mangel 1984, 1986), s would be an integer denoting foraging group size, and a reasonable parameter specification would be

$$\lambda_n = n\lambda \ , \quad Z_n = Z/n \ , \quad \alpha_n(X) = \alpha \tag{3}$$

Over any time interval that does not involve breeding activities, a proxy for expected fitness is the probability of survival. Let $p(x,t)$ denote the maximum probability of survival time t, given $X(0) = x$. It can be shown that $p(x,t)$ satisfies the dynamic programming equation (Mangel 1985):

$$p_t + \alpha p_x = \max_s E\{\lambda_s p(x + z_s, t) - \lambda_s p(x,t)\} \tag{4}$$

with initial condition

$$p(x,0) = \begin{cases} 1 & \text{for } x > 0 \\ 0 & \text{for } x = 0 \end{cases} \tag{5}$$

where $x = 0$ is assumed to correspond to starvation of the forager. This partial differential-functional-integral equation will not be discussed further here. Instead we pass immediately to a discrete-time Markovian framework which offers the dual advantages of simplicity and flexibility.

With $X(t)$ still representing the energy state, Eq. (1) is now replaced by

$$X(t + 1) = X(t) - \alpha_{st} + Z_{st} \tag{6}$$

The distribution of food discoveries Z_{st} is given by

$$Pr(Z_{st} = y_{si}) = \lambda_{si} \ , \quad i = 1, \ldots, i_s \tag{7}$$

We will assume that the energy state is constrained by

$$0 \le X(t) \le C \tag{8}$$

where $X(t) = 0$ implies death by starvation and C is the energy capacity of the forager.

Consider first the specification $p(x,t)$ = maximum probability to survival to period t, given $X(0) = x$. Then

$$p(x,) = \begin{cases} 1 & \text{for } x > 0 \\ 0 & \text{for } x \le 0 \end{cases} \tag{9}$$

and an elementary argument establishes the dynamic programming equation

$$p(x,t+1) = \max_s \Sigma_i \lambda_{si} p(\langle x - \alpha_s + y_{is} \rangle_C, t) \tag{10}$$

where $\langle a \rangle_C = \min(a,C)$. Equation (10) is easily solved by numerical iteration.

In order to indicate the ease with which this basic Markov decision process model can be adapted to other situations, suppose that strategy s involves a risk of predation β_s. Then (10) becomes

$$p(x,t+1) = \max_s (1-\beta_s)\Sigma_i \lambda_{is} p(\langle x - \alpha_s + y_{is} \rangle_C, t) \tag{11}$$

Similarly, suppose that fitness is given in terms of a terminal payoff function $\phi(x)$ --for example, ϕ could itself be determined from a separate MDP model of the breeding period. Then the initial (really, terminal) condition (9) is replaced by

$$p(x,0) = \phi(x) \tag{12}$$

Other situations may involve reformulation of the state equation (6) or the dynamic programming equation (11); in most cases the appropriate changes are obvious from the desired model structure. Several examples are given in Mangel and Clark (1986). Two examples utilizing real data will be described in the following.

Lion Foraging Groups. The optimal size of groups of lions foraging in the Serengeti has been studied by Caraco and Wolf (1975), based on data of Schaller (1972). The probability of a successful chase, for zebra or wildebeest, as a function of group size, is shown in Fig. 1. Multiplying by e/n (where e = average edible biomass per prey) gives the curve of expected edible biomass per lion per day shown in Fig. 2 .

According to to these curves, lions would maximize their expected food intake per chase by hunting in groups of size $n = 2$. The

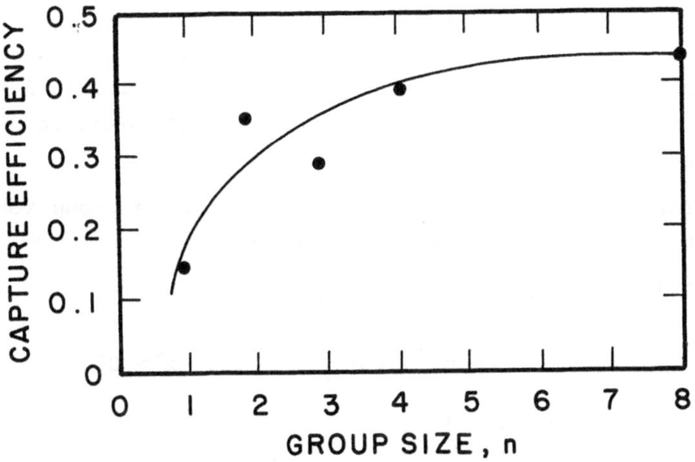

Figure 1. Probability of successful chase of zebra or wildebeest as a function of lion group size (Caraco and Wolf 1975).

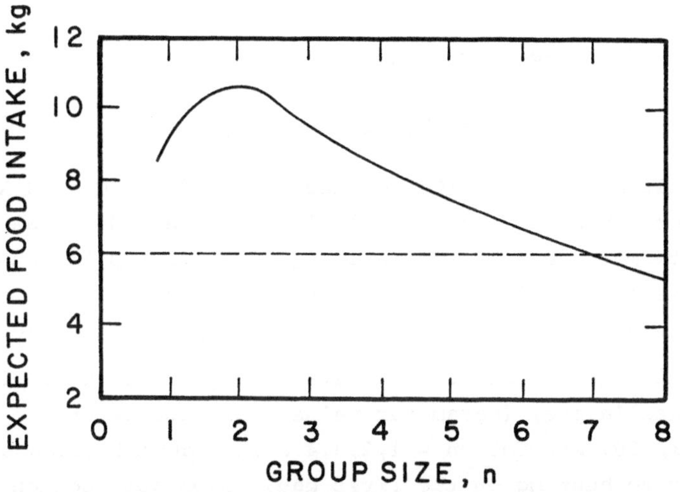

Figure 2. Mean edible prey biomass per lion per chase as a function of lion group sizes preying on zebra. (Caraco and Wolf 1975).

average observed group size was $\bar{n} = 7.3$.

Several hypotheses have been advanced to explain the apparent discrepancy between the apparent optimum and the observations for the case of zebra (and wildebeest). Caraco and Wolf (1975) suggested that other ecological factors, such as breeding success and competition with hyenas, would lead to larger optimal group sizes. Rodman (1981) suggested that kin selection would also favor larger group sizes. Clark and Mangel (1984) suggested that competition between lions could lead to supra-optimal group sizes. None of these hypotheses fully explain the observed group sizes, however.

Our Markovian model can be applied directly to the group foraging problem (Clark 1986). Specifically, the model is

$$p(x,t+1) = \max_n [\lambda_n p(\langle x - \alpha + e/n \rangle_C, t) + (1-\lambda_n) p(x-\alpha, t)] \qquad (\text{for } x > 0) \tag{13}$$

$$p(x,0) = \begin{cases} 1 & \text{for } x > 0 \\ 0 & \text{for } x \le 0 \end{cases} \tag{14}$$

in which the symbols have the following interpretation

t = number of days until end of nonbreeding season
x = individual lion's gut contents (kg)
C = gut capacity (kg)
e = edible biomass of prey (kg)
α = daily food requirement (kg)
n = group size
λ_n = successful kill probability, per day.

Parameter values used were (Caraco and Wolf 1975): $C = 30$ kg, $\alpha = 6$ kg, $e = 50$ kg for zebra (the actual calculations treated e as a random variable with mean $\bar{e} = 50$ kg), and $\lambda_n = 1 - (1 - p_n)^3$, where p_n are the per-chase probabilities given in Fig. 1. Thus we assume that lions can undertake up to three chases per day, stopping after the first kill.

The numerical results for $t = 30$ days are given in Table 1 for both zebra and gazelle prey (parameter values for gazelle are $e = 15$ kg, $p_n = .15, .28, .30, .31$ for $n = 1,2,3,4$). The optimal group sizes n^* of course refer to hunting on the first day, given gut contents x. In fact the same dependence upon x holds for all $t \geq 10$ days; for $t < 10$ days horizon effects alter the relationship between x and n^*.

Note that for zebra prey n^* is always larger than the "optimal" value $n^* = 2$ suggested from Fig. 2. The reason for this is fairly clear when x is large: a successful chase with $n = 2$ lions would provide each lion with 25 kg of zebra meat, far in excess of its capacity. By hunting in a larger group, each lion increases its chances of obtaining its daily requirement. When x is low it is better to hunt in a smaller group, but it is not generally optimal to try to make up the entire deficit $(C + \alpha - x)$ in a single day.

	zebra		gazelle	
x	n^*	p(x,30)	n^*	p(x,30)
5	4	.71	1	.09
10	4	.91	1	.17
15	4	.96	1	.23
20	4	.98	1	.29
25	5	.98	1	.34
30	6	.99	2	.36

Table 1. Optimal hunting groups sizes n^* and probability of survival $p(x,30)$ for lions hunting zebra and gazelle in the Serengti as functions of initial gut contents x (Clark 1986).

The Markovian model not only provides improved predictions of the observed size of groups of lions chasing zebra (groups larger than 6 were not allowed in our calculation), but also provides the additional prediction that hungry lions will hunt in smaller groups. It turns out to be quite important for lions to adjust group size to the current state (gut contents): for example, a lion that always hunts in a group of size 6 would still have a 96% chance of surviving 30 days if $x = 30$ kg, but only a 38% chance when $x = 5$ kg.

The same heuristic reasoning applies to the results for Thomson's gazelle. With $x = 30$ kg a lion can meet its daily need with half a 15 kg gazelle, but not a third; when $x \le 28$ kg, however, half a gazelle is not enough, and the optimal group size becomes $n^* = 1$. (Caraco and Wolf did not comment on the fact that in some cases lions hunted gazelle in apparently less than optimal sized groups. The Markovian model explains this observation also.)

The maximum daily expected food intake per lion is 14.5 kg for zebra prey and 4.7 kg for gazelle. (We have not considered mixed hunting.) The analysis of Caraco (1981) would thus indicate that lions should be risk averse (large groups) in hunting zebra, and risk prone while hunting gazelle. This prescription agrees qualitatively with our Markovian predictions.

<u>Mean-variance Tradeoffs</u>. Many species of animals forage in groups. In some cases group foraging may be simply an aggregative response to food

concentrations, while in other cases it appears to be primarily an anti-predator strategy (Pulliam and Caraco 1984). But in cases where food patches are larger than the capacity of individual animals, group foraging can also significantly increase average feeding rates (Clark and Mangel 1986). Furthermore, by reducing the variance in food intake, group foraging can increase survival probabilities, even possibly at the expense of mean feeding rates (Oster and Wilson 1978, Real 1980, Caraco 1981, Stephens and Charnov 1982).

The question arises whether the compromise between mean and variance is necessary. Could variance be reduced without sacrificing mean food intake? Not unless some form of communal sharing of food is possible. Although we do not believe that lions participate in communal sharing of their kills, it is interesting to speculate on the potential benefits of such behavior.

Consider a pride of 12 lions feeding on zebra. The main daily individual consumption, and its coefficient of variation, are given in Table 2 for various hunting group sizes n. The last row shows the case $n = 2$ with communal sharing.

Group size	Sharing	Mean daily consumption	Coeff. of variation
2	no	14.0 kg	0.89
4	no	8.8 kg	0.65
6	no	6.8 kg	0.48
2	yes	14.0 kg	0.36

Table 2. Mean daily individual consumption and coefficient of variation for members of a pride of 12 lions hunting zebra in groups of various sizes n.

Sharing clearly has the potential for significant reduction of variance, while maintaining a high mean. An optimization model based on communal sharing will be reported elsewhere.

Parasitic Wasps. Animals forage for things other than food, including mates, and in the case of insects, oviposition sites. Charnov and Skinner (1984,1985) have discussed oviposition strategy of parasitoid wasps, using a model analogous to the model developed by Lack (1947) for studying clutch size in birds. According to Lack's model, fitness is a dome-shaped function of the number of egges laid in a given clutch. Consequently fitness is maximized at some intermediate clutch size C^*. Charnov and Skinner estimated C^* for two species of parasitoid wasps, as a function of host volume V. For the case of *Nasonia vitripennis*, they obtained roughly a linear relationship $C^*(V) = kV$.

The data, however, disagreed with the predictions: wasps laid clutches of all sizes $C \le C^*(V)$, while almost completely avoiding clutch sizes larger than $C^*(V)$. In order to explain the discrepancy, Charnov and Skinner suggested that average fitness per host might not be an appropriate measure of a mother's total fitness. If $W(C)$ denotes the contribution to fitness from a single clutch size C, consider the objective of maximizing the gain in fitness per unit time:

$$\max_{C} \frac{W(C)}{r(C) + t_s} \tag{15}$$

where $r(C)$ is handling time, and t_s search time. If $r(C)$ is increasing in C, then the solution of (15) is smaller than the value of C^* which maximizes $W(C)$.

The revised model explains why observed values of C are less than C^* --indeed, C^* becomes an error by this argument. But it fails to explain the spread in the data, which covered the range $[0,C^*]$ more or less evenly. The following simple Markovian model (Mangel 1986; see also Iwasa et al 1984) explains everything.

Assume that an insect is born with a fixed capacity R for lifetime egg production (other cases can be modelled with similar results). Let X_t denote the number of eggs remaining to be laid at time t, and let $W(C)$ be the expected contribution to fitness from a clutch of size C. Assume W concave. Let λ denote the probability of finding a host per time period (Mangel's model allows a range of host sizes). If $p(x,t)$ denotes maximum expected total fitness over t periods, given $X(0)=x$, we have

$$p(x,0) = \lambda \max_{0 \le C \le x} W(C) \tag{16}$$

$$p(x,t+1) = \lambda \max_{0 \le C \le x} [W(C) + p(x - C,t)] + (1 - \lambda)p(x,t) \tag{17}$$

Since $p(x,t)$ is nondecreasing in x, it follows immediately from (17) that the number of eggs laid in any time period is $\le C^*$, and $< C^*$ if p is increasing. Furthermore, the stochastic nature of the discovery process implies that the number of eggs laid will vary over time. In particular, the longer the remaining time t, the fewer eggs should be laid per host (clearly for $t = \infty$ only *one* egg should be laid per host).

Simple as it is, this Markovian model provides a completely different picture of oviposition strategy from the static average model of Charnov and Skinner. Many other realistic details could easily be included in the Markovian model, such as: probability of predation, simultaneous foraging strategy, continuous egg production depending on energy state, and multiple host types (including preparasitized hosts). In the latter case, it can prove to be optimal to lay *no* eggs on an inferior host (even if $W > 0$).

While it might be possible, with difficulty, to model such decision problems in a static framework, we are convinced that the predictions would be next to worthless. Nature is not static, and models that assume otherwise miss much of nature's richness.

References

Caraco, T. 1981. Risk sensitivity and foraging groups. Ecology 62: 527-531.

Caraco, T., and Wolf, L.L. 1975. Ecological determinants of group sizes of foraging lions. Amer. Nat. 109: 343-352.

Charnov, E.L. 1976. Optimal foraging: the marginal value theorem. Theor. Pop. Biol. 9: 129-136.

Charnov, E.L., and Skinner, S.W. 1984. Evolution of host selection and clutch size in parasitoid wasps. Florida Entomol. 67: 5-21.

Charnov, E.L., and Skinner, S.W. 1985. Complementary approaches to the understanding of parasitoid oviposition decisions. Environ. Entomol. 14: 383-391.

Clark, C.W. 1986. The lazy, adaptable lions: a Markovian model of group foraging. Anim. Behav. (to appear).

Clark, C.W., and Mangel, M. 1984. Foraging and flocking strategies: information in an uncertain environment. Amer. Nat. 123: 626-641.

Clark, C.W., and Mangel, M. 1986. The evolutionary advantages of group foraging. Theor. Pop. Biol. (in press).

Feller, W. 1971. An Introduction to Probability Theory and Its Applications, Vol. 2. Wiley, New York.

Fretwell, S.D. 1972. Populations in a Seasonal Environment. Princeton Univ. Press, Princeton, NJ.

Green, R.F. 1980. Bayesian birds: a simple example of Oaten's stochastic model of optimal foraging. Theor. Pop. Biol. 18: 244-256.

Houston, A.I., and McNamara, J.M. 1985. The choice of two prey types that minimises the probability of starvation. Behav. Ecol. Sociobiol. 17: 135-141.

Iwasa, Y., Higashi, M., and Yamamura, N. 1981. Prey distribution as a factor determining the choice of optimal foraging strategy. Amer. Nat. 117: 710-723.

Iwasa, Y., Suzuki, Y., and Matsuda, H. 1984. The theory of oviposition strategy of parasitoids I. Effect of mortality and limited egg number. Theor. Pop. Biol. 25: 205-227.

Krebs, J.R., Stephens, D.W., and Sutherland, W.J. 1983. Perspectives in optimal foraging, in Brush, A.H., and Clark, G.A. (eds.), Perspectives in Ornithology. Cambridge Univ. Press, Cambridge, pp. 165-221.

Lack, D. 1947. The significance of clutch size. Ibis 89: 302-352.

MacArthur, R.H., and Pianka, E.R. 1966. On the optimal use of a patchy environment. Amer. Nat. 100: 603-609.

Mangel, M. 1985. Decision and Control in Uncertain Resource Systems. Academic Press, New York, 255 pp.

Mangel, M. 1986. Oviposition site selection and clutch size in parasitic insects. Anim. Behav. (in press).

Mangel, M., and Clark, C.W. 1986. Unified foraging theory. Ecology (in press).

McFarland, D. (ed.) 1982. Functional Ontogeny. Pitman, London.

McFarland, D., and Houston, A.I. 1981. Quantitative Ethology: the State Space Approach. Pitman, London.

Oaten, A. 1977. Optimal foraging in patches: a case for stochasticity. Theor. Pop. Biol. 12: 263-285.

Oster, G.F., and Wilson, E.O. 1978. Caste and Ecology in the Social Insects. Princeton Univ. Press, Princeton, NJ.

Pulliam, H.R., and Caraco, T. 1984. Living in groups: is there an optimal group size? Ch. 5 in Krebs, J.R., and Davies, N.B. (eds.), Behavioral Ecology: An Evolutionary Approach. Sinauer, Sunderland, MA, pp. 37-64.

Real, L.A. 1980. On uncertainty and the law of diminishing returns in evolution and behavior. In Staddon, J.E.R. (ed.). Limits to Action: the Allocation of Individual Behavior. Academic Press, New York, pp. 37-64.

Rodman, P.S. 1981. Inclusive fitness and group size, with a reconsideration of group sizes of lions and wolves. Amer. Nat. 118: 275-283.

Schaller, G.B. 1972. The Serengeti Lion. Univ. Chicago Press, Chicago, IL.

Schoener, T.W. 1971. Theory of feeding strategies. Ann. Rev. Ecol. System. 2: 369-404.

Stephens, D.W., and Charnov, E.L. 1982. Optimal foraging: some simple stochastic models. Behav. Ecol. Sociobiol. 10: 251-263.

The Theory of Population Dynamics: Back to First Principles

Lev R. Ginzburg

Department of Ecology & Evolution
State University of New York at Stony Brook
Stony Brook, New York 11794, U.S.A.

Models of population dynamics are based on the population size as the complete descriptor of the dynamic state. We can write this central assumption of traditional theory as follows:

$$\frac{1}{N}\frac{dN}{dt} = f\ (E)$$

where N is the population size;

$\frac{1}{N}\frac{dN}{dt}$ is the relative growth rate (average number of surviving offspring per parent per unit of time);

$f\ (E)$ is a function of the environment with the understanding that population size itself might be one of the environmental parameters.

To construct a community model, we write such equations for each of the populations in the ecosystem. The environment, E, for a given population includes sizes of other populations, and that is how we obtain the system of equations designed to describe the dynamics. The differences between models are due to different assumptions concerning the functions f.

Let me call "constant" all environments in which a given population increases (or decreases) exponentially. There is no external way to determine such a constancy. One can, in principle, do it by enumerating: food is sufficient, oxygen is present in necessary quantities, temperature is constant, etc. But there is no need to enumerate the separate causes. Environmental <u>invariability for a given population</u> from the population--dynamics point of view can be established by examining the population itself, since only if the population size is constant or changes exponentially is the environment invariable.

I will write down the Malthusian law $N\ (t) = N_0e^{rt}$ in a slightly unusual form:

$$\frac{d^2}{dt^2}\ (\ell nN) = 0.$$

Paraphrasing Hutchinson (1975): Populations preserve exponential growth unless they do not. Although sounding tautological, this bears, in my opinion, a meaning analogous to Newton's first law, describing what happens when "nothing happens in the environment". In the suggested form, the equation has no parameters and requires two initial conditions: population size and the growth rate. In this paper I will try to reconstruct the principles of population dynamics based on an extended notion of the dynamic state.

Let us start by considering the statics. On a very gross scale, we can say that population sizes that are observed in ecosystems are dictated by the environment, so that if the environment is changed, the sizes change accordingly. This statement ignores dynamics completely but can be accepted on a very crude level as a reasonable approximation to reality. Of course, even in a static case, interrelations between species can be taken into account, and changing the environment for one of the populations will change the sizes of other populations involved in the network of mutual interactions.

The next step is to consider disturbances and to assume that birth and death rates of the species, rather than their abundances, are defined by the environment, and that changes of the environment cause the growth rates to deviate from zero. This will lead us to the traditional rate equations.

Accepting exponential growth as a nondisturbed motion and assuming that deviations of the environment cause changes in the growth rates is another step in the same direction. I find it totally impossible, on purely logical grounds, to distinguish between the last two ways of describing dynamics. When a growing population depletes the food supply, is it the growth rate that is itself changing as a function of the declining supply or is it the change in the growth rate that is created by the food supply being depleted? Either view seems acceptable. The difference between the two views is, however, quite substantial in certain cases.

With the new view, the growth rate becomes another independent variable which should be supplied in addition to the dynamic equations as an extra initial condition. The question of why a given population in a given environment has a certain growth rate is never addressed. The explanations concentrate only on how growth rate will change given that it is set a certain initial value. An attempt to construct such a second order equation benefits from its generality (it will describe the essence of the interaction and be valid in a variety of environments) at the expense of not addressing a universe of questions which are reasonable but too hard to answer. Most of us are so used to the rate-based equations that it takes an effort to see the actual impossibility of rejecting the proposed step outright.

The most striking differences between the two views appear in the description of predator-prey interactions, which will be considered next.

Let us mentally place a pair of interacting predator and prey populations in an ideal unlimited environment. By unlimited, I imply the absence of density-dependence of any sort so that both populations grow (or die) exponentially in the absence of interaction. Although it might be hard to imagine such an environment where all resources expand freely following population growth, this is a necessary step to take in order to see the essence of my argument. Adding limitations will, of course, stop exponential expansion of our populations and provide a degree of necessary realism.

Let me clarify further the concept of an unlimited environment. Sometimes, even when a population grows exponentially, it is said that the growth is limited by a particular substrate that is consumed by the growing population. We mean that increasing the concentration of the substrate in the medium will increase the population growth rate. In this sense, growth is called "limited" by the concentration. This is not the sense in which I use this term. I will call all environments where a particular population grows exponentially unlimited. Thus, for a given population, one can have an infinite variety of unlimited environments characterized by different growth rates. For our purposes limitations will appear only when a resource is being used up in the process of population growth and, thus, the growth itself deviates from exponential. A chemostat with bacteria as prey, and protozoa as predators is the closest effective analogy to the abstract environment I have described.

The proposed model has the following form:

$$\frac{d^2}{dt^2}(\ell n\ N_1) = -G\left(\frac{N_1}{N_2}\right)$$

$$\frac{d^2}{dt^2}(\ell n\ N_2) = F\left(\frac{N_1}{N_2}\right) \qquad [1]$$

Here rates of change of Malthusian parameters, G and F, are affected by the ratio of population sizes. The model is invariant with respect to the proportional changes of both population sizes. Moreover, it is invariant with respect to an arbitrary "Malthusian transformation": that is, multiplication of both sizes by the same exponential function. Initial growth rates, r_{10} and r_{20} should be supplied in addition to initial population sizes, and they reflect the properties of the particular unlimited environment. The right-hand sides of Equations (1) are qualitatively shown on Fig. 1.

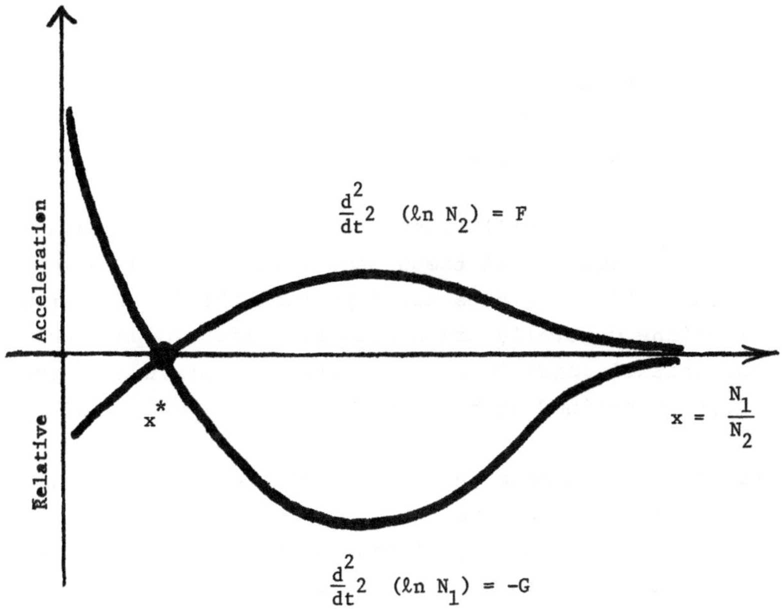

Fig. 1

Accelerations of the ℓnN (rates of change of the Malthusian parameter) as a function of relative abundance of prey and predator according to the proposed model; x^* is the equilibrium ratio of two populations.

Consider the "forces" acting on two populations as shown on Fig. 1. If the initial relative abundance,

$$x = {}^{N_1}/_{N_2}$$

is close to the balanced growth ratio, and $r_{20} - r_{10}$ is relatively small, we can linearize our equations around the equilibrium ratio and obtain $x(t)$ oscillating around the mean value, x^*, while both populations grow (on the average) exponentially with the exponent depending on the intial growth rates. If the initial position is imbalanced favoring predators, ${}^{N_1}/_{N_2} \ll x^*$ and $r_{10} < r_{20}$, we may obtain extinction of prey or predator or both. The precise sequence of events in this case will depend on the absolute values of N_{10} and N_{20}. If, on the contrary, ${}^{N_1}/_{N_2} \gg x^*$, then the prey may effectively "escape" predation and $\frac{N_1}{N_2} \to \infty$, so that the prey will asymptotically attain its independent growth rate while the predator will grow (or decline) with a lower exponent.

Let us summarize all the properties of the proposed model that are different from those of traditional rate based models:

* The outcome of the interaction depends on both initial sizes of the populations and their growth rates.

** In the case of coexistence, oscillations are superimposed on balanced exponential growth. The frequency of these oscillations at the neighborhood of the equilibrium ratio, x^*, does not depend on the initial growth rates. Since oscillations generally are nonlinear, slight dependence of the period on the initial conditions is expected. Amplitude of the oscillations is always determined by both initial sizes and initial growth rates.

*** The balanced ratio of abundances in the case of coexistence, x^*, is independent of the initial growth rates.

Ideal environments in which we mentally placed our pair of interacting prey and predator populations do not exist. Populations do not grow exponentially. Limiting factors, or constraints, reduce and finally stop the growth at some point. This is not to say that considering an unlimited environment was not a useful exercise. It helped to clarify our thinking about the dynamics of predation, and now we will attempt to add realism by placing the system in a limited environment. Let us start with one population and consider the analogy of the logistic equation:

$$\frac{d}{dt} (\ell n\ N) = r - \phi\ (N).$$

We can rewrite it as:

$$\frac{d^2}{dt^2} (\ell n\ N) = - \frac{d\phi}{dN} \frac{dN}{dt}$$

This equation is nothing but a derivative of the logistic-like equation. It is a result of the presence of a constraint in the form of a limited resource. The force, generated by the constraint can be called <u>a reaction of the constraint</u>. It is a growth-rate dependent, "friction-like" force which decelerates the growth and limits it to the carring capacity of the environment, defined by the standard equation

$$r - \phi\ (N) = 0$$

Note that here r reflects all other properties of the environment besides the limiting factor and $\phi(N)$ describes the shape of the density-dependence induced by the presence of the constraint.

Clearly, if this limitation is the only factor influencing growth, there is no sense in differentiating the logistic-like equation once again in order to obtain the force acting on the acceleration of population size. The old model is fully satisfactory by itself. The new form, however, is absolutely necessary if we want to consider the joint action of predation in combination with the limiting factors.

For simplicity, let us assume that in a predator-prey system, only the prey population is affected by the growth limiting resource. We will then have

$$\frac{d^2}{dt^2}(\ell n\, N_1) = -\,G\left(N_1/N_2\right) - \frac{d\phi}{dN_1}\frac{dN_1}{dt}$$

$$\frac{d^2}{dt^2}(\ell n\, N_2) = F\left(N_1/N_2\right)$$

Note, that an important assumption of additivity is made here. The rate of change of the Malthusian parameter is <u>a sum</u> of the rate induced by predation and the rate induced by the declining resource.

This model produces a stable equilibrium around the ratio x^* with damped oscillations if the intensity of density-dependence is low (for low values of $\frac{d\phi}{dN}$) and a nonoscillatory stability when the intensity of density-dependence is high. Note that even with a non-oscillatory stability an overshoot of the equilibrium level is a possibility.

These conclusions seem very similar to the ones obtained from the Lotka-Volterra equations when density-dependence is included. There are, however, important differences. Changing both r_{10} and r_{20}, i.e., non-interactive factors of the environment, will affect the equilibrium values of both prey and predator in our model but will preserve the ratio at x^*. In the Lotka-Volterra model the equilibrium level of prey is defined only by r_2 and not r_1 whereas the equilibrium level of the predator depends on r_2 and r_1. The ratio of prey to predator is not preserved in the Lotka-Volterra model, whereas in our model it is invariant to the strength of limiting factors and other noninteractive properties of the environment. Only the absolute abundance level reached by the system depends on the initial growth rates.

The behavior of the unrestricted predator-prey system in linear approximation is analogous to the simple mechanical oscillator moving uniformly without friction and oscillating with amplitude defined by relative initial position and speed. The behavior of the predator-prey system with density-dependent factors is analogous to the mechanical oscillator with some level of friction. Oscillations are damped or not present at

all (if friction is strong enough) but the relative equilibrium position is not influenced by friction. Absolute final position, on the other hand, depends on the initial speeds and the intensity of friction.
Away from local approximation, there is more complexity to our model. The "friction" intensity is actually position-dependent and the "spring" is strongly nonlinear. The essential features of the analogy are, however, valid. The nonlinearity of "the spring" allows for the predator's extinction in fundamentally the same way as for the non-restricted case. The existence of an extinction threshold allows for extinction of the prey before the predator, but this was also the case with the classical models.
Questions for experimental or even field verification that are raised by this model are whether the equilibrium ratio of prey to predators is the same in different environments, and whether the frequencies of their oscillations are the same when the strength of density-dependence is the same.
It is unfortunate that we are attempting to build a theory of population dynamics without a reasonably complete understanding of statics. As can be seen from the analysis above, understanding static relationships holds the keys to making a choice between competing models of dynamics. In this regard, the work of Sugihara (1984) on the topology of trophic networks may prove to be extremely important. It might happen that the fact that trophic networks are "mechanically stringent," made up of triangles and tetrahedrons, has a much more literal meaning than it seemed even to the author of this analogy. If trophic relationships act like "springs" connecting the equilibrium ratios of preys and predators, the stability of the systems itself might impose the property of the "mechanical stringency" on the network of trophic relationships.

Discussion

Beginning with the classical work of Gause (1934), much unsuccessful experience has accumulated in the attempts to verify the Lotka-Volterra predator-prey models and their analogues. This is very much in contradiction with the equations themselves which predict cyclic behavior and even stable coexistence in the presence of limiting factors. (Note that the frequency of oscillations predicted by the Lotka-Volterra model is strongly dependent on the population's growth rates--in complete contradictions to the model proposed here.) At the same time, experiments with competition have been much more successful at realizing model-predicted behaviors. Competitive exclusion as well as coexistence of competitors have been observed experimentally in the laboratory in a reasonably good agreement with traditional models.

I submit that the failure of experimental verification for the Lotka-Volterra predator-prey model is plausibly a consequence of using a wrong model. In fact, if a model of the sort of proposed here is accepted, it is easy to see why it was so difficult to maintain a predator-prey system in the laboratory without adopting special tricks like environmental heterogeneity or immigration. A very delicate balance of initial sizes <u>and initial growth rates</u> is needed to insure coexistence. According to the new model, the best approach is to start with the balanced ratio and create a minimal difference in the initial growth rate, so that the system is minimally deviant from its equilibrium state. This can be achieved by adjusting the properties of the environment in which the pair of populations is placed. These conditions were definitely not satisfied in the attempted experiments. The expectation of stable coexistence independent of initial sizes and (most importantly) growth rates which was based on the Lotka-Volterra equations might be responsible for the failure of these experiments. Repeating such experiments with the new model in mind would be a valuable exercise.

A series of experiments where prey and predator were successfully maintained (oscillating usually stopped after 6-7 cycles) came from the work with continuous cultures by Dr. Fredrickson (1972) and his colleagues at the University of Minnesota. The oscillations were the result of a three-way interaction, for instance glucose feeding of bacteria which are prey for the protozoa. Although glucose is fed to a chemostat with a constant rate, it fluctuates widely in a counterphrase to the consumer (prey) abundance. However, the glucose concentration is always so much above the Michaelis constant that it is reasonable to assume that the Malthusian parameter of prey is saturated. One can, therefore, interpret the results of an example of purely predator-prey oscillatory interaction.

Changing the flow rate in chemostat experiments, as well as changing the quality of food supplied for the prey, is a very good way to see the effect of initial growth rates on the fate of the interaction. If the proposed model provides a better description of the system, amplitudes of oscillations should be very sensitive to the changes in flow rate while both frequency of oscillations and mean relative abundance should not be sensitive to such changes. In the set of results reported in Tsuchiya et al (1972), the flow rate was reduced twice and four times between the three experiments. Neither the frequency of oscillations, nor average ratio changed significantly in these experiments, although amplitude of oscillations was higher for higher flow rates. This is precisely what would have been predicted based on the new model and is in complete contradiction with traditional models.

Attempts to introduce second order differential equations in population dynamics modelling have been made in the past. First independent publications by Clark (1971) and Ginzburg (1972, in Russian) were quite formal and immature approaches to the problem. The other two publications are a response to Clarke's work (Innes, 1972) and another paper (Yee, 1980) with the analysis and classification of types of behavior in a single population second order model. The interpretation of "forces" acting in population dynamics suggested by Yee is quite different from the one proposed in this paper. Another approach attempting to interpret adaptation due to natural selection as a cause of "inertia" in population's response to environmental changes was made by Ginzburg (1980) and reiterated in Ginzburg (1982). The position towards acceleration based models advocated in the current paper is radically different from all of the above mentioned approaches.

The proposed approach in summary is to consider predation as producing effects on the acceleration of the population growth (changes in Malthusian parameters) rather than on rates. The analysis of "forces" thus defined leads to a description of the coexisting prey-predator system as a stable system oscillating around a particular equilibrium ratio. Density-dependent factors in the proposed model are interpreted as "reactions of constraints". They are responsible for stopping the exponential growth of an otherwise unlimited system, and they influence the frequency and the rate of decay of oscillations.

Experimental work that is needed to reject or support the proposed view should be centered on checking the invariance of the ratio and the oscillation frequencies in prey-predator systems with respect to initial growth rates. These characteristics vary under traditional models as a function of r_1 and r_2 (in different directions for different models) and stay unchanged under the proposed model.

I would like to end by encouraging such experimentation. This will not only help to judge the validity of the proposed view but help to make an intelligent choice between different kinds of population dynamics models which have remained in the literature for 60 years without adequate verification.

This paper is contribution number 572 in Ecology and Evolution at the State University of New York at Stony Brook. This work was supported by the grant R81-1275 from the Office of Exploratory Research of the U.S. Environmental Protection Agency.

REFERENCES

Clark, G.P. 1971. The Second Derivative in Population Modelling. Ecology, 52:606-613.

Gause, G.F. 1934. The Struggle for Existence. Hafner Publishing Co., New York.

Ginzburg, L.R. 1972. The Analysis of the "Free Motion" and "Force" Concepts in Population Theory. In. Coll. "Studies in Theoretical Genetics", V.A. Ratner, ed., Novosibirsk (in Russian).

Ginzburg, L.R. 1980. Ecological Implications of Natural Selection. Proceedings of the Vito Volterra Symposium on Mathematical Models in Biology, 1979. Lecture Notes in Biomathematics, V. 39, Springer-Verlag.

Ginzburg, L.R. 1982. Theory of Natural Selection and Population Growth. Benjamin/Cummings Publishing Company, Menlo Park, California.

Hutchinson, G.E. 1975. Variation on a theme by Robert MacArthur. In. Coll. "Ecology and Evolution of Communities:, pp. 492-521, Harvard University Press. Cambridge, Massachusetts.

Innis, G. 1972. The Second Derivative and Population Modelling: Another View. Ecology, 53:720-723.

Sugihara, G. 1984. Graph Theory, Homology and Food Webs, Proceedings of Symposia in Applied Mathematics, Vol. 30, pp. 83-101.

Tsuchiya, H.M., J.F. Drake, J.L. Jost and A.G. Fredrickson. 1972. Predator-Prey Interactions of Dictyostelium discoideum and Escherichia coli in Continuous Culture, Journal of Bacteriology, Vol. 110, pp. 1147-1153.

Yee, J. 1980. A. Nonlinear, Second-Order Population Model. Theoretical Population Biology, 18:175-191.

ECOLOGICAL AND EVOLUTIONARY ASPECTS OF DISPERSAL

Simon A. Levin
Section of Ecology and Systematics
and
Ecosystems Research Center
Cornell University, Ithaca, New York, 14853

I. INTRODUCTION

One of the most fascinating challenges in ecology involves the statistical description of movement, and the understanding of population distributions in terms of the behavior of individuals. Knowledge of plant and animal dispersal patterns is fundamental to an understanding of basic and applied issues ranging from the evolution of life history traits to the spread of genetically engineered organisms. The spatial and temporal structure of environment, and its effects upon the movements of individuals, are central issues in ecological and evolutionary theory. Seed and pollen dispersal, together with germination of dormant seeds and released growth of understory plants, play important roles in secondary successional patterns of forest communities following disturbance. Dispersal is key to the maintenance of gene flow among populations, the dynamics of pest outbreaks, the recovery of disturbed areas, and the optimal spatial design of agricultural systems and natural reserves.

Yet, in 1951, in his classic paper on dispersal, J.G. Skellam lamented the lack of progress on the quantification of dispersal, and attributed it to the relative reluctance on the part of biologists to formulate problems in terms of simplified, abstract models. As he lamented, the necessary mathematical techniques had been available for nearly a century, and were utilized by Maxwell in developing the kinetic theory of gases.

The biological theory, which had seen scattered developments in the early part of this century (e.g., Pearson and Blakeman 1906, Brownlee 1911), received a boost in the 1930's and 1940's from the investigations of population geneticists such as Fisher (1937) and Dobzhansky and Wright (e.g., 1943), and from related theoretical developments by Kolmogorov et al. (1937); but it was not until

Skellam's 1951 paper that ecologists seriously became interested in the problem. Even then, it took another quarter-century before substantive contact was made between theory and application.

Today, inspired to large extent by Skellam's work, the theory of dispersal represents an active area of investigation. It is one of the success stories of mathematical ecology in that it is testable, quantifiable, and modifiable, and thereby provides a bridge between theory and experiment.

II. EVOLUTIONARY ASPECTS

The evolutionary aspects of dispersal have received even less careful attention than has description of ecological patterns. This is largely because the fascination with optimization arguments in life history theory led theorists into a brick wall when they considered the problem of dispersal, which could only be understood within the context of frequency dependence.

Hamilton and May (1977) dealt with this problem elegantly, by considering the advantages to dispersal for annual plants living in a renewable and stable environment. In such an environment, no microsite (defined as supporting a single adult) is intrinsically any better than any other; given the inherent costs of dispersal, especially the risk of death before reaching a safe site, why disperse at all? A pure optimization argument might suggest that the process, being costly, should be avoided.

Yet an entirely different picture emerges when one considers different genotypes in competition: a disperser, for all that it loses along the way, invariably outcompetes and eliminates a non-disperser. Furthermore, one finds that there is a single type that can outcompete all others, namely that which disperses a fraction of its seeds equal to the reciprocal of 1+m, where m is the probability of death due to dispersal. Thus, if half of the dispersers are lost before reaching a safe site, the evolutionarily stable strategy is to disperse 2/3 of one's seeds. Even as the probability of death approaches unity, a plant still should disperse half of its seeds.

How are these arguments modified when a more general class of environments and possible strategies are considered? Dispersal, dormancy, diapause, iteroparity, and vegetative spread all can be viewed as responses to local unpredictability, environmental deterioration, and the negative consequences of such biological interactions as competition, predation, and parasitism. Levin et al. (1984) introduce a more general model to deal with the evolution of dispersal and its interaction with dormancy, and demonstrate that the degree of dispersal increases with environmental uncertainty and with environmental unfavorability. Delayed germination (dormancy) reduces the selection pressure for dispersal, because the existence of a seed bank has the effect of reducing environmental variability. The basic model is discussed elsewhere in this volume, where results are given for the more general problem of the joint evolution of dispersal and delayed germination (Cohen and Levin, 1986).

III. WAVES OF ADVANCE

The objective of mathematical models of the spread of invading organisms is to capture the central aspects of spread in terms of a few, easily interpretable parameters, and to make prediction feasible. The simplest diffusion models have a number of disadvantages in that they ignore the higher order moments of the redistribution of individuals, and thus do not allow for "great leaps forward" (e.g., Mollison 1977). However, they have the advantages of being testable and easily quantifiable, and in many cases provide remarkably good descriptions of the process of spread (Levin 1981).

There are a wide variety of models that one can use to describe spread, and these can be classified on the basis of when and how often movement occurs in the lifetime of an individual; the lengths and distribution of individual steps; and the principal modes of transport (e.g., active, passive by winds and currents, or vectored by other organisms). The underlying assumption of most simple models is that spread can be understood as the cumulative outcome of a great many small, not necessarily independent, steps. That is, the scale of observation is assumed to be many times larger than the average step length.

At the scale of the individual step, movement has both a deterministic component, as determined by the mean step length and average direction, and a stochastic component, as expressed in variances. As already mentioned, it is in theory possible to deal with higher-order moments, but these add a level of complexity to such models that makes it difficult or impossible to verify or quantify them on the basis of data. Thus, one usually settles for the incorporation of advection and diffusion, represented respectively by the first- and second-order terms in the equation

$$\partial n/\partial t = \partial^2(Dn)/\partial x^2 + \partial^2(Dn)/\partial y^2 - u\partial n/\partial x - v\partial n/\partial y, \qquad [1]$$

in which $n=n(x,y,t)$ denotes the population density at the point (x,y), D is a measure of diffusion (and could be generalized to vary in the x and y directions), and u and v denote mean advective effects. The resultant model is of the same type that appears in diffusion approximations to stochastic processes, with the mean being given by the first-order term, and the variance being expressed by the second-order term. As with stochastic processes, in which much debate revolves around the choice of an appropriate limiting calculus, debate concerning the correct form may ensue in movement models when, for example, rates of movement are dependent on local environmental conditions. Depending on which points of view one takes, one obtains somewhat different forms for the appropriate coefficients. However, such problems do not arise for the simplest versions, in which coefficients are constant.

The general form [1] considers both diffusion and advection, but one often is concerned with special cases of these models in which one or the other of the basic processes dominates. For example, a great deal of the theoretical literature has concerned itself exclusively with the diffusion process, ignoring net advective effects. This is the case for Skellam's examination of the rates of spread of muskrat introduced into Europe, and it applies as well to most subsequent attempts to apply the methodology. The usual approach, as described by Kareiva (1983), is to measure the rate of increase of the variance of the distribution of individuals released at a point. According to the basic theory (pure diffusion with constant coefficients), the variance should increase linearly with time at a rate equal to twice the diffusion coefficient. In situations where one can mark individuals and either follow their movements or at least obtain a time sequence describing the rate of increase of variance, it is possible to derive good estimates of the diffusion coefficient. More sophisticated methods for estimating the parameters in diffusion and advection models are being developed by Banks and Kareiva, using the method of splines; these are reported elsewhere in this volume.

Kareiva's study of the foraging behavior of phytophagous insects showed that when the diffusion coefficient was estimated in the way described above and then used to generate a sequence of probability distributions to be compared with data, agreement was excellent. These studies indicated that, to a first approximation, simple diffusion models provided excellent descriptions of the spread of the populations. In those minority of cases where there was significant disagreement between the predictions of the constant coefficient model and the data, a model in which habitat dependent movements were considered would be sufficient to explain the discrepancies.

The application of diffusion models to invasion problems had its earliest appearance in Fisher's analysis of the rate of advance of advantageous alleles (1937), and in work the same year by the mathematicians Kolmogorov, Petrovsky, and Piscunov. Fisher, considering the case of selection operating on two alleles at a single autosomal locus, proposed the model

$$\partial P/\partial t = rP(1-P) + D\, \partial^2 P/\partial x^2 \qquad [2]$$

where P is the frequency of the advantageous allele. Fisher conjectured that an advancing wave would relax asymptotically to a front, with the characteristic speed of advance $c=2(rD)^{1/2}$ (see Levin 1986). Kolmogorov et al. (1937) confirmed Fisher's result, at least for most reasonable patterns of release.

As discussed in Levin (1986), these results have been those of most interest to biologists, probably because of their inherent simplicity. Equation [2] adds a logistic growth term to the basic diffusion model, and Fisher (1937) and Kolmogorov et al. (1937) also considered more general growth laws. Skellam applied models of this form to the spread of invading species, and obtained similar asymptotic behavior. Even for the linear growth model, for which there is no true front, an apparent front is observed if one includes a detection threshold into the model. Most interestingly, the speed of advance is the same in this case as for the logistic, and is independent of where the threshold level is set.

To apply these models, one would like data concerning:

(1) the intrinsic rate of natural increase of the population of interest (as measured by r);

(2) the distribution of dispersal distances, or at least estimates of its means and variances in the various dimensions (as measured by the advection and diffusion coefficients);

(3) the rates of advance of fronts.

Often, one can do no better than obtain data concerning (3) above. In these cases, one can still examine the degree to which the data are in agreement with the prediction of an asymptotic front advancing with constant speed. In those cases where one observes great leaps forward, it is clear that the basic model is inadequate, although it may describe an important part of the process of spread. More often, one will see patterns of constant increase for long periods of time, followed by rapid changes as the population enters new habitat. In these cases, the model plays the role of suggesting which parameters bear close examination; i.e., it suggests that the change in habitat has modified either r or D, and thereby suggests further biological investigations.

Ideally, one would like to have independent information concerning (1) and (2) above, and to use this information to generate predictions concerning rates of spread. For such applications, one needs to obtain life history data for the species in question, sufficient to estimate r for the variety of habitats that the population is likely to encounter. Similarly, one needs to know from direct or indirect methods what the patterns of individual movements are, and whether the assumption of approximate normality in the distribution of movements is admissible. Specifically, one needs estimates of the mean and variance of the distribution. When simultaneous information on (3) is available, this provides a check on the validity of the underlying model; studies directed at such validation are reported elsewhere (Andow et al. ms).

IV. CONCLUSION

In this paper, I have tried to introduce some of the ecological and evolutionary problems associated with dispersal. Other issues, such as the maintenance and generation of spatial patterns, especially clines, are discussed elsewhere (Levin 1976, 1978, 1979; Nagylaki 1978). Hopefully this discussion will suffice to expose the central ideas, and to introduce one of the richest areas in mathematical biology, one where theory and experiment can proceed in close partnership. The understanding of the evolutionary forces shaping dispersal patterns, and the interplay between dispersal and other mechanisms for dealing with uncertain environments, remain tantalizing and fundamental problems in population biology. Moreover, the need to develop quantitative methods for the description and prediction of population movements is of surpassing importance, and gains in urgency as we face environmental problems associated with invasions of exotic species, and with disturbance and recovery in natural systems. For such challenges, mathematical approaches are essential and already are having influence; but the need for better methods, ones tested against the demands of experimental data, make the subject of dispersal a gold mine for theoretical investigations.

ACKNOWLEDGEMENTS

This work was supported in part by NSF grant DMS-840-6472, by EPA Cooperative Agreement CR 811060, and by Hatch and McIntire-Stennis grants to Simon A. Levin. This paper is also Ecosystems Research Center report ERC-114. The views expressed are those of Simon A. Levin, and do not necessarily reflect those of the sponsoring agencies.

REFERENCES CITED

Andow, D., P. Kareiva, S.A. Levin, and A. Okubo. 1986. Spread of invading organisms (ms.).

Brownlee, J. 1911. The mathematical theory of random migration and epidemic distribution. Proc. Roy. Soc. Edinburgh, 31:262-289.

Cohen, D. and S.A. Levin. 1986. The interaction between dispersal and dormancy strategies in varying and heterogeneous environments. This volume.

Dobzhansky, T. and S. Wright. 1943. Genetics of natural populations. X. Dispersion rates in *Drosophila pseudoobscura*. Genetics 28:304-340.

Fisher, R.A. 1937. The wave of advance of advantageous genes. Ann. Eugen. London 7:355-369.

Hamilton, W.D. and R.M. May. 1977. Dispersal in stable habitats. Nature 269:578-581.

Kareiva, P. 1983. Local movements in herbivorous insects: applying a passive diffusion model to mark-recapture field experiments. Oecologia 57:322-324.

Kolmogorov, A., I. Petrovsky, and N. Piscunov. 1937. Étude de l'équations de la diffusion avec croissance de la quantité de la matière et son application a un problème biologique. Bull. Univ. Moscou, Ser. Internation., Sec. A,1, #6, 1-25.

Levin, S.A. 1976. Population dynamic models in heterogeneous environments. Annu. Rev. Ecol. Syst. 7:287-310.

Levin, S.A. 1978. Population models and community structure in heterogeneous environments. pp. 439-476 in S.A. Levin (ed.), Studies in Mathematical Biology II: Populations and Communities. Mathematical Association of America, Washington, D.C.

Levin, S.A. 1979. Non-uniform stable solutions to reaction-diffusion equations: Applications to ecological pattern formation. pp. 210-222 in H. Haken (ed.), Pattern formation by dynamic systems and pattern recognition. Springer-Verlag, Heidelberg.

Levin, S.A. 1981. The role of theoretical ecology in the description and understanding of populations in heterogeneous environments. Amer. Zool. 21:865-875.

Levin, S.A. 1986. Random walk models of movement and their implications. pp. 143-154 in Hallam, T. and Levin, S.A. (eds), Mathematical Ecology: An Introduction. Springer-Verlag, Heidelberg.

Levin, S.A., D. Cohen, and A. Hastings. 1984. Dispersal strategies in patchy environments. Theor. Pop. Biol. 26:165-191.

Mollison, D. 1977. Spatial contact models for ecological and epidemic spread. J. Roy. Stat. Soc. B 39, 283-326.

Nagylaki, T. 1978. The geographical structure of populations. pp. 588-624 in S.A. Levin (ed.), Studies in Mathematical Biology II: Populations and Communities. Mathematical Association of America, Washington, D.C.

Okubo, A. 1980. Diffusion and Ecological Problems: Mathematical Models. Biomathematics 10. Springer-Verlag, Heidelberg. 254 + xiii pp.

Pearson, K. and J. Blakeman. 1906. Mathematical contributions to the theory of evolution - XV. A mathematical theory of random migration. Draper's Company Research Mem. Biometric Series III. Dept. Appl. Math. Univ. College, Univ. London.

Skellam, J.G. 1951. Random dispersal in theoretical populations. Biometrika 38:196-218.

The Speeds of Traveling Frontal Waves in Heterogeneous Environments

Nanako Shigesada, *Kohkichi Kawasaki and Ei Teramoto

Department of Biophysics, Kyoto University
Kyoto 606, Japan

*Science and Engineering Research Institute
Doshisha University, Kyoto 602, Japan

1. Introduction.

Since Fisher's pioneering work (Fisher 1937), many studies on traveling waves in a growing population have been performed. The model proposed by Fisher consists of a diffusion equation with a logistic growth term:

$$u_t = du_{xx} + (\varepsilon - u)u \quad \text{for } x \in (-\infty,\infty).$$

Here $u(x,t)$ denotes the population density at position x and time t, and d and ε are diffusivity and intrinsic growth rate, respectively. It has been shown from this equation that, starting from a localized distribution, the population evolves into a propagating wave of constant speed, $2\sqrt{\varepsilon d}$.

Following Fisher's work, this equation and its extensions have been analyzed by, among others, Kolmogorov, Petrovskii and Piskunov(1937), Aronson and Weinberger(1975), Hadeler and Rothe(1975), Murray(1977), Fife(1979), Okubo(1980), Dunbar(1983), Novick-Cohen and Segel(1984). The results of these analyses have been applied to explain such phenomena as the spread of pest species, epidemic diseases, chemotactic bacteria, and seed dispersal. Most of these models have dealt with traveling waves propagating in homogeneous environments. Therefore, the diffusivity coefficient d and intrinsic growth rate ε are treated as constants. However, organisms generally diffuse through heterogeneous environments which may be described as a mosaic of patches. In such a case, diffusivity and intrinsic growth rate are no longer constant, but, rather, they vary in space.

Here we consider a heterogeneous environment consisting of two patch types, "favorable" and "less favorable", arranged alternately along a one-dimensional axis. The population is assumed to grow logistically and to be subject to random dispersal in a manner similar to Fisher's treatment. Thus the equation for population density $u(x,t)$ in dimensionless units is similar to the above equation, except that d and ε are now functions of x instead of constants:

$$\frac{\partial}{\partial t}u = \frac{\partial}{\partial x}(d(x)\frac{\partial}{\partial x}u) + (\varepsilon(x) - u)u \quad \text{for } x \in (-\infty,\infty). \quad (1)$$

Here we assume $d(x)$ and $\varepsilon(x)$ to be given by stepwise functions as follows:

$$\begin{aligned} d(x) &= 1, & \varepsilon(x) &= 1 & &\text{for } x_{2m} < x < x_{2m+1}, \\ d(x) &= d(>0), & \varepsilon(x) &= \varepsilon(\leq 1) & &\text{for } x_{2m+1} < x < x_{2m+2}, \end{aligned} \quad m=0,\pm 1,\pm 2,\ldots (2)$$

where $x_{i+1}=x_i+\ell_i$ and $x_0=0$, with ℓ_i denoting the width of the i-th patch. Since we use dimensionless units, $d(x)$ and $\varepsilon(x)$ can be taken as unity in favorable patches ($x_{2m}<x<x_{2m+1}$ for $m=0,\pm 1,\pm 2,..$), whereas in less-favorable patches ($x_{2m+1}<x<x_{2m+2}$ for $m=0,\pm 1,\pm 2,..$) these parameters are assigned values d and ε, respectively. Note that ε should be less than 1 but can be either positive or negative.

The initial distribution is assumed to be localized at the center of the x axis. At boundaries between adjacent patches, x_i ($i=0,\pm 1,\pm 2,..$), we require the density $u(x,t)$ and the flux $d(x)u_x$ to be continuous.

We shall first consider periodically varying environments and then proceed to analyze the case of irregularly varying environments.

2. Periodically Varying Environments

We start with a periodically varying environment in which $\ell_{2m}=\ell_f$ (const.) and $\ell_{2m+1}=\ell_u$(const.) for all m so that $\varepsilon(x)$ and $d(x)$ are given by periodic step functions as shown in Fig. 1. Thus the favorable and less-favorable patches have constant widths, ℓ_f and ℓ_u, respectively, and the spatial period of the environment is given by $\ell_f+\ell_u$, which we denote by ℓ^*. Therefore our model in dimensionless units includes four independent parameters d, ε, ℓ_f and ℓ_u.

Eq.(1) always possesses the trivial solution $u(x)\equiv u^0=0$, which represents the absence of the species. Further, it may possess

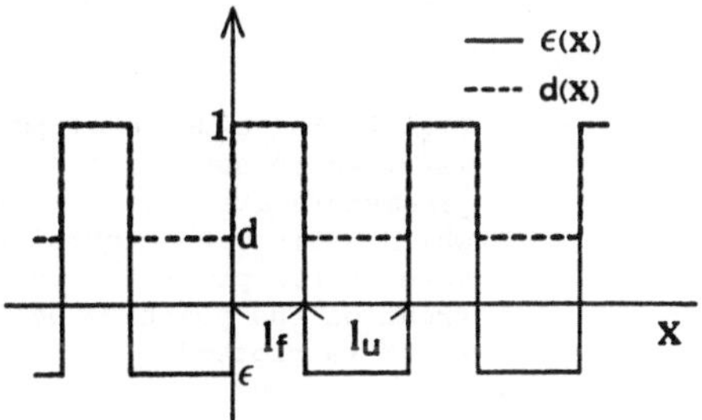

Fig. 1. Spatial patterns of diffusivity $d(x)$ and intrinsic growth rate $\varepsilon(x)$ in a periodic environment.

another positive stationary solution $u^s(x)$ which is periodic with spatial period ℓ^* $(u^s(x) = u^s(x+\ell^*))$. We may conjecture that if the trivial solution u^o is dynamically unstable while $u^s(x)$ is stable, then the transition from u^o to u^s take places in a spatially mediated way. That is, if a few organisms are introduced into a local area, they will grow into an invasive propagating wave.

To examine this problem more closely, we first study the stability of the trivial solution. By linearizing (1) about u^o and denoting the function satisfying the linearized equation by $v(x,t)$, we obtain

$$v_t = (d(x)v_x)_x + \varepsilon(x)v, \qquad x \in (-\infty,\infty). \tag{3}$$

We can show that the trivial solution of (3), $v=0$, is dynamically

$$\text{stable,} \qquad \text{if } \ell_f < 2\mathrm{Tan}^{-1}[\sqrt{-\varepsilon d}\, \tanh(\sqrt{-\varepsilon/d}\, \ell_u/2)], \tag{4-a}$$

$$\text{unstable,} \qquad \text{if } \ell_f > 2\mathrm{Tan}^{-1}[\sqrt{-\varepsilon d}\, \tanh(\sqrt{-\varepsilon/d}\, \ell_u/2)]. \tag{4-b}$$

We have also found by numerical computation of Eq.(1) that when the parameters satisfy inequality (4-a), any population starting from a localized distribution will eventually become extinct. On the other hand, when (4-b) is satisfied, the population always evolves into a propagating wave as shown in Fig. 2. This propagating wave is periodic in the sense that at each point x, $u(x,t)$ monotonically increases with time, and after each time interval t^*, the spatial pattern shifts to the right by the spatial period ℓ^*. This property is described by the following equations:

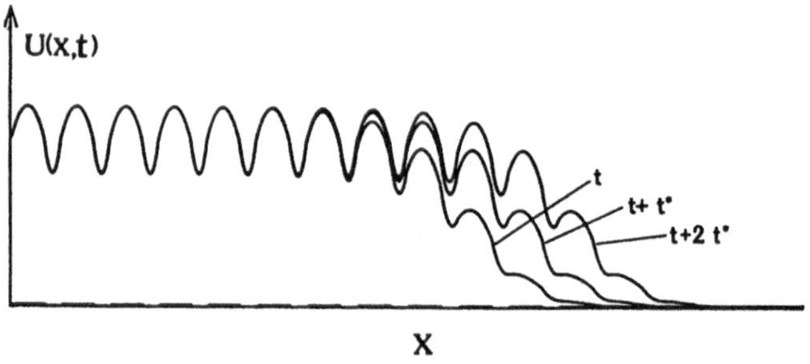

Fig. 2. The time development of a traveling periodic wave, which satisfies $u(x,t)=u(x+\ell^*,t+t^*)$. The parameters are chosen as $d=0.5$, $\varepsilon=-0.5$, $\ell_f=2$, $\ell_u=1$, which satisfy the invasion condition (14-b). Although wave fronts propagate in both directions, only the wave front on the positive half plane at times t, $t+t^*$ and $t+2t^*$ is plotted.

$$\begin{aligned} u(x,t) &= u(x+\ell^*, t+t^*) \\ u(x,t) &\to 0 \qquad \text{for } x \to \infty \\ u(x,t) &\to u^s(x) \quad \text{for } x \to -\infty. \end{aligned} \tag{5}$$

Hereafter, we shall call a propagating wave satisfying conditions (5) a "traveling periodic wave". If we define $c=\ell^*/t^*$, c can be regarded as the speed of the traveling periodic wave. It should be noted that the shape of the traveling periodic wave is not constant but changes continually. However after each time t^*, the spatial pattern is superimposable. In this sense, the traveling periodic wave should be distinguished from the traveling monotonic wave in a homogeneous environment, as was demonstrated by Fisher.

The velocity c can be calculated analytically by assuming that the leading edge of a traveling periodic wave is given by the product of two functions, f and g, as follows:

$$v(x,t) = f(z)g(x), \qquad z = x-ct \qquad (c = \ell^*/t^*) \tag{6}$$

By substituting (6) into (3) and solving the equations for f and g under condition (5), we derive the so-called "dispersion relation" (for details, see Shigesada, Kawasaki and Teramoto (1986)):

$$\begin{aligned} &\cosh s(\ell_f+\ell_u) \\ &= \cosh q_1\ell_f \cosh q_2\ell_u + \frac{q_1^2+(dq_2)^2}{2dq_1q_2} \sinh q_1\ell_f \sinh q_2\ell_u \end{aligned} \tag{7}$$

where

$$q_1 = \sqrt{cs-1}, \qquad q_2 = \sqrt{(cs-\varepsilon)/d}.$$

s is the damping coefficient of the leading edge of the traveling periodic wave. From Eq.(7), c is calculated as a function of s, and it can be seen that c has a minimum, c_{min}. Actual wave speeds should be larger than $c_{min}(\varepsilon,d,\ell_f,\ell_u)$.

From this analysis, we cannot determine the speed $c(\geq c_{min})$ ultimately attained by a stable traveling periodic wave. However, our computer calculation indicates that, for any set of parameters, the speed of a traveling periodic wave always approaches precisely the value c_{min} derived here.

Further analysis of the dispersion relation shows the following inequality to hold:

$$c_{min}(\varepsilon,d,\ell_f,\ell_u) \geq 2\sqrt{\langle\varepsilon\rangle_a \langle d\rangle_h} \qquad \text{for } \langle\varepsilon\rangle_a > 0, \tag{8}$$

where $\langle\varepsilon\rangle_a$ and $\langle d\rangle_h$ are the arithmetic and harmonic means, respectively, defined as follows:

$$\langle\varepsilon\rangle_a \equiv \frac{1}{\ell^*}\int_0^{\ell^*} \varepsilon(x)dx = \frac{\ell_f+\varepsilon\ell_u}{\ell_f+\ell_u}, \tag{9}$$

$$\langle d\rangle_h \equiv \frac{1}{\frac{1}{\ell^*}\int_0^{\ell^*}\frac{1}{d(x)}dx} = \frac{\ell_f+\ell_u}{\ell_f+\ell_u/d}\,.$$

The equality sign of (8) holds at the limit ℓ_f, $\ell_u \to 0$. It should be pointed out here that the right-hand side of (8) has the same form as the speed of the traveling monotonic wave in a homogeneous environment obtained in Fisher's model, except that intrinsic growth rate and diffusivity are replaced by the mean values $\langle\varepsilon\rangle_a$ and $\langle d\rangle_h$, respectively.

In Fig. 3, the stable speeds of traveling periodic waves, c_{min}, is plotted as a function of ε for various fixed values of d. c_{min} monotonically decreases with decreasing ε. After c_{min} reaches zero, it remains at zero thereafter, so that the population never evolves into a traveling periodic wave. As diffusivity d increases, c_{min} also increases when ε is positive, while if ε is smaller than some critical negative value, c_{min} decreases with diffusivity, d.

3. Irregularly Varying Environments

We now extend our analysis to irregularly varying environments. We assume that favorable and less-favorable patches are alternately arranged as expressed by Eq.(2) in which the widths of the favorable and less-favorable patches, ℓ_{2m}, ℓ_{2m+1} (m=0,±1,±2,...), are assigned random values uniformly in the intervals $\ell_f-\sigma_f<\ell_{2m}<\ell_f+\sigma_f$ and $\ell_u-\sigma_u<\ell_{2m+1}<\ell_u+\sigma_u$, respectively. Hereafter, we shall call ℓ_f (ℓ_u) and σ_f (σ_u) the "mean patch size" and the "variance" of favorable patches (less-favorable patches), respectively. When σ_f and σ_u tend to 0, the irregular environment reduces to the periodic one discussed in the previous section. To explore the temporal development of spatial

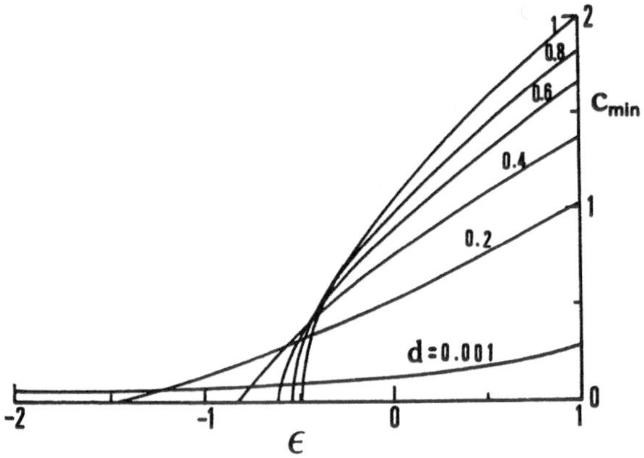

Fig. 3. The speed of a traveling periodic wave, c_{min}, as a function of ε for various fixed values of d. ℓ_f=1; ℓ_u=2.

patterns in this model, we have run computer simulations for various sets of parameter values. One simulation result is illustrated in Fig. 4. Although the spatial pattern exhibits some irregularity due to the random arrangement of patches, it can be seen that any localized initial distribution evolves into a frontal wave whose edge is moving to the right. Here we define the average speed for such a wave as follows. Let $x^*(t)$ denote the position where the wave front reaches a certain value, say ξ. That is, $x^*(t)=\max\{x|u(x,t)=\xi\}$ (see Fig. 4). The speed of $x^*(t)$, $dx^*(t)/dt$, exhibits an irregular behavior fluctuating around some constant value. However, if we introduce a time-averaged speed of dx^*/dt defined by the following equation

$$\bar{c} = \lim_{t\to\infty} \frac{1}{t}\int_0^t \frac{dx^*}{dt'}\, dt', \tag{10}$$

computer simulation shows that $\bar{c}$ takes a constant value, irrespective of the value of ξ.

Fig. 5 illustrates the dependence of time-averaged speeds $\bar{c}$ on σ_f when the ratio σ_u/σ_f is fixed at 1. Although $\bar{c}$ monotonically increases with $\sigma_f(=\sigma_u)$, the change is so gradual that we can consider $\bar{c}$ to be nearly equal to the wave speed for $\sigma_f=\sigma_u=0$ (*i.e.*, the wave speed in the periodic environment which was obtained analytically from dispersion relation (7)).

Fig. 6 shows the dependence of the time-averaged speed $\bar{c}$ on the mean patch size ℓ_f when the ratio ℓ_u/ℓ_f is fixed at 3/2 and $\sigma_f=\sigma_u=0.5\ell_f$ (dots in figure). The solid curve represents the wave speed for $\sigma_f=\sigma_u=0$, with other parameters remaining the same. At ℓ_f, $\ell_u \simeq 0$,

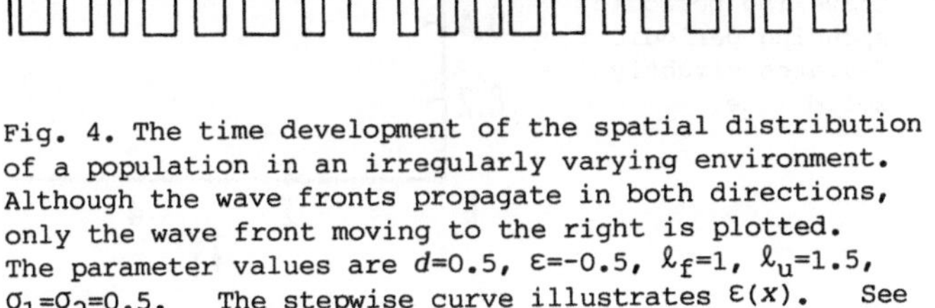

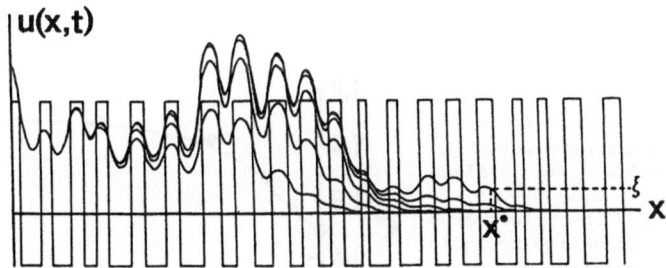

Fig. 4. The time development of the spatial distribution of a population in an irregularly varying environment. Although the wave fronts propagate in both directions, only the wave front moving to the right is plotted. The parameter values are d=0.5, ε=-0.5, ℓ_f=1, ℓ_u=1.5, $\sigma_1=\sigma_2$=0.5. The stepwise curve illustrates $\varepsilon(x)$. See the text concerning ξ and x^*.

the dots falls on the solid curve. However, as $\ell_f (=\frac{2}{3}\ell_u)$ increases, the dots deviate slightly upwards from the solid line. This tendency can be seen generally for various sets of the parameters. In particular, we can mathematically prove, by using the method of averaging, that in the limit of $\ell_f, \ell_u \to 0$, the time-averaged speed is given by the following equation:

$$\bar{c} \simeq 2\sqrt{\langle\varepsilon\rangle_a \langle d\rangle_h} \qquad \text{for } \ell_f, \ell_u \ll 1, \tag{11}$$

where $\langle\varepsilon\rangle_a$ and $\langle d\rangle_h$ are the arithmetic and harmonic means, respectively, defined as follows:

$$\langle\varepsilon\rangle_a \equiv \lim_{L\to\infty} \frac{1}{2L}\int_{-L}^{L} \varepsilon(x)dx = \frac{\ell_f+\varepsilon\ell_u}{\ell_f+\ell_u},$$
$$\langle d\rangle_h \equiv \lim_{L\to\infty} \frac{1}{\frac{1}{2L}\int_{-L}^{L} \frac{1}{d(x)}dx} = \frac{\ell_f+\ell_u}{\ell_f+\ell_u/d}. \tag{12}$$

Fig. 5. The time-averaged speed $\bar{c}$ as a function of σ_f(dots). The ratio σ_u/σ_f is fixed at 1. d=0.5; ε=-0.5; ℓ_f=2; ℓ_u=3. The broken line indicates the speed of the traveling periodic wave in the corresponding periodic environment.

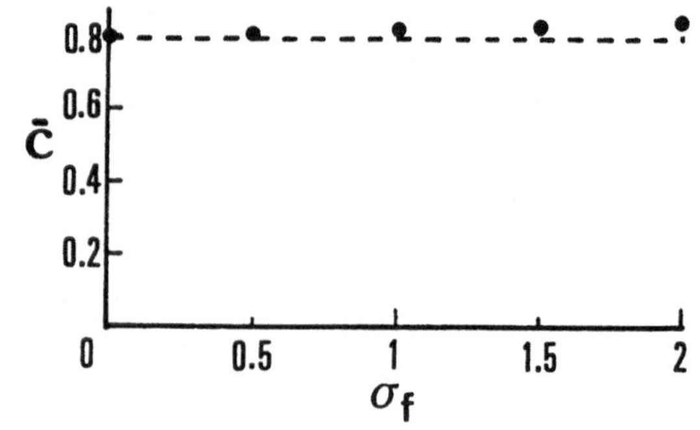

Fig. 6. The time-averaged speed $\bar{c}$ as a function of the mean patch size ℓ_f. The ratio ℓ_u/ℓ_f is fixed at 1.5. d=0.5; ε=-0.5; $\sigma_f=\sigma_u$= 0.5ℓ_f. The solid line illustrates the speed of the traveling periodic wave in the corresponding periodic environment. $\bar{c}$ deviates slightly upwards from the solid line.

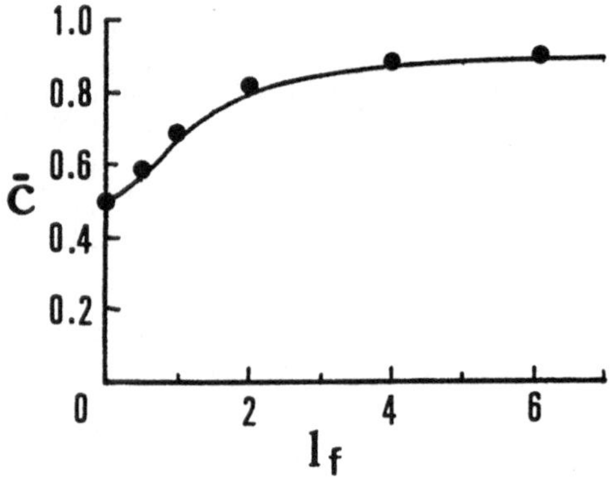

See Appendix for the derivation of Eq.(11); this derivation was originally suggested by Othmer(pers. comm.). Since the values of $\langle\varepsilon\rangle_a$ and $\langle d\rangle_h$ as given by (12) for an irregular environment are the same as those given by (9) for the periodic environment, we can see that, as $\ell_f, \ell_u \to 0$, the time-averaged speed c coincides with the right hand side of Eq.(8).

Thus we can conclude that (1) the time-averaged speed in an irregularly varying environment as defined here can be approximated by the speed of a traveling periodic wave in a corresponding periodic environment and (2) the approximation becomes more precise as ℓ_f and ℓ_u tend to zero.

Acknowledgment: The authors would like to thank Dr. H.G. Othmer for his valuable suggestions and comments. They also wish to thank Dr. L.A. Segel and Dr. L. Ricciardi for their encouragement and interest in this work. This work was supported by a Grant-in-Aid for Special Project Research on Biological Aspects of Optimal Strategy and Social Structure from the Japan Ministry of Education, Science and Culture.

References

Aronson, D.G. and Weinberger, H.F. 1975. Nonlinear diffusion in population genetics, combustion and nerve propagation, in: Proceedings of the Tulane Program in Partial Differential Equations and Related Topics, Lecture Notes in Mathematics **446**, Springer, Berlin, 5-49.

Dunbar, S.R. 1983. Travelling wave solutions of diffusive Lotka-Volterra equations, J. Math. Biology **17**, 11-32.

Fife, P.C. 1979. "Mathematical Aspects of Reacting and Diffusing Systems", Lecture Notes in Biomathematics **28**. Springer-Verlag, Berlin-Heidelberg-New York.

Fisher, R.A. 1937. The wave of advance of advantageous genes, Ann. of Eugenics **7**, 255-369.

Hadeler, K.P. and Rothe, F. 1975. Travelling fronts in nonlinear diffusion equations, J. Math. Biol. **2**, 251-263.

Kolmogorov, A.N., Petrovskii, I.G. and Piskunov, N.S. 1937. A study of the equation of diffusion with increase in the quantity of matter, and its application to a biological problem, Bjul. Moskovskovo Gos. Univ. **17**, 1-72.

Murray, J.D. 1977. "Lectures on Nonlinear Differential-Equation Models in Biology". Clarendon Press, Oxford.

Novick-Cohen, A. and Segel, L.A. 1984. A gradually slowing travelling band of chemotactic bacteria. J. Math. Biology **19**, 125-132.

Okubo, A. 1980. "Diffusion and Ecological Problems: Mathematical Models". Berlin-Heidelberg-New York: Springer Verlag.

Othmer, H.G. 1983. A Continuum Model for Coupled Cells. J. Math. Biology **17**, 351-369.

Shigesada, N., Kawasaki, K. and Teramoto, E. 1986. Traveling Periodic Waves in Heterogeneous Environments. Theor. Popul. Biology **30**, 143-160.

Appendix

We rewrite the governing equation for the irregularly varying environment defined in Section 3:

$$u_t = (d(x)u_x)_x + (\varepsilon(x) - u)u, \tag{A-1}$$

where $d(x)$ and $\varepsilon(x)$ are given by (2), and the widths ℓ_{2m} and ℓ_{2m+1} respectively of the favorable and less-favorable patches are assigned random values uniformly in the intervals $\ell_f-\sigma_f<\ell_{2m}<\ell_f+\sigma_f$ and $\ell_u-\sigma_u<\ell_{2m+1}<\ell_u+\sigma_u$, respectively.

Suppose that the average patch sizes, ℓ_f and ℓ_u, are very small. Then the solution of (A-1) varies slowly in space compared with the rapid change of the environmental parameters $\varepsilon(x)$ and $d(x)$. Othmer (1983) has analyzed an equation similar to (A-1), with periodically varying parameters, which describes diffusion in a coupled cellular system, and he obtained a slowly varying solution by using the method of averaging. Although our model (A-1) involves randomly varying parameters $\varepsilon(x)$ and $d(x)$, we can apply his method with a slight modification. In the following we briefly illustrate the procedure to derive Eq.(11) presented in the text.

Suppose the average period of the environment $\ell^*=\ell_f+\ell_u$ to be sufficiently small and let $\ell^*\equiv\delta<<1$. We introduce the following two space variables:

$$\begin{aligned} X_0 &= x \quad \text{(slow space variable)}, \\ X_1 &= x/\delta \quad \text{(fast space variable)}. \end{aligned} \tag{A-2}$$

Since $d(x)$ and $\varepsilon(x)$ are rapidly varying functions, we put

$$d(x) = d(X_1), \qquad \varepsilon(x) = \varepsilon(X_1).$$

We assume the solution of (A-1) to be given by a function of these two variables, $u(x,t)=u(X_0,X_1,t;\delta)$ and seek an expansion of the following form:

$$u(X_0,X_1,t;\delta) = u_0(X_0,X_1,t)+\delta u_1(X_0,X_1,t)+\delta^2u_2(X_0,X_1,t)+ \dots , \tag{A-3}$$

where u_k, $\partial u_k/\partial X_0$, $\partial u_k/\partial X_1$ $(k=1,2)$ are bounded for all X_0 and X_1 and

$$\lim_{L\to\infty} \frac{1}{2L}\int_{-L}^{L} u_k dX_1 = 0 \qquad \text{for } k=1,2. \tag{A-4}$$

Using the chain rule, differentiation with respect to space is transformed according to

$$\frac{\partial}{\partial x} = \frac{\partial}{\partial X_0} + \frac{1}{\delta}\frac{\partial}{\partial X_1}. \tag{A-5}$$

Upon inserting (A-3) and (A-5) into (A-1), and matching powers of δ, we obtain

$$\frac{\partial}{\partial X_1} d \frac{\partial}{\partial X_1} u_0 = 0 \ , \tag{A-6}$$

$$\frac{\partial}{\partial X_1} d \frac{\partial}{\partial X_0} u_0 + \frac{\partial}{\partial X_0} d \frac{\partial}{\partial X_1} u_0 + \frac{\partial}{\partial X_1} d \frac{\partial}{\partial X_1} u_1 = 0, \tag{A-7}$$

$$\frac{\partial}{\partial t} u_0 = \frac{\partial}{\partial X_0} d \frac{\partial}{\partial X_0} u_0 + \frac{\partial}{\partial X_1} d \frac{\partial}{\partial X_0} u_1 + \frac{\partial}{\partial X_0} d \frac{\partial}{\partial X_1} u_1 + \frac{\partial}{\partial X_1} d \frac{\partial}{\partial X_0} u_2 + (\varepsilon(X_1) - u_0) u_0 . \tag{A-8}$$

By solving (A-6) with the restriction that u_0 is bounded for all X_1, we see that u_0 is independent of X_1. Substituting $\partial u_0/\partial X_1 = 0$ into (A-7) and integrating (A-7) with respect to X_1, we obtain the general solution of (A-7) as follows:

$$u_1(X_0, X_1, t) = a_1(X_0, t) \int_0^{X_1} \frac{1}{d} dX_1 - X_1 \frac{\partial}{\partial X_0} u_0 + a_2(X_0, t) \ . \tag{A-9}$$

a_1 and a_2 can be determined by use of the restrictions that $\lim_{L\to\infty} [\{u_1(X_0, L, t) - u_1(X_0, -L, t)\}/2L] = 0$ and (A-4) for $k=1$. Thus we have

$$u_1 = [\int_0^{X_1} (\frac{1}{d} - \langle\frac{1}{d}\rangle) dX_1 - \langle\int_0^{X_1} (\frac{1}{d} - \langle\frac{1}{d}\rangle) dX_1\rangle] \frac{\partial}{\partial X_0} u_0 / \langle\frac{1}{d}\rangle \ , \tag{A-10}$$

where we use the notation

$$\langle f(X_1)\rangle = \lim_{L\to\infty} \frac{1}{2L} \int_{-L}^{L} f(X_1) dX_1$$

for the spatial average of $f(X_1)$. Substituting (A-10) into (A-8) and taking the spatial average of (A-8) under the conditions $\partial u_k/\partial X_0 < \infty$ $(k=1,2)$ for all X_1, we obtain the equation of the slowly varying solution as follows:

$$\frac{\partial}{\partial t} u_0 = \langle d\rangle_h \frac{\partial^2}{\partial X_0^2} u_0 + (\langle\varepsilon\rangle_a - u_0) u_0, \tag{A-11}$$

where

$$\langle d\rangle_h = \frac{1}{\langle 1/d\rangle}, \qquad \langle\varepsilon\rangle_a = \langle\varepsilon\rangle.$$

(A-11) is Fisher's equation with a constant diffusion coefficient $\langle d\rangle_h$ and intrinsic growth rate $\langle\varepsilon\rangle_a$, and hence the speed of the traveling wave is given by

$$\bar{c} \simeq 2\sqrt{\langle\varepsilon\rangle_a \langle d\rangle_h} \quad \text{for } \ell_f, \ell_u \ll 1,$$

which agrees with Eq.(11) in the text.

QUANTITATIVE MODELING OF GROWTH AND DISPERSAL IN POPULATION MODELS

H.T. Banks and K.A. Murphy
Lefschetz Center for Dynamical Systems
Division of Applied Mathematics
Brown University
Providence, RI 02912

INTRODUCTION

In this presentation we discuss methods for inverse or parameter estimation problems which can be employed as quantitative modeling techniques in models for distributed (spatially, age, size, etc.) biological systems. In this context they may be useful in attempts to understand, elaborate on, or further refine details of specific mechanisms for dispersal, growth, interaction, etc. in wide classes of models. We have also used these techniques in a number of biologically related problems [1] such as bioturbation [12], [14], [15] and climatology [19]. In addition to an overview of ideas underlying these techniques, we shall present here brief discussions and some findings on two specific biological problems for which we are currently using them successfully.

A typical inverse problem entails some given or hypothesized dynamical model with "parameters" q (often temporally and/or "spatially" or even state dependent) and "states" $u(t,x;q)$, $0 \leq t \leq T$, $x \in \Omega$, which depend on the parameters through a dynamical system of equations. One has observations or data $\hat{u}_{ij}$ for $u(t_i,x_j;q)$ and wishes to choose, from some admissible parameter set Q, parameters $\bar{q}$ so as to give a best fit of the model to the data. For example, we might have a hypothesized model for transport

$$\frac{\partial u}{\partial t} + \frac{\partial}{\partial x}\left[V(t,x,u)u\right] = \frac{\partial}{\partial x}\left[D(t,x,u)\frac{\partial u}{\partial x}\right] + F(t,x,u) \tag{1}$$

with initial and boundary conditions also possibly depending on the unknown parameters $q=(V,D,F)$. Given data $\hat{u}_{ij}$, we seek to minimize a fit-to-data criterion such as a least-squares

$$J(q) = \sum_{i,j} |u(t_i,x_j;q) - \hat{u}_{ij}|^2 \tag{2}$$

over a specified class Q of functions (V,D,F) so as to obtain a best estimate $\bar{q} = (\bar{V},\bar{D},\bar{F})$. In addition to obtaining estimates for q, usually one desires to analyze in some way the "goodness" of the model in describing the phenomena one is modeling. We shall elaborate on some related questions in this regard below.

The methods we discuss briefly here can be powerful modeling tools when carefully and correctly used. Some of the novel features of our recent efforts include the capabilities for estimation of (i) state or density dependent dispersal coefficients such as D above in (1), (ii) system nonlinearities such as F in (1), and (iii) boundary parameters in both simple and not-so-simple boundary conditions (we give an example in the discussions on size dependent models below). Furthermore, there are a number of modeling related questions that one might hope to address from a theoretical or computational (or both) viewpoint with the aid of these techniques. These include:

(a) Experimental design [3], [4]. [5]: What is the appropriate data required to support analysis of a particular model or mechanism? E.g., How many time vs. spatial observations must be made, or what type of initial data is needed to study movement patterns?

(b) Robustness of model parameters [1], [3], [8], [16], [19]: Do the problem formulation and the methods enjoy certain stability properties? E.g., Do the parameter estimates and the estimation methods depend continuously on observation noise, initial data, amounts of data available, problem constraints, etc.?

(c) Identifiability [18], [25]: Is the map from the parameter space to the observation sufficiently well-behaved so that the methods can produce unique estimates?

(d) Model comparison [3], [5], [19]: Can one make evaluations regarding the importance and type of mechanisms needed to model given phenomena? E.g., Which is more important in particular transport phenomena: convection, diffusion, nonlinear effects, dynamic (time varying) vs. heterogeneous (spatially varying) terms or coefficients? What level of refinement in modeling terms can be supported by the experimental design and data? Do model refinements yield statistically significant improvements in explanation of the data?

CONCEPTUAL CONSIDERATIONS

We next outline briefly certain ideas related to the problems and methods that are the focus of this presentation. At the same time we shall indicate some questions that may arise in either theoretical or computational aspects of investigations using the methods. These discussions can be made precise and mathematically rigorous, but for the sake of brevity, we shall not do that here.

For the purposes of illustration, we return to the problem of minimizing the functional of (2) subject to the system (1) relating the states and parameters. Such problems lead to the need for optimization techniques for constrained problems that are infinite dimensional in nature. One has a system with states $(t,x)\to u(t,x)$ in some infinite dimensional function space X and parameters $(t,x)\to q(t,x)$, or $(t,x,u)\to q(t,x,u)$ if the parameters are state dependent, in some infinite dimensional function space Q. These problems can be concisely stated in a theoretical framework using either the theory of semigroups or evolution operators, or the theory of sesquilinear forms in Hilbert spaces. We won't pursue the details here, but refer the reader to [2], [7], [9], [13], [15].

In any case, this leads to the recognition that in order to effectively develop computational techniques, one must introduce approximation schemes for the state and parameter spaces. That is, one needs families of finite dimensional spaces X^N and Q^M (such as finite elements, splines, spectral families) such that X^N approximates X well as $N\to\infty$ and Q^M approximates Q as $M\to\infty$. (We shall (imprecisely) write this as $X^N\to X$, $Q^M\to Q$ or simply $N\to\infty$, $M\to\infty$, in the discussions below.) One then must develop schemes to solve the approximate problems obtained when u in J of (2) is replaced by the approximate states $u^N\in X^N$ satisfying some equation approximating (1). Minimization is carried out over Q^M yielding approximate best-fit parameters $\bar{q}^{N,M}$. Thus, the algorithms we have developed and used (e.g. see [3], [5], [8] for details) entail iterative optimization techniques combined with appropriately chosen approximation schemes based on families X^N, Q^M.

Among the important questions associated with these approximation ideas are those of method convergence and method stability. In the first, one must argue that $\bar{q}^{N,M}\to\bar{q}$ as $X^N\to X$, $Q^M\to Q$, where $\bar{q}$ is a solution to the original problem involving (1) and (2). That is, one must assure fidelity of the estimates under sufficiently accurate approximation of state and parameter spaces. The concept of stability is related to a continuous dependence of the estimates on the observed data, X^N, and Q^M. More precisely, if $\bar{q}^{N,M}(\hat{u})$ denotes solutions to the approximate problems corresponding to state space X^N, parameter space Q^M and data $\hat{u}$, and if $\{\hat{u}^K\}$ is a sequence of data with $\hat{u}^K\to\hat{u}$, then one desires to guarantee that $\bar{q}^{N,M}(\hat{u}^K)\to\bar{q}(\hat{u})$ as $N,M,K\to\infty$, where $\bar{q}(\hat{u})$ is a solution of the original problem with data $\hat{u}$. That is, fidelity of the estimates will hold with sufficiently small noise in the observations as well as sufficiently accurate approximations of the state and parameter spaces. For further discussions see [1], [18].

One can develop a general theory to guarantee theoretically and computationally well-behaved algorithms based on the ideas we have used in a wide class of problems including the examples discussed below. The arguments rely heavily on ideas from functional analysis, approximation theory and compactness. We refer the reader to [1], [4], [8], [15], [18], [25] for further elaboration and details. We only note here that fundamental to all these convergence and stability results is the establishment that $u^N(t,x;q^M)\to u(t,x;q)$ in some sense (i.e., in an appropriate X-topology) whenever $q^M\to q$ in an appropriate sense (i.e., in a Q-topology). For further discussion of mathematical ideas, and implementation and testing of the methods, we refer to the presentations in [3], [5] in addition to those references cited above. Here we discuss several projects in which these methods are playing a fundamental role and outline some new results in two areas.

INSECT DISPERSAL/GROWTH MODELS

We have, in collaboration with P. Kareiva (U. Washington), considered a number of aspects of insect movement and growth. In several cases our quantitative methods have proved useful in planning the experiments as well as actually investigating various models. Among the investigations we have pursued are:

(i) quantifying "initial disturbance" effects in dispersal rates for flea beetle movement in mark-recapture experiments in cultivated collard patches [3], [5], [21], [22], [23];

(ii) studying the effects of density-dependent dispersal rates, nonlinear growth, interaction, and predation in multiple species models such as those for ladybug-aphid-goldenrod experiments [10], [24];

(iii) quantifying "preferred direction" components in cabbage root fly movement in two-dimensional domains [6], [31].

In regard to the investigations of (ii), we note that the methods can be used effectively to estimate the shape of density-dependent dispersal coefficients D and nonlinear growth terms f in models of the form

$$\frac{\partial u}{\partial t} = \frac{\partial}{\partial x}\left[D(t,x,u)\,\frac{\partial u}{\partial x} \right] + f(u) \tag{3}$$

where $D(t,x,\cdot)$ has the form depicted in Figure 1.

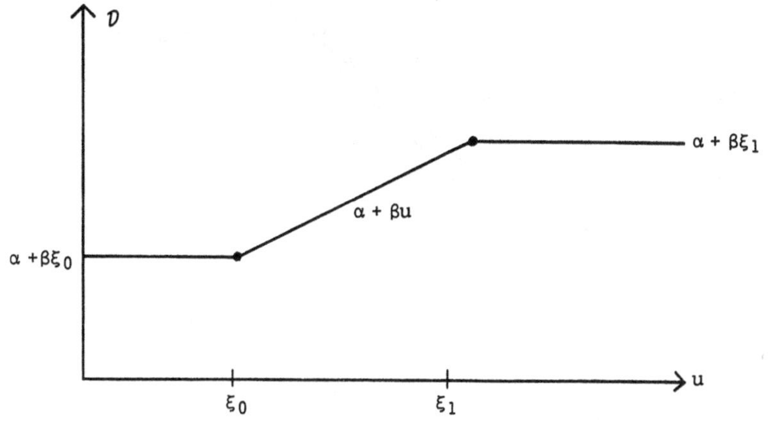

Figure 1

Such dispersal coefficients represent a rate that is bounded below and above (basal and saturation limits) and depends linearly on the density between these bounds. Problems with density-dependent dispersal have received attention elsewhere [28], [29] (see also [26] for further discussions regarding the importance of such problems).

Before using our estimation or inverse techniques in problems with experimental data, we carry out a rather careful testing of the methods with "synthetic" data on numerous examples. This procedure involves a series of tests using "data" generated (with noise) from a system with known (prechosen) parameters to ascertain the ability to recover the parameters from given sets of "data". For detailed explanations of this procedure, see for example [3], [5]. This testing is also combined with attempts to establish convergence and stability results for the methods. For problems involving systems of the type (3), such results are given in [1], [10], [11]. We present here results from two of the numerical tests we performed.

Example 1: We seek to estimate $D = D(u)$, i.e., ξ_0, ξ_1, α, β in Figure 1, in the system

$$\frac{\partial u}{\partial t} = \frac{\partial}{\partial x}\left[D(u)\frac{\partial u}{\partial x}\right] + 2u^2 - \frac{1}{2}u^3 + F(t,x),\ u(t,0) = u(t,1) = 0,\ u(0,x) = 6x(1-x),$$

where F is known (computed analytically so that $u(t,x) = 6x(1-x)(1+t^2)$ is a solution corresponding to a "true" parameter D^* with ξ_0=.5, ξ_1=1.6, α=3, β=1). "Data" corresponding to observations at (t_i,x_j), t_i=0,.5,1,x_j=.1,.2,...,.9, were used for the inverse procedure. Results for estimation with N=6 and N=14 (cubic splines were used for the state approximations - see [10] for details) along with the initial estimate D^0 and true value D^* are depicted in Figure 2.

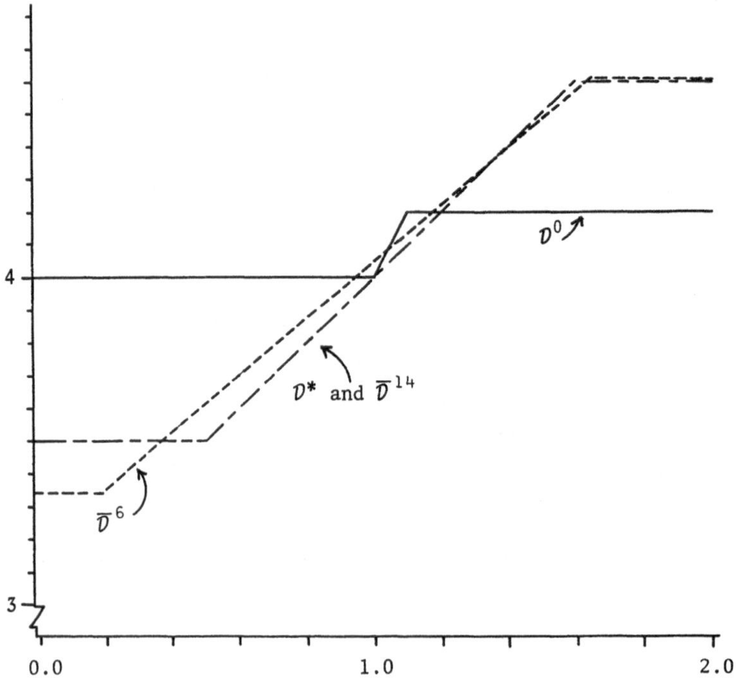

Figure 2

Example 2: We seek to estimate q and the nonlinearity f (we do not make any a priori parametrization or shape assumption on f) in

$$\frac{\partial u}{\partial t} = q\,\frac{\partial^2 u}{\partial x^2} + f(u) + F(t,x),\ u(t,0) = u(t,1) = 0,\ u(0,x) = 6x(1-x),$$

where again F is computed so that $u(t,x) = 6x(1-x)(1+t^2)$ is a solution corresponding to true values q^*= 3.0, $f^*(u) = 2u^2 - \frac{1}{2}u^3$. Results for the simultaneous estimation of f and q with state approximation N=6 (cubic splines with mesh h= $\frac{1}{6}$) and parameter approximations (linear splines) for f with mesh size h=.65 (see [11] for further details) are presented in Figure 3. These results for the initial estimate f^0, the converged value $\bar{f}^{6,.65}$, and f^* correspond to initial guess q^0=1.0 and converged value $\bar{q}^6$=2.9993.

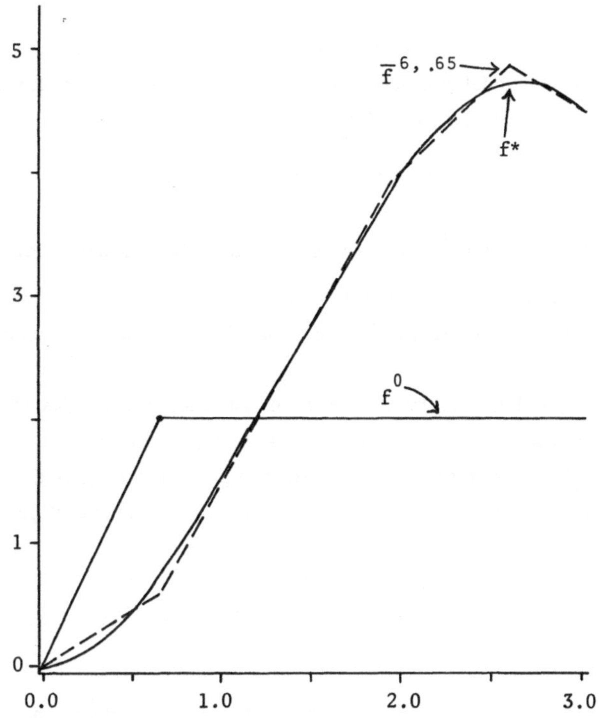

Figure 3

SIZE DEPENDENT GROWTH MODELS

We are currently using our parameter estimation techniques in investigations that entail size dependent population growth models. Data from experiments with mosquito fish populations in rice paddies have motivated our collaborative efforts with L. Botsford (U. California, Davis). While the basic problems we are considering are control problems for the mosquito fish/mosquito populations, a substantial effort is required in developing the underlying dynamics (i.e., growth models). For a more detailed description of the modeling and control problems, we refer to [17], [30].

A simple version of the basic modeling problem entails estimation of $q=(g,m,b)$ in the system

$$\frac{\partial u}{\partial t} + \frac{\partial}{\partial x}(gu) = -mu \quad x_0 \leqslant x \leqslant x_1,\ t>0,\ u(0,x) = \Phi(x),\ g(x_0)u(t,x_0) = \int_{x_0}^{x_1} b(t,\xi)u(t,\xi)d\xi, \qquad (4)$$

where g=growth rate, m=mortality rate, and b=fecundity are in general dependent on time t and size x, with x_0, x_1 the minimum and maximum observable sizes, respectively. Data for the system generally consists of observations that yield values $\hat{u}(t_i,x)$, $x_0 \leqslant x \leqslant x_1$, so that a distributed least-squares criterion, e.g.,

$$J(q) = \sum_i |u(t_i,\cdot;q) - \hat{u}(t_i,\cdot)|^2_{L^2}$$

is appropriate.

Since a convergence theory for the estimation problem has not appeared elsewhere, we sketch one approach to this problem. This approach is the analogue to that given in [1], [10], [11] for the insect dispersal model problems. We first rewrite (4) in variational or weak form. We seek $u(t)\epsilon H^0(x_0,x_1)$ satisfying for all $\phi\epsilon H^1(x_0,x_1)$

$$<u_t,\phi> + <mu,\phi> - <gu,D\phi> - \phi(x_0)R(t,u) = 0, \quad u(0,\cdot) = \Phi \tag{5}$$

where $R(t,\psi) \equiv \int_{x_0}^{x_1} b(t,\xi)\psi(\xi)d\xi$, $D = \frac{\partial}{\partial x}$, and $<,>$ is the usual $L^2=H^0$ inner product. For brevity, we assume that the parameters g and m depend only on size x. The ideas we present here can be readily modified to treat theoretically and computationally the more general case where g and m also are time dependent.

We assume that (g,m,b) are to be chosen from function spaces $G\times M\times B\subset H^1(x_0,x_1)\times H^0(x_0,x_1)\times H^0([0,T]\times[x_0,x_1])$ containing only nonnegative functions g,m,b where the functions in G also satisfy $g(x_1)=0$ and $g(x_0)\geq\nu_1$ for some positive constant ν_1. For the approximating systems (see [4], [10]) we assume that subspaces $Z^N\subset H^1(x_0,x_1)$ are chosen and let $u^N(t)\epsilon Z^N$ denote solutions to

$$<u_t^N,\psi> + <mu^N,\psi> - <gu^N,D\psi> - \psi(x_0)R(t,u^N)=0, \text{ for all } \psi\epsilon Z^N, \ u^N(0)=P^N\Phi, \tag{6}$$

where P^N is the orthogonal projection of $H^0(x_0,x_1)$ onto Z^N. We assume that the subspaces Z^N satisfy:

(H_1) For $\phi\epsilon H^1$ we have $P^N\phi\to\phi$ in H^1 while for $\phi\epsilon H^0$ we have $P^N\phi\to\phi$ in H^0.

A number of the commonly used approximating families (piecewise linear, cubic splines [27]) satisfy this hypothesis. The ideas here can be slightly modified (see the remarks in [1]) to also include spectral families (such as Legendre polynomials - see [20]) in the state approximation schemes for which the convergence theory presented here is valid.

We further assume that G, M, B are compact in H^1, H^0, H^0 respectively, and that compact approximation families G^M, M^M, B^M for G, M, B respectively have been chosen satisfying:

(H_2) For $g^M\epsilon G^M$, $g^M(x_1)=0$, $g^M(x_0)\geq\nu_1$, $|Dg^M|_\infty\leq\mu_1$, and $G^M=i_1^M(G)$ where for each $g\epsilon G$, $i_1^M(g)\to g$ in H^1 with the convergence uniform in $g\epsilon G$;

(H_3) For $m^M\epsilon M^M$, $|m^M|_\infty\leq\mu_2$, and $M^M=i_2^M(M)$ where for each $m\epsilon M$, $i_2^M(m)\to m$ in H^0 with the convergence uniform in $m\epsilon M$;

(H_4) For $b^M\epsilon B^M$, $|b^M(t,\cdot)|\leq\mu_3$, and $B^M=i_3^M(B)$ where for each $b\epsilon B$, $i_3^M(b)\to b$ in H^0 with the convergence uniform in $b\epsilon B$.

We next remark that to give a convergence theory it suffices (see [8]) to argue that $u^N(t;q^M)\to u(t;q)$ in H^0 for each t as $N,M\to\infty$ whenever $\{q^M\}$ is an arbitrary sequence with $q^M\epsilon G^M\times M^M\times B^M$ and $q^M\to q\epsilon G\times M\times B$. Indeed, it suffices to give these arguments in the form $u^N(t;q^N)\to u(t;q)$ whenever $\{q^N\}$ is arbitrary in $G\times M\times B$ with $q^N\to q$ in $G\times M\times B$ (in the $H^1\times H^0\times H^0$ topology in this case). We sketch the arguments; let $\{q^N\}$ be arbitrary with $q^N\to q$ in $G\times M\times B$ and let $u^N(q^N)$, $u(q)$ be the solutions to (6), (5), respectively, corresponding to $q^N=(g^N,m^N,b^N)$,

$q=(g,m,b)$ respectively. From (H_1), it suffices to argue that $z^N(t) \equiv u^N(t;q^N) - P^N u(t;q) \to 0$ in $H^0(x_0,x_1)$ for each t in (0,T).

Letting $R^N(t,\psi) \equiv \int_{x_0}^{x_1} b^N(t,\xi)\psi(\xi)d\xi$, we have from (5) and (6) that for all $\phi \in Z^N$

$$<(u^N-P^Nu)_t,\phi> = <u_t^N - u_t + (u-P^Nu)_t,\phi>$$

$$= <(I-P^N)u_t,\phi> + <mu-m^Nu^N,\phi> + <g^Nu^N - gu,D\phi> + \phi(x_0)\Delta R^N$$

where $\Delta R^N \equiv R^N(t,u^N)-R(t,u)$. With z^N as defined above we have $z^N(0)=0$ and

$$<z_t^N,\phi> = <(I-P^N)u_t,\phi> + <mu-m^NP^Nu,\phi> - <m^Nz^N,\phi>$$

$$+ <g^NP^Nu-gu,Du> + <g^Nz^N,D\phi> + \phi(x_0)\Delta R^N.$$

Choosing $\phi=z^N$ in this identity we find

$$\frac{1}{2}\frac{d}{dt}|z^N|^2 + <m^Nz^N,z^N> - <g^Nz^N,Dz^N>$$

$$= <(I-P^N)u_t,z^N> + <mu-m^NP^Nu,z^N> + <g^NP^Nu-gu,Dz^N> + z^N(x_0)\Delta R^N.$$

Recalling that $g^N(x_1) = g(x_1) = 0$, with integration by parts we find

$$<g^Nz^N,Dz^N> = -\frac{1}{2} <Dg^Nz^N,z^N> - \frac{1}{2} g^N(x_0)\, z^N(x_0)^2$$

and

$$<g^NP^Nu-gu,Dz^N> = - <D(g^NP^Nu-gu),z^N> - [g^NP^Nu-gu](x_0)z^N(x_0).$$

Hence, we have

$$\frac{1}{2}\frac{d}{dt}|z^N|^2 + <m^Nz^N,z^N> + \frac{1}{2} <Dg^Nz^N,z^N> + \frac{1}{2} g^N(x_0)z^N(x_0)^2$$

$$= <(I-P^N)u_t,z^N> + <mu-m^NP^Nu,z^N> - <D(g^NP^Nu-gu),z^N> \tag{7}$$

$$+ z^N(x_0)\,[(gu-g^NP^Nu)(x_0) + \Delta R^N].$$

Using the bounds from (H_2), (H_3) we find

$$<m^Nz^N,z^N> + \frac{1}{2} <Dg^Nz^N,z^N> + \frac{1}{2} g^N(x_0)z^N(x_0)^2 \geq \mu|z^N|^2 + \nu z^N(x_0)^2$$

for positive constants μ, ν. Standard inequalities imply that the right side of the equality (7) is less than or equal to

$$\frac{1}{4\epsilon}\left\{|(I-P^N)u_t|^2 + |mu-m^NP^Nu|^2 + |D(g^NP^Nu-gu)|^2\right\}$$

$$+ 3\epsilon|z^N|^2 + \epsilon|z^N(x_0)|^2 + \frac{1}{4\epsilon}\left\{2|(gu-g^NP^Nu)(x_0)|^2 + 2|\Delta R^N|^2\right\},$$

where $\epsilon>0$ is arbitrarily chosen. From the bounds of (H_4) we have that $|R^N(t,z^N)| \leq k|z^N|$ and defining $\delta^N(t) \equiv \mathcal{R}^N(t,P^Nu-u) + \mathcal{R}^N(t,u) - \mathcal{R}(t,u)$ we may conclude that

$$|\Delta \mathcal{R}^N| \leq k|z^N| + |\delta^N(t)|.$$

Using these inequalities in (7) we obtain

$$\frac{1}{2}\frac{d}{dt}|z^N|^2 + \left[\mu-3\epsilon-\frac{k^2}{\epsilon}\right]|z^N|^2 + (\nu-\epsilon)|z^N(x_0)|^2 \leq h^N(t) \tag{8}$$

where

$$h^N(t) \equiv \frac{1}{4\epsilon}\Big\{|(I-P^N)u_t|^2 + |mu-m^NP^Nu|^2 + |D(g^NP^Nu - gu)|^2$$
$$+ 2|(gu-g^NP^Nu)(x_0)|^2 + 4|\delta^N(t)|^2\Big\}.$$

Choosing $\epsilon=\nu$ and using $z^N(0)=0$, we may apply Gronwall's lemma to conclude that it suffices to argue that $h^N(t)\to 0$ in $L^1(0,T)$ to obtain the desired results. Under sufficient smoothness for u(t), this follows readily from (H_1) - (H_4).

While we are still testing our methods for use with models such as (4), our initial findings are quite positive. We present one of our simple test examples.

Example 3: We seek to estimate g in (4) with m=2, $b(x)=12x\sqrt{1-x}$, $\Phi(x)=\sqrt{2}\sqrt{1-x}$, $x_0=0$, $x_1=1$. The solution corresponding to $g^*(x)=2(1-x)$ is given by $u(t,x)=e^t\sqrt{2}\sqrt{1-x}$. Cubic splines ($Z^N$ of dimension N+3) were used for the state approximations while linear splines (Q^M of dimension M) were used in parameter approximations $g^M(x) = \Sigma\, \alpha_i\, \ell_i\,(x)$, where the sum is from i=0 to i=M-1 and ℓ_i is the usual "hat" function basis element [27] with support in ((i-1)/M, (i+1)/M). For M=4, the true value g^* corresponds to coefficients $g^*\sim(\alpha_0^*,\alpha_1^*,\alpha_2^*,\alpha_3^*) = (2.0,1.5,1.0,.5)$. Using data at eleven points in time and space each and initial guess (1,1,1,1), we obtained the converged values (1.998,1.498,1.000,.496) for α-coefficients in the representation for $\bar{g}^{32,4}$, i.e., with N=32, M=4. The graphs of g^* and $\bar{g}^{32,4}$ are not distinguishable using ordinary plotting devices and hence we do not present them.

COMPUTATIONAL CONSIDERATIONS

The problems on which we have focused in this presentation are computationally intensive. Even simple examples such as those presented above can require from 10^2 to 10^4 seconds on an IBM 3081 and we are now using the ideas discussed here in research problems for which use of such a sequential machine would require rather prohibitive computational expenditures. The necessary software packages must deal with reasonably large vector/tensor systems and involve many repetitive routine calculations. Therefore, a substantial part of our research efforts over the last year have entailed development of ideas, algorithms, and software to take advantage of emerging computer architectures involving parallel and vector computational capabilities. Use of such architectures (in our research programs at Brown University and ICASE, we are currently employing a CRAY X-MP - a widely known vector machine - and a STAR ST-100 array processor with parallel features) has substantially enhanced our efforts to investigate some of the research questions in modeling outlined above. For example, results for problems of the type given in Example 1 typically require from 10^3 to 10^4 seconds on the IBM 3081, but when the algorithms and corresponding software are modified to take advantage of the arithmetic speed and vector capabilities of the CRAY, we can carry out the same computational runs in 50 to 200 seconds on a CRAY X-MP.

Research machines such as the CRAY are, through NSF and other research sponsoring agencies, becoming readily accessible to many scientists in the U.S. We recognize that the machines we are presently using for these techniques and methods are not widely available to the world-wide biological research community. However, we firmly believe that the current revolution in computer hardware development has important implications for the community. We note that many of the high speed, parallel and vector features of large, expensive research array processors such as the FPS-164 and STAR ST-100 and research vector machines such as the CRAY X-MP (and 1-S) and CYBER 205 are rapidly becoming available in small, relatively inexpensive desk-top configurations. A number of attached array processor units and boards are now available for use with personal computers such as the IBM PC XT. Recently, new high speed chips (e.g., INTEL 80386 - a 32 bit, 4 MFLOPS chip) have been announced and will be available in PC's in 1987. We believe that the wide availability of "desk-top CRAY" capability discussed in the computer science community is only several years away. If we can successfully develop ideas further along the lines of those discussed above, the potential for a significant impact on biological modeling and research is enormous.

ACKNOWLEDGEMENTS:

This research was supported in part under NSF Grant MCS-8504316, AFOSR Contract 84-0398 and ARO Contract DAAG-29-83-K-0029. Part of the research was carried out while the authors were visiting scientists at ICASE, NASA Langley Research Center, Hampton, Virginia, which is operated under NASA Contract NAS1-17070.

REFERENCES

[1] H.T. Banks, On a variational approach to some parameter estimation problems, Proc. Int. Conf. on Control Theory for Distributed Parameter Systems and Applications (Vorau, Austria, July 9-14, 1984); LCDS Report #85-14, Brown University, May 1985.

[2] H.T. Banks, J.M. Crowley and K. Kunisch, Cubic spline approximation techniques for parameter estimation in distributed systems, IEEE Trans. Auto. Control, AC-28, (1983), 773-786.

[3] H.T. Banks and P. Kareiva, Parameter estimation techniques for transport equations with applications to population dispersal and tissue bulk flow models, J. Math. Biology, 17 , (1983), 253-273.

[4] H.T. Banks, P. Kareiva and P.K. Lamm, Estimation techniques for transport equations, Mathematics in Biology and Medicine, Proceedings, Bari 1983, (V. Capasso et al. Eds.), Springer Lecture Notes in Biomath., 57 , (1985), 428-438.

[5] H.T. Banks, P.M. Kareiva and P.K. Lamm, Modeling insect dispersal and estimating parameters when mark-release techniques may cause initial disturbances, J. Math. Biol., 22, (1985), 259-277.

[6] H.T. Banks, P. Kareiva and L. Zia, Analyzing field studies of insect dispersal with two-dimensional transport equations, to appear.

[7] H.T. Banks and K. Kunisch, An approximation theory for nonlinear partial differential equations with applications to identification and control, SIAM J. Control and Opt. 20, (1982), 815-849.

[8] H.T. Banks and P. Daniel Lamm, Estimation of variable coefficients in parabolic distributed systems, LCDS Report #82-22, Brown University, Sept. 1982; IEEE Trans. Auto. Control, 30 , (1985), 386-398.

[9] H.T. Banks and K.A. Murphy, Estimation of coefficients and boundary parameters in hyperbolic systems, LCDS Report #84-5, Brown University, February, 1984; SIAM J. Control and Opt., to appear.

[10] H.T. Banks and K.A. Murphy, Estimation of parameters in nonlinear distributed systems, Proc. 23rd IEEE Conference on Decision and Control, Las Vegas, (Dec. 12-14, 1984), 257-261.

[11] H.T. Banks and K.A. Murphy, Estimation of nonlinearities in parabolic models for growth, predation and dispersal of populations, to appear.

[12] H.T. Banks and I.G. Rosen, Fully discrete approximation methods for the estimation of parabolic systems and boundary parameters, LCDS Report #84-19, Brown University, May 1984; Acta. Applic. Math., to appear.

[13] H.T. Banks and I.G. Rosen, A Galerkin method for the estimation of parameters in hybrid systems governing the vibration of flexible beams with tip bodies, CSDL Report R-1724, June 1984; C.S. Draper Labs, Cambridge, MA.

[14] H.T. Banks and I.G. Rosen, Approximation methods for the solution of inverse problems in lake and sea sediment core analysis, Proc. 24th Conf. on Dec. and Control, (Dec. 11-13, 1985), Ft. Lauderdale, 732-736.

[15] H.T. Banks and I.G. Rosen, Numerical schemes for the estimation of functional parameters in distributed models for mixing mechanisms in lake and sea sediment cores, LCDS Report 85-27, Brown University, October 1985.

[16] H.T. Banks and D. Iles, A comparison of stability and convergence properties of techniques for inverse problems, LCDS Report #86-3, Brown University, January 1986.

[17] L. Botsford, B. Vandracek, T. Wainwright, A. Linden, R. Kope, D. Reed, and J.J. Cech, Jr., Population development of the mosquitofish, Gambusia Affinis, in rice fields, preprint.

[18] F. Colonius and K. Kunisch, Stability for parameter estimation in two point boundary value problems, Inst. fur Math. Bericht No. 50-1984, Tech. Univ. Graz, October, 1984.

[19] F. Dexter, H.T. Banks and T. Webb, Modeling Holocene changes in the location and abundance of beach populations in eastern North America, J. Biogeography, to be submitted.

[20] D. Gottlieb and S. Orszag, Numerical Analysis of Spectral Methods: Theory and Applications, SIAM, Philadelphia, 1977.

[21] Kareiva, P., Experimental and mathematical analysis of movement: quantifying the influence of plant spacing and quality on foraging discrimination, Ecol. Mon. 52 (1982) 261-282.

[22] Kareiva, P., Local movement in herbivorous insects: applying a passive diffusion model to mark-recapture field experiments, Oecologia (Berl.) 57, (1983) 322-327.

[23] Kareiva, P., Influence of vegetation texture on herbivore populations: resource concentration as herbivore movement. In Denno & McClure (eds.) Variable Plants and Herbivores in Natural and Managed Systems. New York: Academic Press (1983).

[24] Kareiva, P., Predator-prey dynamics in spatially-structured populations: manipulating dispersal in a coccinellid-aphid interaction. Springer Lect. Notes in Biomathematics, 54, (1984) 368-389.

[25] K. Kunisch and L.W. White, Identifiability under approximation for an elliptic boundary value problem, SIAM J. Control and Opt., to appear.

[26] A. Okubo, Diffusion and Ecological Problems: Mathematical Models, Springer-Verlag, New York, 1980.

[27] M.H. Schultz, Spline Analysis, Prentice Hall, Englewood Cliffs, 1973.

[28] N. Shigesada, K. Kawasaki and E. Teramoto, Spatial segregation of interacting species, J. Theor. Biol. 79, (1979), 83-99.

[29] N. Shigesada and E. Teramoto, A consideration on the theory of environmental density, Japanese J. Ecol., 28 (1978), 1-8.

[30] T.C. Wainwright, R.G. Kope, L.W. Botsford, and J.J. Cech, Jr., Implications of laboratory mosquitofish experiments for population development in rice fields, Proc. 52nd Conf. Calif. Mosquito and Vector Control Assoc. (Jan. 29-Feb. 1, 1984).

[31] L.L. Zia, Parameter Estimation Techniques for Two-dimensional Transport Equations with Application to Models of Insect Dispersal, Ph.D. Thesis, Brown University, May 1985.

THE INTERACTION BETWEEN DISPERSAL AND DORMANCY STRATEGIES IN VARYING AND HETEROGENEOUS ENVIRONMENTS

Dan Cohen
Dept. of Botany, The Hebrew University of Jerusalem,
Jerusalem, Israel

Simon A. Levin
Section of Ecology and Systematics, Cornell University
Ithaca, New York 14853, U.S.A.

INTRODUCTION

Natural environments are heterogeneous in space and time. This heterogeneity favors the evolution of mechanisms such as dispersal of seeds or other propagules, because dispersal allows escape in space from locally unfavorable conditions and, on the average, exploitation of ones temporarily more favorable (Levin 1976, Motro 1982). Similarly, delayed germination may be advantageous because it allows seeds to avoid exposure to unfavorable conditions and, on the average, to exploit more favorable ones (Cohen 1966, Templeton and Levin 1979).

It is intuitively clear that dispersal and dormancy are alternative strategies for dealing with and exploiting the spatial and temporal variability of the environment (Venable and Lawlor 1980; Levin, Cohen, and Hastings 1984; Klinkhamer et al. 1985). One should therefore expect the evolved level of dispersal to be a decreasing function of dormancy, and vice versa. Dispersing seeds average the environment over a variety of local patches, reducing variability and reducing selection for dormancy. Similarly, with delayed germination, the seeds produced in any one year are exposed to a weighted average of environmental conditions as they germinate over a number of years. In either case, the overall variation in yield is reduced, and this selects for averaging mechanisms such as dormancy and dispersal.

A better understanding of the similarities and differences between the two alternatives, and the conditions under which one or the other strategy may be expected to dominate, requires a careful analysis of quantitative models of population growth in varying environments.

Simplified models for the analysis of complex life history strategies have proved to be very valuable tools, because they provide insight and understanding and identify critical parameters and tendencies.

We (Levin, Cohen, and Hastings 1984) introduced a model to permit examination of dispersal and dormancy strategies. In this paper, we investigate the effects of the spatial and temporal structure of the environment, and of the relative costs of the two strategies. An important aspect is the characterization of the conditions determining the joint optimal strategy for dispersal and dormancy.

THE BASIC MODEL

We use the growth model presented by Levin, Cohen and Hastings (1984) for the seed population of an annual plant in a patchy environment. This model rests upon the basic growth equation for the seed population (before germination) in any one patch j,

$$S^j_{t+1}=S^j_t[GY^j_t(1-D)+(1-G)V] +ADG\ (1/L)\sum_{i=1}^{i=L} Y^i_t S^i_t \ , \qquad [1]$$

and supposes that an equation of this type applies for each genotype. Here G is the constant annual germination fraction and D is the constant dispersal fraction of the seeds; G and D are genotype dependent, but all other parameters are the same for all genotypes. A is the fraction of dispersing seeds that succeed in reaching a habitable patch, V is the survival of dormant nongerminating seeds, and the summation is taken over all L patches because seeds are dispersed uniformly over all patches. Y is a density dependent yield function.

In this paper, for simplicity we assume Y has the form $Y(Z)=K/Z$, where Z is the total density of all the adult plants (of all types) in the patch, and the random variable K is the total seed yield of the patch. Such a saturation yield function is a good representation of the seed yield in plant populations over a wide range of densities (Harper 1977). More general growth functions are treated by Levin, Cohen and Hastings (1984), but the "saturating" case K/Z provides a good first approximation to the more realistic and more common hyperbolic relation. We assume that K is independently distributed among the patches. Within any one patch, K may vary independently among years, or it may show a positive or a negative temporal correlation.

We define an optimal strategy D* for dispersal, or G* for germination, to be an evolutionarily stable strategy, that is, a type that once established cannot be invaded by any rare mutant with a different strategy. For details and a more rigorous definition, see Levin, Cohen, and Hastings (1984). We disregard details of the genetic system, and consider alternative strategies as if they were competing asexual clones.

METHODS

We approximated optimal strategies by numerical simulation of a multi-species version of Equation [1], with one or two very rare (initial frequency=0.0001) mutants competing against a single resident type. The rare types differed from the resident type by ± 0.01 in either G or D. Simulations were usually run with 20 patches, but test runs with a larger number of patches showed no substantial differences. The basic growth equation was iterated for several hundred generations, usually 200-300, until an ESS strategy could be identified clearly. Longer runs of up to 1000 generations were sometimes necessary when the changes in the densities of the rare types were small or inconclusive.

Optimal strategies were evaluated for different levels of the parameters A and V, and for different degrees of environmental variation and of temporal correlation among years. We evaluated D* as a function of G, and G* as a function of D. The joint optimal strategy, if it existed, lay at the intersection of the D* and the G* curves; but such a cross-point does not necessarily represent a stable strategy.

Approximate analytical expressions for dispersal ESSs were derived as in Levin, Cohen and Hastings (1984).

I. THE EFFECT OF DORMANCY ON OPTIMAL DISPERSAL

In general, the presence of a seed bank in the soil reduces the variability of the densities of germinating seeds in the patches.

This reduces the advantage of dispersing seeds relative to nondispersing seeds, and therefore is expected to decrease the level of optimal dispersal in the population. Note that the density of the seed bank in any one patch is a weighted average of the seed densities in the previous years, discounted annually by the fraction $H=(1-G)V$ of surviving nongerminating seeds.

For the case of full germination ($G=1$), the ESS is obtained (for the saturating yield function) by equating the one-year expected yields of dispersing and non-dispersing seeds (Levin, Cohen, and Hastings 1984). If $G < 1$, one can start from such calculations as an approximation, but without formal justification or estimates of error. Equating the one-year yields of dispersers and non-dispersers should bias the estimates against dispersal, since it undervalues the advantages of arriving in younger habitats.

Dispersing seeds sample all environments equally; hence their average one-year yield per seed is

$$E^D = A[G\bar{K}E(S_t^{-1}) + H], \qquad [2]$$

where $\bar{K}$ is the expectation of K and $H = (1-G)V$ is the annual residual factor for nongerminating seeds in the soil.

For nondispersing seeds, environments are sampled with a distribution reflecting the seed distribution; hence, the one-year expectation is

$$E^{ND} = GE(K_{t-1}\, S_t^{-1}) + H \,. \qquad [3]$$

In both [2] and [3], S_t is defined by the recurrence equation

$$S_t = ADG\bar{K} + (1-D)GK_{t-1} + (1-G)VS_{t-1} \,. \qquad [4]$$

Thus, for t large, and $H < 1$,

$$S_t \sim ADG\bar{K}/(1-H) + (1-D)G \sum_{r=1}^{\infty} K_{t-r} H^{r-1}, \qquad [5]$$

from which $E(S_t^{-1})$ and $E(K_t S_t^{-1})$ can be evaluated numerically to any desired accuracy. Set

$$U_t = \sum_{r=o}^{\infty} K_{t-r-1}\, H^r,$$

which is a measure of the underground seed pool. The approximate solution for D^* is obtained by setting $E^D = E^{ND}$. In particular, this gives $D^* = 0$ if A is less than the critical value

$$A' = (H + E(K_{t-1}/U_t))/(H + E(K_{t-1})E(1/U_t)) \qquad [6]$$

$$= 1 + \mathrm{cov}(K_{t-1}, U_t^{-1})/(H + \bar{K}E(U_t^{-1})).$$

When H=0 (no dormant pool), $U_t=K_{t-1}$ and so

$$A' = 1/E(K_t)E(1/K_t) = \hat{K}/\bar{K},$$

where $\hat{K}$ is the harmonic mean of K (see Levin, Cohen, and Hastings 1984). As H increases, the dormant pool increases and reduces environmental variability; this should reduce the selective advantage of dispersal, and increase the dispersal threshold A'. This seems consistent with [6], which suggests that A' increases with H and tends to 1 as H tends to 1. It remains to demonstrate this rigorously.

More generally, these discussions suggest (but this is not proven) that D^* is a decreasing function of H (that is, of the size of the dormant pool), and that G and V affect D^* primarily through the product $H=(1-G)V$. These considerations motivate the numerical investigations reported next.

SIMULATIONS AND ANALYTICAL RESULTS

Some selected results are shown in Figure 1. The results obtained by simulations and by the analytical approximation (dotted) are shown together, for comparison.

In each case, D^* is an increasing function of the germination fraction G, and a decreasing function of the survival of nongerminating seeds V. Figure 1 shows that the factor $H=(1-G)V$ explains the effects of G and V on optimal dispersal; this agrees with the theoretical predictions. We also see that the critical A' (below which $D^* = 0$) increases when H increases, and that $D^*=1$ at A=1 for all levels of H, as suggested by the theoretical analysis. The detailed approximation breaks down over a range of the parameters, for which the one-year yield is not an adequate measure of the relative advantages of dispersal. It remains to investigate these cases in more detail.

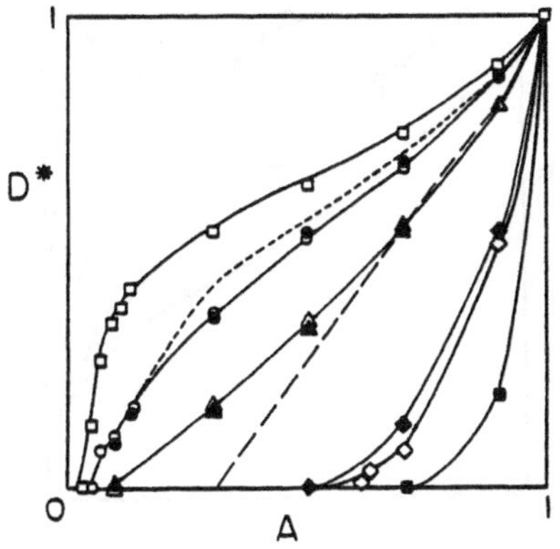

Figure 1. Optimal dispersal D* as a function of the effectiveness of dispersal A, for different levels of the residual factor for dormant seeds, $H=(1-G)V$. $K_1=1$, $K_2=100$. $P(K_1)=0.5$. Here and in later figures, K_1 and K_2 refer to the possible environmental types; subscripts do not refer to years.

G	1	.01	.8	.01	.4	.01	.2	.01
V		.1	.5	.3	.5	.7	.9	.9
H	0	.1	.1	.3	.3	.7	.72	.9
Symbol	□	○	●	△	▲	◆	◇	■

II. THE EFFECT OF DISPERSAL ON THE OPTIMAL DORMANCY

The examination of dormancy (see Cohen, 1966) yields complementary results. In particular, in the absence of dispersal, the optimal germination fraction G* without dispersal is given implicitly by

$$1/V=E(S_t/S_{t+1}) \qquad [7]$$

(Ellner, 1985). Analysis suggests that G* is a decreasing function of the seed survival V, and that $G^*=1$ is reached at the critical level $V'=\hat{K}/\bar{K}$, which is identical to the critical level of A' for $D^*=0$ without dormancy (Ellner, 1985).

The inclusion of dispersal in the germination model changes its structure completely. We cannot derive an analytical expression for

G*, and therefore present here the results of extensive simulations from which we obtained the optimal germination fraction over a range of parameters, and at different levels of dispersal (Figure 2).

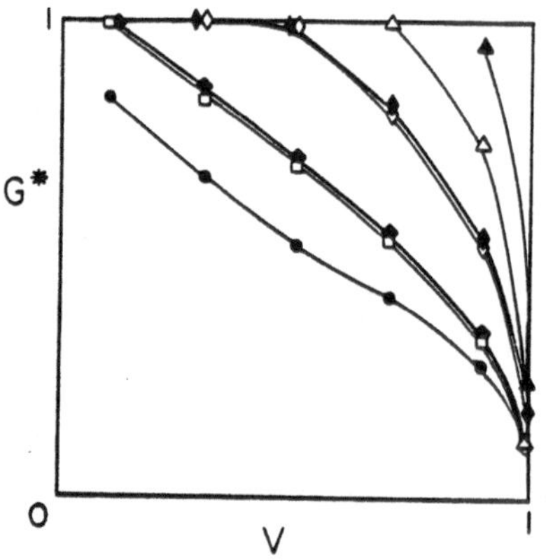

Figure 2. Optimal germination fraction G* as a function of the survival of dormant seeds V, at different levels of the effective dispersal fraction F=AD/(1-D). The environment is as in Fig. 1.

D	.01	.4	.2	.6	.8	.6	.8
A	.1	.1	.3	.3	.7	.9	.9
F=AD/(1-D)	.001	.066	.075	.45	.47	1.35	3.6
Symbol	●	□	◆	◇	♦	△	▲

The results demonstrate that G* is an increasing function of dispersal D, and of the effectiveness of dispersal A. A and D appear to affect optimal germination primarily through the factor F=AD/(1-D) representing the effective dispersal fraction of the seeds. Also, the critical level V' at which G*=1 increases as D and A increase. As in the absence of dispersal, G* approaches zero as V approaches 1.

THE JOINT OPTIMAL STRATEGY FOR DISPERSAL AND DORMANCY

When both dispersal and dormancy are subject to selection, the joint optimal strategy is obtained as the intersection of the curves of D*(G) and G*(D), as shown in Figures 3, 4 and 5. A graphical analysis of the evolutionary stability of the intersection point suggests that it may be a stable equilibrium when there is no temporal autocorrelation (Fig. 3a) in environmental variation; however when the environment varies periodically, the intersection points are unstable, and there are two competing boundary equilibria (Fig. 3b). Under some circumstances, the two boundary strategies can coexist in a joint equilibrium, while under other conditions a single type wins.

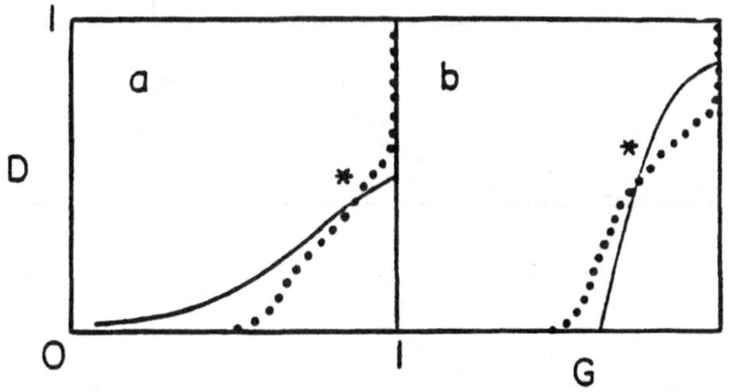

Figure 3. D*(G) and G*(D) in random and periodic environments. D*(G)= continuous curves, G*(D)= dotted curves. K_1=1, K_2=100.

a. Random variation, P(K_1)=0.33. A=0.5, V=0.7. The intersection of the curves represents an evolutionary stable joint D*, G* strategy.

b. Unsynchronized periodic variation. A=0.5, V=0.6. K_1 occurs every third year. The intersection is an unstable saddle point, and there are two evolutionary stable boundary strategies.

In Figures 3-6, the scale of both D and G is between 0 and 1 in each of the GxD squares. Optimal D, G strategies are indicated by star symbols in the figures.

The dependence of D*(G) and G*(D), and of the joint optimal strategy, on A and V and on environmental variation is shown in Figure 4. The ratio between optimal dispersal and optimal dormancy depends on the ratio between the effectiveness of dispersal A, and the survival of dormant seeds V. A higher value of A relative to V favors higher dispersal and lower dormancy, and vice versa.

When A is large relative to V, the optimal joint strategy is to specialize in dispersal with zero dormancy, i.e., G*=1. On the other hand, when V is large relative to A, the optimal joint strategy is to specialize in dormancy with zero dispersal.

The optimal levels of both dispersal D and dormancy (1-G) increase when both A and V increase. This can be seen in Fig. 4 by following the increase of both A and V along the diagonal.

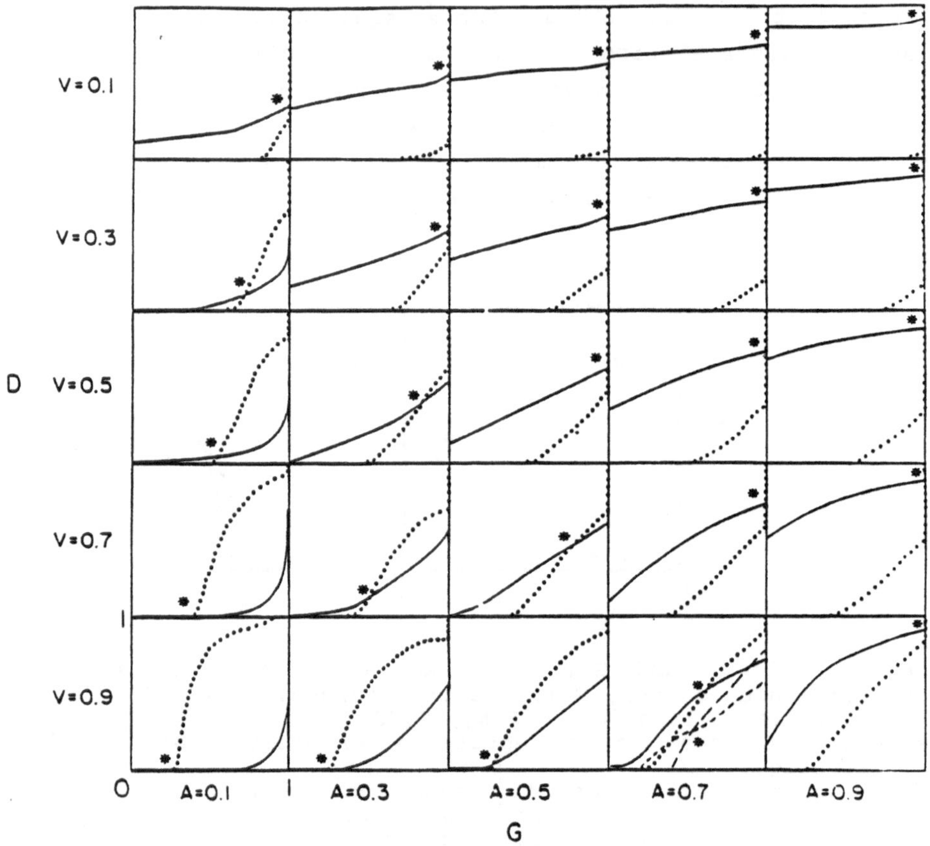

Figure 4. D*(G) and G*(D) at different combinations of A and V in a randomly varing environment. D*(G)= continuous curves, G*(D)= dotted curves. The stable joint equilibrium points are indicated by a * sign. K_1=1, K_2=100. $P(K_1)$=0.5. In the particular case of A=0.7, and V=0.9, optimal curves are also shown for a reduced amplitude of environmental variation, i.e., K_1=1, K_2=10, with the result that both optimal dispersal and optimal dormancy are reduced.

The optimal levels of dispersal and dormancy decrease as the variability of the environment decreases, i.e., K_2/K_1=10 instead of 100, for any given level of A and V (see A=0.7, V=0.9 in Fig. 4). This is as expected by the qualitative arguments presented before.

When bad years are very bad and good conditions occur relatively rarely, the success of dispersing seeds is determined largely by A times the probability of good years, while dormant seeds have to survive for the average interval between good years, and the surviving fraction is H=(1-G)V raised to the power of the duration of the interval. Thus for equal levels of A and V, dispersal rather than dormancy is expected to be the favored strategy under these conditions of rare good years. On the other hand, when bad years are rare, in which case dormancy should be favored over dispersal.

Figure 5 shows the optimal strategy for fixed levels of A and V for different probabilities of the occurrence of the good environment (K_2=100). As expected, when good years are rare, dispersal is favored over dormancy, while the opposite is the case when good years are common.

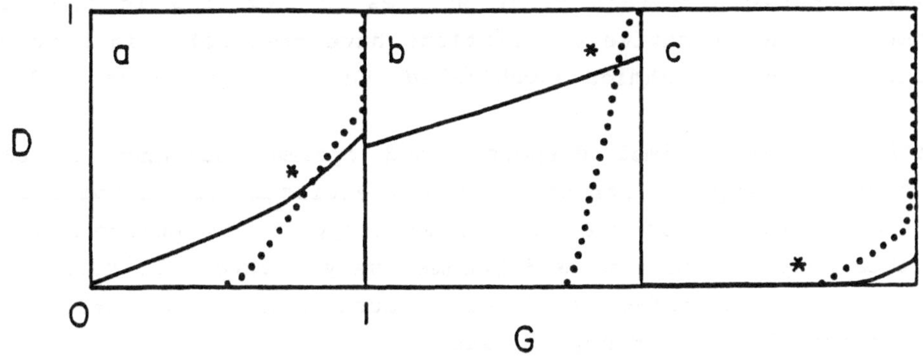

Figure 5. The joint optimal strategy in different temporal structures. A=0.3, V=0.5, K_1=1, K_2=100. a. Equal probabilities for good and bad years, i.e., $P(K_1)$=0.5; b. High probability for bad years, i.e., $P(K_1)$=0.9; c. High probability for good years, i.e., $P(K_1)$=0.1.

DISCUSSION

Our analyses and simulations lead to several conclusions:

1. Optimal dispersal is a decreasing function of the size of the seed bank, as represented by the annual residual factor of dormant seeds H=(1-G)V. When this factor increases, it decreases the variability among years of the seed densities in any single patch, and thus selects for a lower level of optimal dispersal.

2. Optimal dormancy is a decreasing function of dispersal and of the effectiveness of dispersal A. A and D exert their influence primarily through the factor F=AD/(1-D), the fraction of the overall annual yield attributable to the input of dispersing seeds. This input reduces local variability, and therefore higher inputs select for lower levels of dormancy.

3. The ratio between optimal dispersal and optimal dormancy in the joint optimal strategy is affected by the ratio between the effectiveness of dispersal A and the survival of dormant seeds V. Plant species or families should differ in their levels of A and V. Thus, in any given environment, some species will have high A/V ratios and consequently adopt strategies of high dispersal and low dormancy, while others will have low A/V ratios and adopt strategies of low dispersal and high dormancy. Therefore, the distribution of dispersal and dormancy may be expected to be negatively correlated when one compares different plant species or families in the same environment. Such negative correlations have been noted in a number of studies (Ellner and Shmida unpublished, Venable and Lawlor 1980).

The ratio between optimal dispersal and optimal dormancy is also affected by the temporal structure of the environment. Dispersal is favored over dormancy as the optimal strategy if the environment is unfavorable most of the time and becomes very favorable relatively infrequently. This effect will also cause a negative correlation between dispersal and dormancy in nature.

It should be noted however that both optimal dispersal and optimal dormancy are expected to increase if the variability of the environment increases when A and V are kept constant, or when both A and V increase simultaneously in the same environment. Comparisons of plant species growing in different environments should therefore show a positive correlation between dispersal and dormancy if A and V are about the same in most species, or if both A and V change considerably but to about the same extent.

The observed overall correlation between dispersal and dormancy in natural systems may therefore be negative or positive or zero, depending on the patterns of variation of A and V relative to the variation in the environment (see Figure 6). The correlation between dispersal and dormancy should be more positive when similar or related species growing in a wide range of habitats are compared. It should be more negative when plants of different orders or families growing in the same habitat are compared.

4. In environments which vary periodically, we find that there is no single joint optimal strategy with intermediate levels of dispersal and dormancy. Although we could not characterize analytically the stability of the equilibria, we used simulations to

investigate stability and found that the only stable equilibria were at the boundaries G*=1 or D*=0. It was found however that two such boundary strategies could coexist in a stable frequency dependent equilibrium, which would be the eventual evolutionary equilibrium reached between many competing mutants with a wide range of dispersal and dormancy levels.

The degree of asynchronous local periodicity in natural environments may be very variable. Some disturbance cycles, for example, have a strong periodic element. If a single joint strategy for dispersal and dormancy in periodic habitats is indeed evolutionary unstable, then this may have important implications. Under these conditions, dispersal and dormancy may be expected to evolve as alternative divergent strategies, even if they start from very similar initial states. These divergent paths may lead to rapid ecological and eventual evolutionary specializations for either dispersal or dormancy as primary strategies.

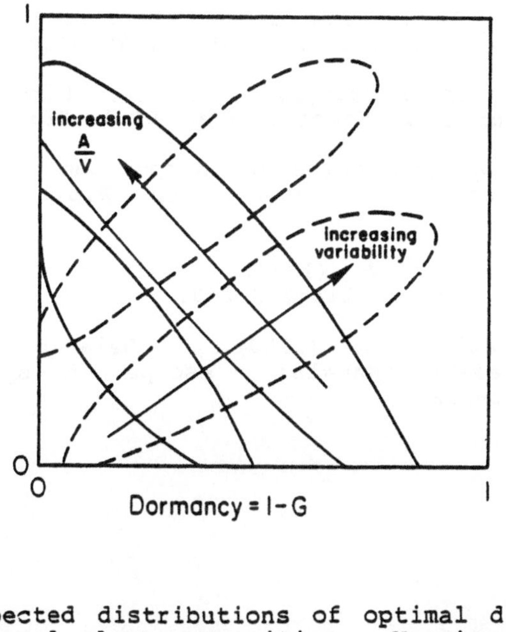

Figure 6. Expected distributions of optimal dispersal and optimal dormancy in natural plant communities. Varying A/V while keeping A and V constant, or changing the ratio between bad and good years while maintaining the overall variability constant, will cause a strong negative correlation between optimal dispersal and optimal dormancy (continuous lines). Variation in the variability of the environment, or the levels of both A and V, will cause a positive correlation between optimal dispersal and optimal dormancy (dotted lines).

ACKNOWLEDGEMENT

This research was supported in part by U.S.-Israel Binational Science Foundation Grant 79/1970 to Dan Cohen and Simon A. Levin; and by Hatch and NSF Grants to Simon A. Levin.

REFERENCES

Cohen, D. (1966). Optimizing reproduction in a randomly varying environment. J. Theoret. Biol. 16:267-282.

Ellner, S.P. (1985). ESS germination strategies in randomly varying environments. II. Reciprocal yield models. Theoret. Pop. Biol. 28:80-116.

Harper, J.L. (1977). Population Biology of Plants. Academic Press. New York. 892 pages.

Klinkhamer, P.G.L., de Jong, T.J., Metz, J.A.J., and Val, J. (1985). Life history tactics of annual organisms: the joint effects of dispersal and delayed germination. Submitted.

Levin, S.A. (1976). Population dynamics models in heterogeneous environments. Ann. Rev. Ecol. Syst. 7:287-310.

Levin, S.A., Cohen, D. and Hastings, A. (1984). Dispersal strategies in patchy environments. Theoret. Pop. Biol. 26:165-191.

Motro, U. (1982). Optimal rates of dispersion and migration in biological populations. Theoret. Pop. Biol. 21:394-411 and 412-429.

Templeton, A.R. and Levin, D.A. (1979). Evolutionary consequences of seed pools. Am. Natur. 114:232-249.

Venable, D.L. and Lawlor, L. (1980). Delayed germination and dispersal in desert annuals: escape in space and time. Oecologia 46:272-282.

SEGREGATION STRUCTURES OF COMPETING SPECIES MEDIATED BY A DIFFUSIVE PREDATOR

Y. Kan-on† and M. Mimura††

†Hiroshima National College of Maritime Technology

††Department of Mathematics, Hiroshima University

It has been suggested that, in some circumstances, predation may tend to increase species diversity in competitive communities, which is the so called *predator-mediated coexistence hypothesis*. Experimental and observational evidence includes studies by, for instance, Paine[1966], Harper[1969] and Connel[1970]. Intuitively speaking, we can argue as follows: consider two prey and one predator species for simplicity. The two prey species are competing so strongly that one would become extinct but for the presence of the predator species, which exerts higher predation pressure on the competitively dominant species. Then coexistence of the two prey species is possible, because competitive pressure is relaxed. From theoretical points of view, Parrish and Saila[1970], May[1971], Fujii[1977], Vance[1978], Hsu[1980], Hutson and Vickers[1983] and Takeuchi and Adachi[1983] studied possibility of predator-mediated coexistence by using Lotka-Volterra equations of one predator and two competing prey. The system takes the form

$$\left\{\begin{array}{l} \dfrac{du_1}{dt} = u_1 (a_1 - b_1u_1 - c_1u_2 - k_1v) \\ \dfrac{du_2}{dt} = u_2 (a_2 - b_2u_1 - c_2u_2 - k_2v) \quad , \quad t > 0 \quad . \\ \dfrac{dv}{dt} = v (- r + \alpha_1k_1u_1 + \alpha_2k_2u_2) \end{array}\right. \tag{1}$$

where $u_1(t)$, $u_2(t)$ and $v(t)$ are the population densities of two prey species and one predator at time $t > 0$. a_1, a_2 and r are the intrinsic growth and death rates, respectively. b_1, b_2 and c_1, c_2 are measures of intraspecific and interspecific competition, respectively. k_1 and k_2 are the predation rates suffered by the prey species. α_1 and α_2 are the conversion factors of the predator. To express (1) in dimensionless variables, we let

$$\bar{t} = a_1 t,\quad \bar{u}_1 = b_1 u_1/a_1,\quad \bar{u}_2 = c_2 u_2/a_1,\quad \bar{v} = k_2 v/a_1,$$
$$\bar{r} = r/a_1,\quad \bar{\alpha}_1 = \alpha_1 k_2/b_1,\quad \bar{\alpha}_2 = \alpha_2 k_2/c_2,\quad k = k_1/k_2,$$
$$a = a_2/a_1,\quad b = b_2/b_1,\quad c = c_1/c_2.$$

Hereafter, the overbars will be omitted. Then (1) becomes

$$\begin{cases} \dfrac{du_1}{dt} = u_1 (1 - u_1 - cu_2 - kv) \\ \dfrac{du_2}{dt} = u_2 (a - bu_1 - u_2 - v) , \quad t > 0 . \\ \dfrac{dv}{dt} = v (- r + \alpha_1 k u_1 + \alpha_2 u_2) \end{cases} \tag{2}$$

System (2) models the dynamics of one predator and two competing prey species which are all spatially uniformly distributed. We define A by

$$A = \begin{pmatrix} 1 & c & k \\ b & 1 & 1 \\ -\alpha_1 k & -\alpha_2 & 0 \end{pmatrix}$$

and state the following two theorems (Mimura and Kan-on[1986]):

Theorem 1 Suppose $|A| \neq 0$. If there is not an interior equilibrium point in $R_+^3 = \{(u_1,u_2,v);u_1,u_2,v>0\}$, then there are equilibrium points on the boundaries of R_+^3, some of which are locally stable. Solutions of (2) approach one of them as t tends to infinity.

Theorem 2 Suppose $|A| = 0$. Let A_{ij} be the (i,j)-th element of the cofactor matrix of A (i,j=1,2,3).

(i) If $A_{i_0 1} + aA_{i_0 2} - rA_{i_0 3} = 0$ for some $i_0 \in \{1,2,3\}$, then the assertion of theorem 1 is still valid.

(ii) If $A_{i1} + aA_{i2} - rA_{i3} = 0$ for any i, then the solution of (2) is represented by

$$u_1(t)^{A_{i1}} u_2(t)^{A_{i2}} v(t)^{A_{i3}} = u_1(0)^{A_{i1}} u_2(0)^{A_{i2}} v(0)^{A_{i3}}, \quad i=1,2,3.$$

Hereafter, in order to simplify the discussion, we fix parameters a, b, c, k, α_1 and α_2 so that

$$|A| > 0 \quad \text{and} \quad a < b,\ 1/c$$

and take r as a free parameter. The latter condition implies that species (u_1) always exists while (u_2) becomes extinct in the absence of a predator. It is easy to prove that (2) has an interior equilibrium point $u^* = (u_1^*, u_2^*, v^*)$ in R_+^3 if $ak > 1$ and

$$0 \le \underline{r}(k) = \alpha_2 \frac{ak-1}{k-c} < r < \alpha_1 \frac{k(ak-1)}{bk-1} = \overline{r}(k) \tag{3}$$

(Figure 1).

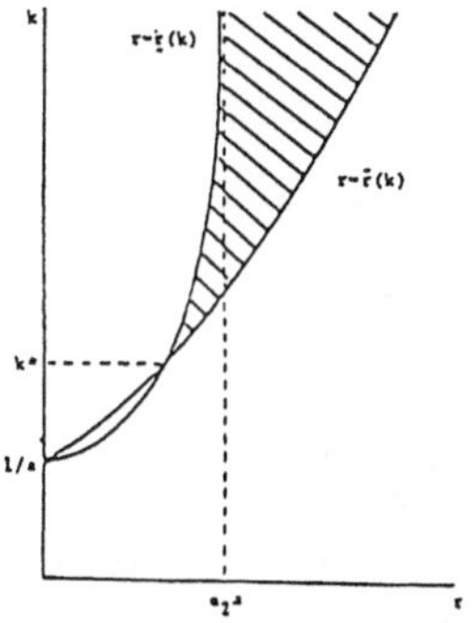

Figure 1 Shaded region is characterized by the existence of the interior equilibrium point E_{+++} in (k,r)-space. k^* is a positive solution of $|A| = 0$.

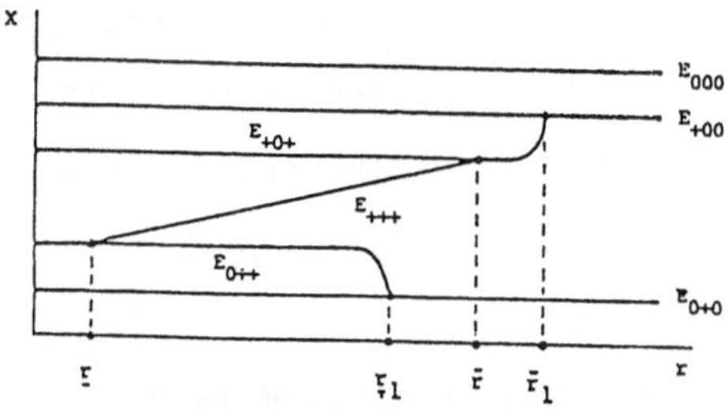

Figure 2. Schematic global bifurcation diagram of equilibrium points with respect to $r \in R_+$. E_{000}, E_{0+0} and E_{+++} denote (0,0,0), (0,a,0) and (u_1^*, u_2^*, v^*), respectively, and E_{++0}, E_{+0+} and E_{0++} also denote the corresponding equilibrium points. X is a certain solution space.

The global picture of the equilibrium points of (2) with respect to $r \in R_+$ is shown in Figure 2. It is known that E_{+++}-branch is connected with E_{0++}-branch at $r = \underline{r}$ and E_{+0+}-branch at $r = \overline{r}$, respectively. By simple calculations, we find that E_{0++}-branch is stable for $r \in (0,\underline{r})$ and unstable for $r \in (\underline{r},\underline{r}_1)$, while E_{+0+}-branch is unstable for $r \in (0,\overline{r})$ and stable for $r \in (\overline{r},\overline{r}_1)$. As for the stability of E_{+++}-branch, we have the following theorem:

Theorem 3 Consider (2) under the assumption of (3). Let r be a free parameter. Then either

(i) E_{+++}-branch is stable

or

(ii) r_* and r^* are two critical values of r such that E_{+++}-branch is unstable for $r \in (r_*,r^*)$ and Hopf bifurcations from $u = u^*$ occur at $r = r_*$ and r^*.

It should be noted that the occurrence of such Hopf bifurcations was already suggested by Fujii[1977], Vance[1978], Kawasaki and Teramoto [private communication] and Takeuchi and Adachi[1933]. Case (ii) in Theorem 3 can be studied more precisely. Let us show one example for some parameter choices. As r is increasing, there occurs a Hopf bifurcation from E_{+++}-branch to stable periodic solution branch, and, as r continues to increase, the periodic solution undergoes cascades

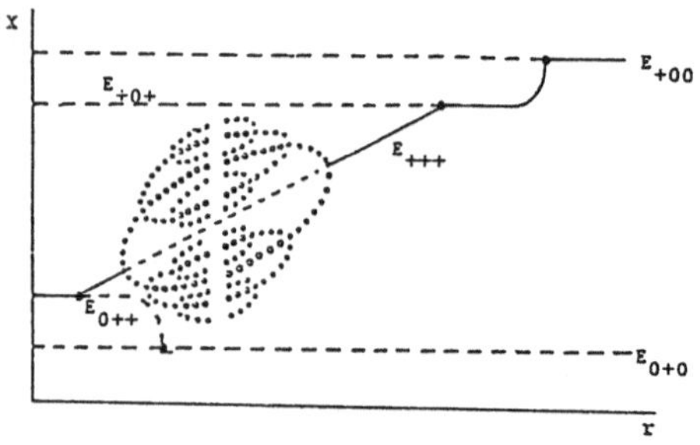

Figure 3 Global picture of equilibrium points and periodic solutions with respect to $r \in R_+$ where a=0.992, b=1.5, c=1.0, α_1= α_2=0.5 and k=10.0.
—— : stable equilibrium branch,
---- : unstable equilibrium bramch,
•••• : stable periodic branch,
oooo : unstable periodic branch.

of period doublings and eventually exhibits apparently chaotic behavior. Finally it condenses back into E_{+++}-branch. The details are stated in Mimura and Kan-on[1986].

Thus Theorem 1 implies that for any fixed $r \in (\underline{r},\overline{r})$, predator-mediated coexistence occurs, while it does not for $r \in (0,\underline{r})$ or $(\overline{r},\infty)$.

Here, we address the question of whether such coexistence is possible for $r \in (0,\underline{r})$ or $(\overline{r},\infty)$ if the three species can migrate solely by *diffusion*. The equations treated here are of the form

$$\begin{cases} \dfrac{\partial u_1}{\partial t} = d_1 \dfrac{\partial^2 u_1}{\partial x^2} + u_1 (a - u_1 - cu_2 - kv) \\ \dfrac{\partial u_2}{\partial t} = d_2 \dfrac{\partial^2 u_2}{\partial x^2} + u_2 (1 - bu_1 - u_2 - v) , \quad t > 0 , \ x \in (0,1). \\ \dfrac{\partial v}{\partial t} = D \dfrac{\partial^2 v}{\partial x^2} + v (- r + \alpha_1 ku_1 + \alpha_2 u_2) \end{cases} \tag{4}$$

where d_1, d_2 and D are diffusion rates of the three species. The boundary and initial conditions are taken to be

$$\frac{\partial u_1}{\partial x} = \frac{\partial u_2}{\partial x} = \frac{\partial v}{\partial x} = 0 \quad , \quad x = 0,1 \tag{5}$$

and

$$u_i(0,x) = u_{0i}(x) \geq 0 \ (i=1.2), \ v(0,x) = v_0(x) \geq 0 \ , \ x \in [0,1] \tag{6}$$

respectively. Let us impose the following conditions:

(H.1) $a < 1/c < b$ and $\alpha_1 = \alpha_2 = \alpha$

The former implies that when the predator is absent, one prey species (u_1) always exists and the other (u_2) becomes extinct even if the species can migrate by diffusion (de Mottoni[1979] and Hsu[preprint]).

(H.2) $0 < \underline{r} < r$ or $\overline{r} > r$

This indicates that either (u_1) and (u_2) becomes extinct in the absence of diffusion (see also Theorem 1).

Our aim is to investigate possibility for coexistence of the competing species under the assumptions (H.1), (H.2). First, we should note that, if all of the diffusion rates are large enough, solutions of (3)-(5) become spatially homogeneous for large time (Conway, Hoff and Smoller[1978]), and, hence, one of the two competing species becomes extinct by (H.2); that is, predator-mediated coexistence never occurs. Then, when at least one of the diffusion rates is not large, is predator-mediated coexistence possible? The results of some numerical experiments appear in Figure 4. When d_1 and d_2 are sufficiently small relative to D, the two competing species coexist in segregated patterns, and they fail to do so otherwise. This indicates that the difference between the diffusion rates of prey and predator is important for the occurrence of coexistence. To study this, we obtain the numerical global structure of *stationary* solutions with respect to ε (we put $d_1 = \varepsilon^2$, $d_2 = \varepsilon^2 d$ with $d = O(1)$ and D is large enough). Figure 5 shows that the appearance of spontanous bifurcation depends upon ε. That is, there is an ε_0 such that two non-uniform stationary solutions exist for $0 < \varepsilon < \varepsilon_0$, while for $\varepsilon_0 < \varepsilon$ does not occur. Moreover, one of the solutions is *stable* and the other is *unstable*. We should remark here that these solution branches do not bifurcate from the trivial branch of equilibrium point $u = u^*$, because $u = u^*$ does not exist when $r > \overline{r}$. The existence and stability of these solutions have not been rigorously proved. However, we note that when ε is sufficiently small, we can apply the singular perturbation techniques of Mimura

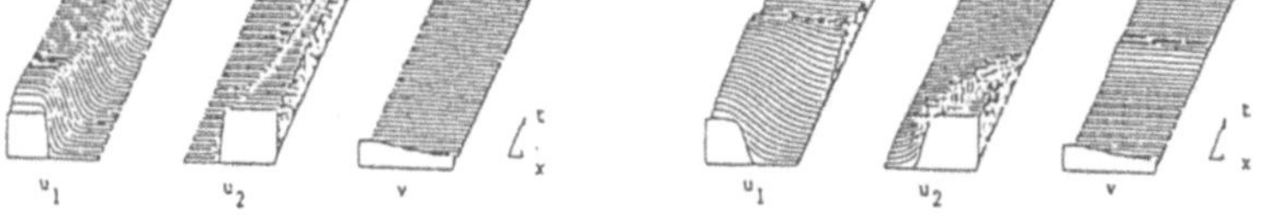

(a) Segregated pattern of solutions where $a=0.95$, $b=1.5$, $c=1.0$, $r=3.1$, $k=10.0$, $\alpha_1=\alpha_2=0.5$, $d_1=d_2=0.001$ and $D=5.0$.

(b) Spatially homogeneous pattern of solutions where a, b, c, r, k, α_1, α_2 and D take the same values as in (a) except for $d_1=d_2=0.1$.

Figure 4

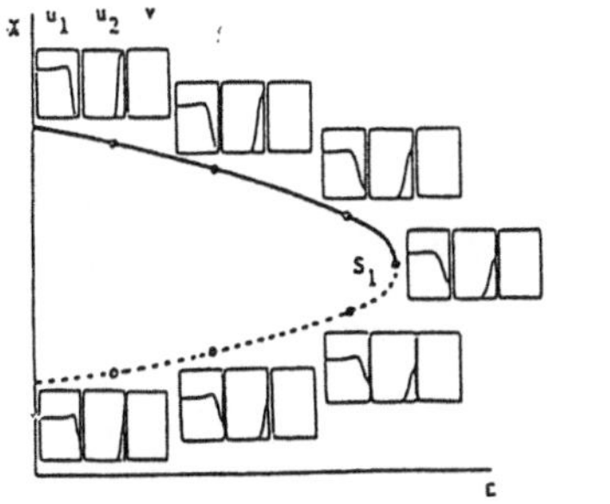

Figure 5 Shapes of u_1, u_2, v on for chosen $r > \bar{r}$. The solid curve is a stable branch and the broken curve is an unstable branch. This is computed with a=0.95, b=1.5, c=1.0, r=3.1, k=10.0, $\alpha_1=\alpha_2=0.5$, d=1.0 and D=5.0.

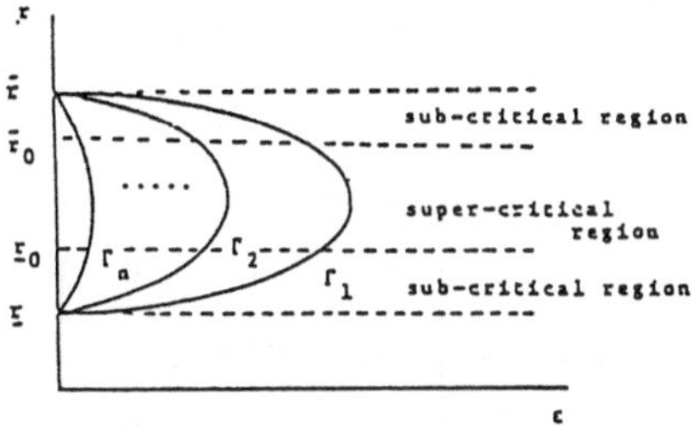

Figure 6 Primary bifurcation curves Γ_n (n=1,2, .., ∞) in (r,ε)-space when D is sufficiently large.

and Fife[1986] to construct approximate solutions to such non-uniform solutions (see Mimura and Kan-on[1986]). This numerical experiment indicates that there exist *stable* non-uniform stationary solutions in which segregated coexistence of the competing species occurs even if r is chosen as $\bar{r} < r$. It is concluded that *a suitable difference in the diffusion rates enchances possibility of coexistence.*

Next, we seek to clarify the reason for which such stable stationary solutions exist for $r > \bar{r}$. To this purpose, we draw a picture of global parametric dependence of ε and r on non-uniform stationary solutions of (3) and (4). The analytic tool used to study such solutions is local bifurcation analysis with respect to ε and r, while a, b, c, k and α are fixed to satisfy (H.1) and (H.2). Furthermore, D is assumed to be large in order to avoid mathematical technicalities. We first fix r and ε as $(r,\varepsilon) \in (\underline{r},\bar{r})\times R_+$, under which a prey coexistence equilibrium point E_{+++} exists. To consider the stability of E_{+++}, we study the linear eigenvalue problem associated with (4) and (5)

$$\begin{cases} \lambda(\varepsilon,r)\Phi = D(\varepsilon)\dfrac{d^2\Phi}{dx^2} + M(r)\Phi , & x \in (0,1) , \\ \dfrac{d\Phi}{dx}(0) = \dfrac{d\Phi}{dx}(1) = 0 , & \end{cases} \tag{7}$$

where

$$\Phi = {}^t(\varphi_1,\varphi_2,\varphi_3) ,$$

$$D(\varepsilon) = \begin{pmatrix} \varepsilon^2 & 0 & 0 \\ 0 & \varepsilon^2 d & 0 \\ 0 & 0 & D \end{pmatrix}$$

and $M(r)$ is the Jacobian matrix at E_{+++} associated with the dynamics of (3). The set $(r,\varepsilon) \in (\underline{r},\overline{r})\times R_+$ where (6) has zero eigenvalues corresponds to primary (stationary) bifurcation points at which E_{+++} loses its stability property. One finds that this set consists of an infinite number of curves $\{\Gamma_n\}(n=1,2,\cdots,\infty)$ by using the eigenfunctions $\{\cos(n\pi x)\}(n=1,2,\cdots,\infty)$ for $r \in (\underline{r},\overline{r})$. For sufficiently large D, $\{\Gamma_n\}$ has the following properties: (i) the $\{\Gamma_n\}$ never intersect each other (there is no double criticality; (ii) the $\{\Gamma_n\}$ are ordered with respect to n, and (iii) Γ_1 is the primary bifurcation curve first appearing when ε decreases (Figure 6). It should be noted that if D is not so large, the $\{\Gamma_n\}$ do not satisfy the above properties and the situation is extremely complicated (see Fujii et al.[1982] for models of one prey and one predator). We arbitarily fix $r_c \in (\underline{r},\overline{r})$ and take ε as a bifurcation parameter. Then, local bifurcation analysis indicates that for any $(r_c,\varepsilon_c) \in \Gamma_n$, there exists $\sigma_0 > 0$ such that (3) has a unique one-parameter family of stationary solutions $(\varepsilon(\sigma),u(\sigma)) \in R_+\times X$ bifurcating from (ε_c,u^*) for $|\sigma| < \sigma_0$, where X is an appropriate solution space such as $(H_N)^3$ with H_N = closure of $\{\cos(n\pi x)\}$ in $H^2(I)$. Furthermore one can see by computations that there are $\underline{r}_0$ and $\overline{r}_0$ such that the primary bifurcation branch from a point on Γ_1, say D_1-branch, is *super-critical* for any $r \in (\underline{r}_0,\overline{r}_0)$, while it is *sub-critical* for any $r \in (\underline{r},\underline{r}_0)$ or $(\overline{r}_0,\overline{r})$. Thus, we know the local structure of stationary

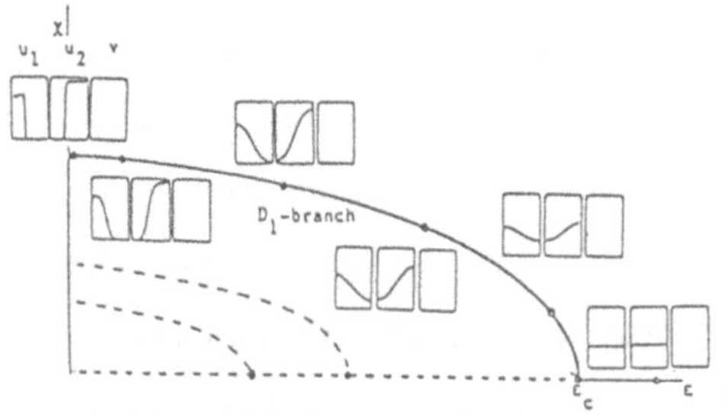

Figure 7 Shapes of u_1, u_2, v on D_1-branch for chosen $r \in (\underline{r}, \overline{r})$. a, b, c, k, α_1, α_2, d and D take the same values as in Figure 5 except for r=2.0.

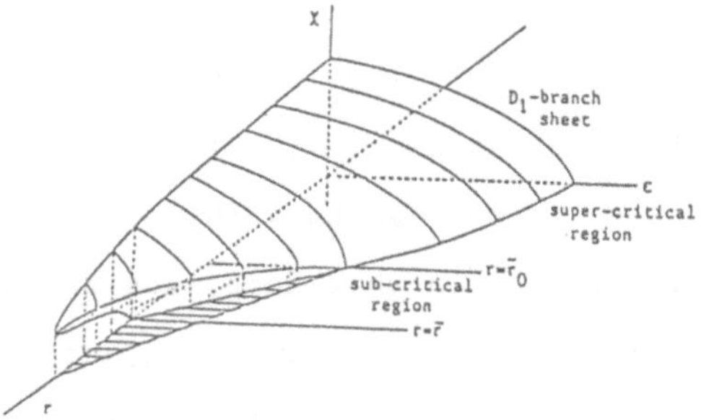

Figure 8 Schematic D_1-branch sheet in (r,ε)-space. The upper sheet is stable and the lower one is unstable.

solutions of (3) near (r_c, ε_c). However, we do not yet possess any information on the global destination of the D_1-branch. Fugure 7 shows a numerically obtained picture of the global behavior of the D_1-branch for fixed $r \in (\underline{r}_0, \overline{r}_0)$. It is observed that the D_1-branch exists in the limit $\varepsilon \downarrow 0$, in which no secondary bifurcation point appears (of course D is taken to be large).

Let us consider the stability of the D_1-branch. First, we recall (ii) of Theorem 3 and fix parameters such that, except for r and ε, (ii) of Theorem 3 holds. Then Figure 7 indicates that the D_1-branch is stable and never changes stability for $\underline{r}_0 < r < \overline{r}_0$ (super-critical case). This implies the well-known *diffusion-induced instability*. On the other hand, if r is increasing ($\overline{r}_0 < r < \overline{r}$), the branch deforms to become sub-critical, and it proceeds to the right and turns back in the limit $\varepsilon \downarrow 0$; that is, the D_1-branch is originally unstable but recovers its stability. If r is continued to increase to $\overline{r}$, then E_{+++} is absorbed by E_{0++} and no bifurcation from E_{+++} occurs. If r further increases beyond $\overline{r}$ ($r > \overline{r}$), the equilibrium point E_{+++}, in which coexistence of the prey species occurs, no longer exists and either E_{+0+} or E_{+00} is a stable equilibrium point. However, the D_1-branch still exists with a limit point S_1. This implies that there are two different types of solution branches in the limit $\varepsilon \downarrow 0$ (Figure 5). Moreover, we find that S_1 is decreasing with r and the D_1-branch disappears for large r (Figure 8). In this case, non-uniform stationary solutions no longer exist. In summary, it is conjectured that when ε is sufficiently small, there exist stable stationary solutions which exhibit sharp spatial segregation between u_1 and u_2 even for some $r \in (\underline{r},\overline{r})$. In other words, suitable differences in the diffusion rates between prey and predator make coexistence of competing prey species possible even if such coexistence cannot occur in the absence of diffusion migration.

References.

J. H. Connell, A predator-prey system in the marine intertidal region. I. Balanus glandula and several predator species of thais, Ecol. Monogr. 40, 1970, 49-78.

E. Conway, D. Hoff and J. Smoller, Large time behavior of solutions of systems of nonlinear reaction-diffusion equations, SIAM J. Appl. Math. 35, 1978, 1-16.

H. Fujii, M. Mimura and Y. Nishiura, A picture of global diagram in ecological interacting and diffusing systems, Physica 5D, 1982, 1-42.

K. Fujii, Complexity-stability relatioship of two-prey-one-predator species system model: Local and global stability, J. Theoret. Bio. 69, 1977, 613-623.

J. L. Harper, The role of predation in vegetational diversity, in Diversity and Stability in Ecological Systems (G.M.Woodwell and H. H. Smith, Eds), Brookhaven National Laboratory, Upton, N. Y., 1969, 48-62.

S. B. Hsu, On general two-species competition model with diffusion, preprint.

S. B. Hsu, Predator-mediated coexistence and extinction, Math. Biosci. 54, 1980, 231-269.

V. Hutson and G. T. Vickers, A criterion for permanent coexistence of species with an application to a two-prey one-predator system, Math. Biosci. 63, 1983, 252-269.

K. Kawasaki and E. Teramoto, Private communication.

R. M. May, Stability in multispecies community models, Math. Biosci. 12, 1971, 59-79.

M. Mimura and P. C. Fife, A 3-component system of competition and diffusion, Hiroshima Math. J. 16, 1986, 189-207.

M. Mimura and Y. Kan-on, Predation-mediated coexistence and segregation structures, in Patterns and Waves (T. Nishida, M. Mimura and H. Fujii, Eds), Kinokuniya/North-Holland, 1986, 129-155.

P. de Mottoni, Qualitative analysis for some quasi-linear parabolic systems, Institute of Math. Polish Academy of Sci. Zam 11/79, 1979.

R. T. Paine, Food web complexity and species diversity, Amer. Natur. 100, 1966, 65-75.

J. D. Parrish and S. B. Saila, Interspecific competition, predation an and species diversity, J. Theoret. Biol. 27, 1970, 207-220.

Y. Takeuchi and N. Adachi, Existence and bifurcation of stable equilibrium in two-prey, one-predator communities, Bull. Math. Biol. 45, 1983, 877-900.

R. R. Vance, Predator and resource partitioning in one predator-two prey model communities, Amer. Natur. 112, 1978, 797-813.

A spatially aggregating population model involving size-distributed dynamics

Masayasu Mimura and Shinya Takigawa
Department of Mathematics
Hiroshima University, Hiroshima 730 Japan

1. Introduction.

Many biological populations tend to aggregate in response to concentration gradients of a chemoattractant secreted by themselves. The present study is motivated by the aggregation observed in *Blattella germanica*. At properly high densities, *B. germanica* individuals grow faster than do isolated ones, and they aggregate so as to maintain such densities (Ishii [1969]). This prominent feature suggests that there is evidently a correlation between the growth rate and population density of individuals.

In order to investigate the dependence of aggregation on the growth rate of individuals, we propose a chemotactic population model incorporating size (length, volume or weight)-distributed dynamics. To model chemotactic movement of individuals, we use the well-known model for aggregation of slime mold amoebae by Keller and Segel [1970]. Also, for the dynamics of size growth, we use a model similar to age-dependent models (see, for instance, Webb [1985]) and cell size distribution models (Diekmann, Heijmans and Thieme [1984]). The one space-dimension model is

$$u_t + \varepsilon[g(r(t,x))(h(z)u)_z + \kappa(z)u] = (\mu u_x - \chi(u,v)v_x)_x, \quad (t,z,x) \in R_+ \times I \times \Omega, \tag{1.1}$$

$$v_t = Dv_{xx} + \int_0^1 \delta(z)u(t,z,x)dz - \beta v, \quad (t,x) \in R_+ \times \Omega, \tag{1.2}$$

together with the boundary conditions

$$g(r(t,x))h(0)u(t,0,x) = \int_0^1 b(z,r(t,x))u(t,z,x)dz, \quad (t,x) \in R_+ \times \Omega, \tag{1.3}$$

$$u_x = 0, \qquad (t,z,x) \in R_+ \times I \times \partial\Omega, \tag{1.4}$$

$$v_x = 0, \qquad (t,x) \in R_+ \times \partial\Omega, \tag{1.5}$$

and the initial conditions

$$u(0,z,x) = \phi(z,x), \quad (z,x) \in \overline{I} \times \overline{\Omega}, \tag{1.6}$$

$$v(0,x) = \psi(x), \qquad x \in \overline{\Omega}, \tag{1.7}$$

where $R_+ = (0,\infty)$, $I = (0,1)$, $\Omega = (0,\ell)$ and $r(t,x) = \int_0^1 a(z)u(t,z,x)dz$. Here u

and v denote the population density of individuals and the concentration of chemoattractant, respectively; εgh, $\varepsilon\kappa$ and b the density-dependent growth rate, the death rate and the birth rate, respectively. h is the intrinsic growth rate of the single individual which satisfies $h(z) > 0$ for $z \in I$ and $h(1) = 0$. It is assumed that the maximal size of an individual is normalized to 1. $g(r)$ is some function of the density r, which affects the growth rate of individuals. Let us show two examples of functional forms of g. In Figure 1, g is monotonically decreasing as the density increases, while in Figure 2, g is humped so that there is an optimal total density at which growth rate is maximal. For both examples, g is assumed to be strictly decreasing for large r because of limited environmental resources. μ and D denote diffusivities of individuals and chemoattractant, respectively. χ measures the chemotactic behavior of the individuals. $\delta(z)$ and β denote the production rate of chemoattractant by individuals of size z and the degradation rate of the chemoattractant, respectively. The system is supposed to be closed so that the boundary conditions on the space are of zero flux. Finally, since spatial movement should be rapid in comparison with growth processes, ε is assumed to be sufficiently small, so that these two processes proceed on totally different time scales.

This article will investigate the asymptotic behavior of solutions to the problem (1.1)-(1.7) by using 2-timing methods (Ei and Mimura [1984]). The result makes clear the relationship between spatially aggregative force and the existence or extinction of individuals for large time.

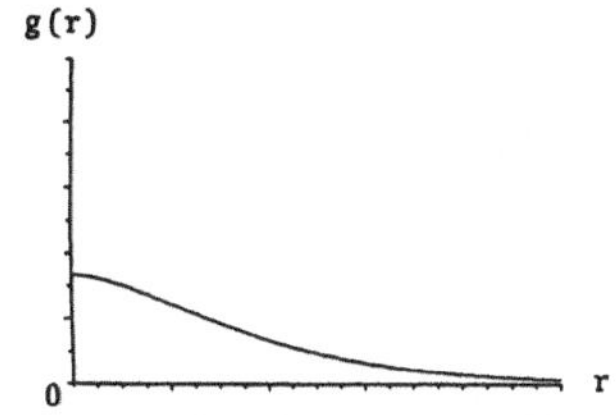

Figure 1. An example of $g(r)$ which is monotonically decreasing.

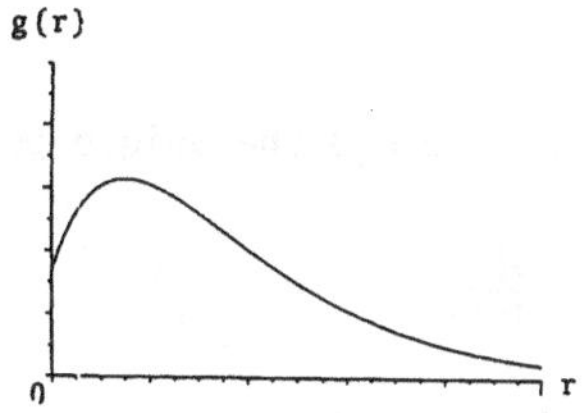

Figure 2. Another example of $g(r)$. Here there is an optimal total density at which the growth rate is maximal.

2. Asymptotic behavior of solutions in the absence of spatial movement.

In this section we consider the case where spatial migration is not present. Then (1.1) (for $\varepsilon = 1$), (1.3) and (1.6) reduce to

$$u_t + g(r(t))(h(z)u)_z + \kappa(z)u = 0, \qquad (t,z) \in R_+ \times I, \tag{2.1}$$

$$g(r(t))h(0)u(t,0) = \int_0^1 b(z,r(t))u(t,z)dz, \qquad t \in R_+, \tag{2.2}$$

and

$$u(0,z) = \phi(z), \qquad z \in \overline{I}, \tag{2.3}$$

respectively, where $r(t) = \int_0^1 a(z)u(t,z)dz$.

We have proved (Mimura and Takigawa, in prep.) the global existence and uniqueness of classical solutions as well as the stability of stationary solutions to the problem (2.1)-(2.3) under appropriate conditions on g, h, κ, a, b and ϕ. Here we only show the results for the asymptotic behavior of solutions to (2.1)-(2.3).

Throughout this paper, we assume that

$$b(z,r) = b(z)g(r), \tag{A.1}$$

$$a(z) \not\equiv 0, \tag{A.2}$$

$$\kappa(z) = \kappa = \text{constant} , \tag{A.3}$$

$g(r)$ is a strictly positive, bounded function on $\overline{R}_+$ and $g(0) = 1$. (A.4)

From (A.1), the boundary condition (2.2) takes a simple form

$$u(t,0) = \int_0^1 \frac{b(z)}{h(0)} u(t,z)dz. \tag{2.4}$$

The first theorem gives a condition under which the problem (2.1)-(2.3) has at least one non-trivial stationary solution.

<u>Theorem 2.1.</u> We assume that

$$\int_0^1 \frac{b(z)}{h(z)} dz > 1 , \tag{2.5}$$

and let κ_0 denote the unique positive number satisfying

$$\int_0^1 \frac{b(z)}{h(z)} \exp(-\kappa_0 \int_0^z \frac{dy}{h(y)})dz = 1 . \tag{2.6}$$

Let $\kappa > 0$ be arbitrarily fixed. Then, a necessary and sufficient condition for the existence of a non-zero stationary solution $\overline{u}(z)$ with $\int_0^1 a(z)\overline{u}(z)dz = \overline{r}$ is that $\overline{r}$ is a positive solution of

$$g(\overline{r}) = \frac{\kappa}{\kappa_0} . \tag{2.7}$$

Furthermore, for each solution $\overline{r} > 0$ of (2.7), $\overline{u}(z)$ is uniquely determined as

$$\overline{u}(z) = \frac{c(\overline{r})}{h(z)} \exp(-\kappa_0 \int_0^z \frac{dy}{h(y)}) , \tag{2.8}$$

where

$$c(\overline{r}) = \frac{\overline{r}}{\int_0^1 \frac{a(z)}{h(z)} \exp(-\kappa_0 \int_0^z \frac{dy}{h(y)})dz} . \tag{2.9}$$

On the other hand, if $\int_0^1 \frac{b(z)}{h(z)} dz \leq 1$, then the unique stationary solution is $u(z) \equiv 0$.

Next, we discuss the stability of stationary solutions. In what follows, a stationary solution $\bar{u}(z)$ is said to be *exponentially asymptotically stable* if there exists some constant $\rho > 0$ such that, for any $\varepsilon > 0$, there exists some constant $\delta > 0$ with $\| u(t,\cdot)-\bar{u}\|_{L^1} \leq \varepsilon e^{-\rho t}$ for $\|\phi-\bar{u}\|_{L^1} < \delta$ and $t \geq 0$.

Theorem 2.2. Assume (2.5). Then the following assertions hold:

(a) The trivial equilibrium $u(z) \equiv 0$ is exponentially asymptotically stable for $\kappa > \kappa_0$. Furthermore, if $\kappa > \kappa_0 g_0$ with $g_0 = \sup_{r\geq 0} g(r)$, then there are positive constants M and ρ such that

$$\| u(t,\cdot)\|_{L^1} \leq Me^{-\rho t}\|\phi\|_{L^1} \tag{2.10}$$

for any initial data ϕ and $t \geq 0$. (2.10) also holds for any $\kappa > 0$ if $\int_0^1 \frac{b(z)}{h(z)}dz \leq 1$. On the other hand, it is unstable for $0 < \kappa < \kappa_0$.

(b) Assume that there is a non-zero stationary solution $\bar{u}(z)$ with $\int_0^1 a(z)\bar{u}(z)dz = \bar{r}$ and that $g(r)$ is differentiable at $r = \bar{r}$. Then, $\bar{u}(z)$ is exponentially asymptotically stable if $g'(\bar{r}) < 0$, while it is unstable if $g'(\bar{r}) > 0$.

The proofs of the above-mentioned theorems are given in [Mimura and Takigawa, in prep.].

From a biological point of view, Theorem 2.2 (b) can be interpreted as follows: Let us define $N(r)$ by

$$N(r) = \int_0^1 \frac{b(z)}{h(z)} \exp\left(-\frac{\kappa}{g(r)} \int_0^z \frac{dy}{h(y)}\right)dz , \tag{2.11}$$

which is called the *net reproductive rate* when the total density is r (see, for instance, Gyllenberg [1982]). Let us assume $g'(\bar{r}) < 0$. Then, since $N(r)$ is monotonically decreasing in some neighborhood of $\bar{r}$ and $N(\bar{r}) = 1$, it follows that, if the total density $r(t)$ becomes slightly larger (smaller) than $\bar{r}$, then the net reproductive rate becomes smaller (larger) than unity and so $r(t)$ returns to $\bar{r}$. The remaining part of Theorem 2.2 can be interpreted similarly.

We apply our results to the cases depicted in Figures 1 and 2. Figures 3 and 4 show the global bifurcation diagram for each case with respect to the death rate κ when (2.5) is assumed. Moreover, for the case of Figure 2, the graphs of non-zero stationary solutions for specific $a(z)$, $b(z)$, $h(z)$ and $g(r)$ are shown in Figure 5. It should be noted that the constant κ_0, determined by (2.6), doesn't depend on the growth function $g(r)$. The two situations are similar for $\kappa > \kappa_0 g_0$ or $0 < \kappa < \kappa_0$; that is, if $\kappa > \kappa_0 g_0$, then any solution tends to zero in $L^1(0,1)$ as $t \to \infty$, but, if $0 < \kappa < \kappa_0$, then there is a stable non-zero stationary solution $\bar{u}(z)$ as well as an unstable one $u(z) \equiv 0$. On the other hand, if κ is chosen as $\kappa_0 < \kappa < \kappa_0 g_0$, the two cases differ greatly. In the first case (Figure 1), any solution tends to zero as $t \to \infty$ — the population becomes extinct. In the second case (Figure 2), there exist two stable stationary solutions, $u \equiv 0$ and $u = \overline{u}(z)$ ($\not\equiv 0$), and an unstable one, $u = \underline{u}(z)$ ($\not\equiv 0$). In fact, if $\phi(z)$ lies in a small neighborhood of $\overline{u}(z)$, the corresponding solution tends to $\overline{u}(z)$ as $t \to \infty$; that is, the population persists for large t. Roughly speaking, the unstable solution $\underline{u}(z)$ acts as a separator between $u \equiv 0$ and $u = \overline{u}(z)$. Thus, we can conclude that a suitable density-dependent growth rate enhances its survival probability.

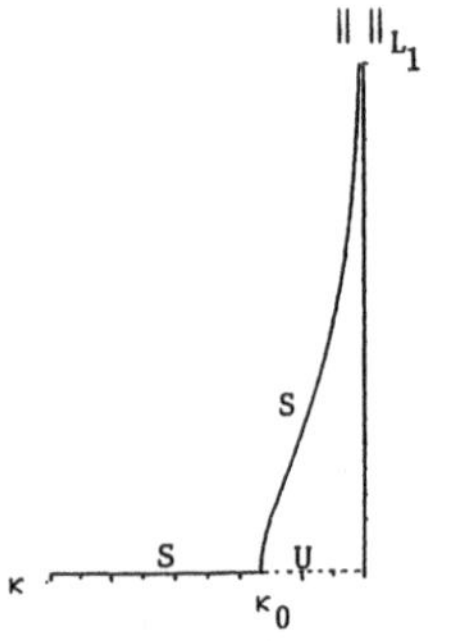

Figure 3. Bifurcation diagram for (2.1)-(2.3) in the case of Figure 1.

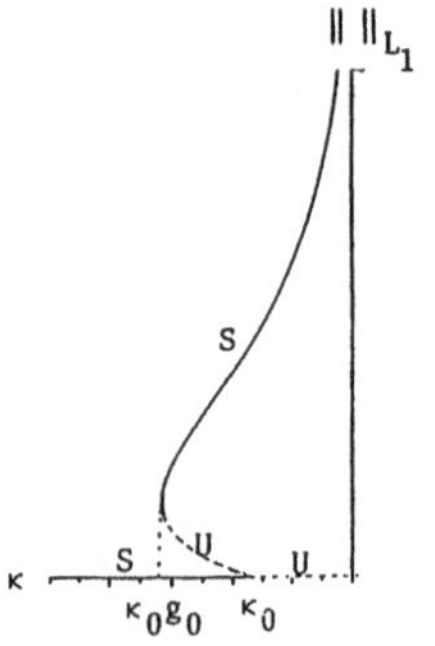

Figure 4. Bifurcation diagram for (2.1)-(2.3) in the case of Figure 2.

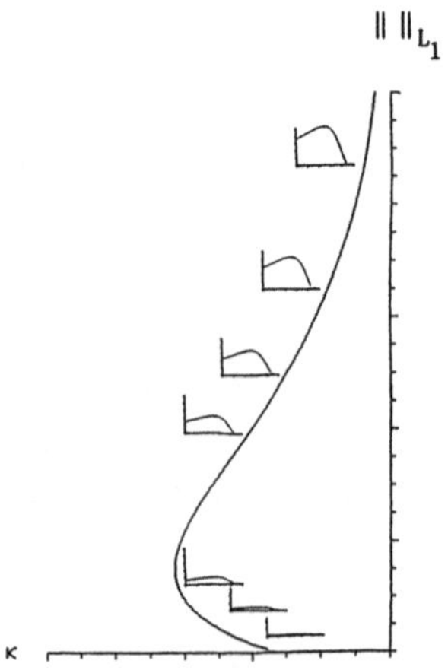

Figure 5. Stationary solutions for various values of κ when $a(z) = b(z) = \frac{5}{2}z$, $g(r) = (1+2r)\exp(-\frac{1}{2}r)$ and $h(z) = (1-z)^2$.

3. Presence of migration in size-distributed dynamics.

Here we assume

$$\mu = 1 \quad \text{and} \quad \chi(u,v) = \chi \cdot u \quad \text{with a positive constant } \chi, \tag{A.5}$$

$$a(z) = \delta(z), \tag{A.6}$$

in addition to (A.1)-(A.4). Since the study of asymptotic behavior of solutions to (1.1)-(1.7) is extremely difficult, we here assume ε to be sufficiently small and use the 2-timing method which is applied to usual reaction-diffusion systems by Ei and Mimura [1984].

We look for a solution (u,v) in the form of

$$\begin{cases} u(t,\tau,z,x;\varepsilon) = u^0(t,\tau,z,x) + \varepsilon u^1(t,\tau,z,x) + O(\varepsilon^2) \\ v(t,\tau,x;\varepsilon) = v^0(t,\tau,x) + \varepsilon v^1(t,\tau,x) + O(\varepsilon^2) , \end{cases} \tag{3.1}$$

where $\tau = \varepsilon t$ is assumed to be a new time scale. Our purpose is to study the asymptotic behavior of (u^0,v^0) in order to obtain information about that of (u,v) for large τ as well as t, by assuming that ε is sufficiently small.

Substituting (3.1) into (1.1)-(1.7) and equating equal powers of ε^0 and ε^1, we obtain

$$\begin{cases} u_t^0 = (u_x^0 - \chi u^0 v_x^0)_x \\ v_t^0 = Dv_{xx}^0 + \int_0^1 a(z)u^0 dz - \beta v^0, \quad (t,x) \in R_+ \times \Omega, \end{cases} \tag{3.2}$$

and

$$\begin{cases} u_\tau^0 + u_t^1 + g(\int_0^1 a(z)u^0 dz)(h(z)u^0)_z + \kappa u^0 = (u_x^1 - \chi u^1 v_x^0 - \chi u^0 v_x^1)_x \\ v_\tau^0 + v_t^1 = Dv_{xx}^1 + \int_0^1 a(z)u^1 dz - \beta v^1 , \end{cases} \tag{3.3}$$

respectively. The boundary conditions on (u^0,v^0) and (u^1,v^1) can be similarly derived.

We first consider the asymptotic behavior of solutions to (3.2) for large time t (we assume τ to be $O(1)$). This is obtained by studying stationary solutions (U^0,V^0) of (3.2) satisfying $\lim_{t\to\infty} (u^0(t,\tau,z,x), v^0(t,\tau,x)) = (U^0(\tau,z,x), V^0(\tau,x))$. For this purpose we consider the stationary problem

$$\begin{cases} 0 = U_x^0 - \chi U^0 V_x^0 & \\ 0 = DV_{xx}^0 + \int_0^1 a(z)U^0 dz - \beta V^0 & , x \in \Omega \\ V_x^0 = 0, \ x \in \partial\Omega. & \end{cases} \tag{3.4}$$

The first equation of (3.4) is written as

$$U^0(\tau,z,x) = C(\tau,z)\exp(\chi V^0(\tau,x))$$

for an arbitrary function $C(\tau,z)$. Integrating the first equation of (3.2) with respect to x, we find

$$\frac{\partial}{\partial t} \int_\Omega u^0(t,\tau,z,x)dx = 0 ;$$

this shows that $\int_\Omega u^0(t,\tau,z,x)dx$ is independent of t, so we write it as $\ell M(\tau,z)$. Using

$$\xi(\tau,x) = \frac{\exp(\chi V^0(\tau,x))}{\int_\Omega \exp(\chi V^0(\tau,x))dx} ,$$

with $\int_\Omega \xi(\tau,x)dx = 1$, we can rewrite U^0 as

$$U^0(\tau,z,x) = \ell M(\tau,z)\xi(\tau,x) . \tag{3.5}$$

It is obvious that $\xi(\tau,x)$ is a solution of the problem

$$\begin{cases} 0 = \xi_x - \chi\xi V^0_x & , x \in \Omega \\ 0 = DV^0_{xx} + \ell\alpha(\tau)\xi - \beta V^0 & \\ V^0_x = 0, \ x \in \partial\Omega & \\ \int_\Omega \xi(\tau,x)dx = 1 , & \end{cases} \tag{3.6}$$

where $\alpha(\tau) = \int_0^1 a(z)M(\tau,z)dz$. (3.6) is known as the stationary version of the Keller-Segel model [Keller and Segel, 1970] which has been studied by a great number of authors (Schaaf[1984], Childress and Percus [1981], for instance). It is shown that a uniform steady state becomes unstable and nonuniform steady states (aggregated patterns) appear when χ or α is large. More precisely speaking, a uniform stationary solution $(\bar{\xi},\bar{V}^0) = (\frac{1}{\ell}, \frac{\alpha}{\beta})$ is stable for $\chi\alpha < \frac{D\pi^2}{\ell^2} + \beta$, while, for $\chi\alpha > \frac{D\pi^2}{\ell^2} + \beta$, it is unstable and there appear stable bifurcating solutions $(\bar{\xi}(x), \bar{V}^0(x))$. Figure 6 shows the bifurcation diagram of stationary solutions with respect to α. Thus we find that stable stationary solutions $\xi(x;\alpha)$ consist of

$$\xi(x;\alpha) = \begin{cases} \frac{1}{\ell} & \text{for} \quad \chi\alpha < \frac{D\pi^2}{\ell^2} + \beta \\ \bar{\xi}(x) & \text{for} \quad \chi\alpha > \frac{D\pi^2}{\ell^2} + \beta. \end{cases} \tag{3.7}$$

It is noted that $\xi(x;\alpha)$ is unique except for the reflection ($\xi(x) = \bar{\xi}(\ell-x)$ is also a solution). Thus we may assume

$$\lim_{t\to\infty} u^0(t,\tau,z,x) = \ell M(\tau,z)\xi(x;\alpha(\tau)) . \tag{3.8}$$

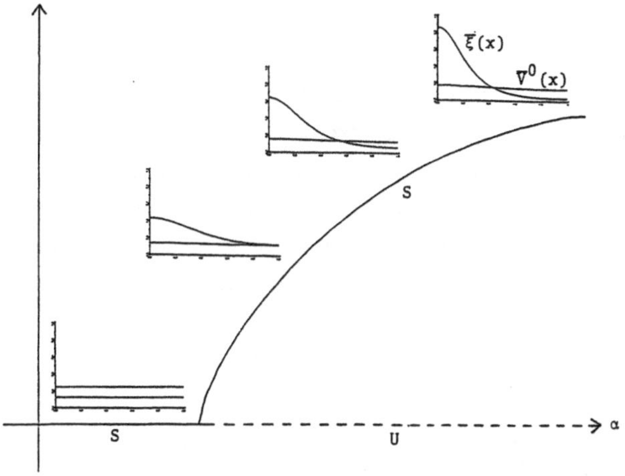

Figure 6. Bifurcation diagram for (3.6) with respect to α and numerically obtained stationary solutions for various values of α when $D = \ell = \beta = 1$ and $\chi = 21$.

We next consider the asymptotic behavior of solutions (u^0, v^0) for large time τ. To do so, we study the behavior of $M(\tau,z)\cdot\xi(x;\alpha(\tau))$ in (3.8). Integrating the first equation of (3.3) with respect to x and assuming

$$\lim_{t\to\infty} \frac{\partial}{\partial t} \int_\Omega u^1(t,\tau,z,x)dx = 0 ,$$

we obtain

$$\begin{cases} M_\tau + \tilde{g}(\alpha(\tau))(h(z)M)_z + \kappa M = 0 \\ M(0,z) = \frac{1}{\ell} \int_\Omega \phi(z,x)dx \equiv M_0(z) \\ M(\tau,0) = \int_0^1 \frac{b(z)}{h(0)} M(\tau,z)dz , \end{cases} \tag{3.9}$$

where $\alpha(\tau) = \int_0^1 a(z)M(\tau,z)dz$ and $\tilde{g}(\alpha) = \int_\Omega g(\alpha\ell\xi(x;\alpha))\xi(x;\alpha)dx$. We note that, by replacing $M(\tau,z)$ and $\tilde{g}(\alpha)$ in (3.9) with $u(t,z)$ and $g(r)$ in (2.1)-(2.3), respectively, (3.9) reduces to the problem (2.1)-(2.3). Thus, by using Theorem 2.2, the stability of stationary solutions to (3.9) can be completely investigated. If, for some stable stationary solution $\overline{M}(z)$ to (3.9), $\lim_{\tau\to\infty} M(\tau,z) = \overline{M}(z)$ can be known, we find that u^0 tends to $\ell\overline{M}(z)\xi(x;\overline{\alpha})$ as t and $\tau \to \infty$, where $\overline{\alpha} = \int_0^1 a(z)\overline{M}(z)dz$. It is shown that $\overline{M}(z)$ and $\xi(x;\overline{\alpha})$ exhibit the size and the spatial distributions of individuals, respectively. However, $\tilde{g}(\alpha)$ cannot be known *a priori* because it depends on $\xi(x;\alpha)$. Of course, when $\chi\alpha < \frac{D\pi^2}{\ell^2} + \beta$, $g(\alpha) = \tilde{g}(\alpha)$. Figure 7 shows the numerically obtained functional forms of $\tilde{g}(\alpha)$ for some specific χ when β is fixed. Since the picture shows that $\tilde{g}(\alpha)$ is still humped even if χ is varied, Theorem 2.2 implies that there is a stable non-zero stationary solution $\overline{M}_\chi(z)$ as well as an unstable one $\underline{M}_\chi(z)$ for suitable κ.

Let us study the difference in the asymptotic behavior of solutions (u^0,v^0) for the specific values $\chi = 0$ (in the absence of chemotactic force) and $\chi = 46$ (in the presence of strong chemotactic force). We fix the death rate κ as $\kappa_* < \kappa < \kappa^*$ and take $M_0(z)$ such that it lies below $\underline{M}_0$ and above $\underline{M}_{46}$ (Figure 8). Then we can infer that, when $\chi = 0$, $\lim_{\tau\to\infty} M(\tau,z) = 0$, while, when $\chi = 46$, $\lim_{\tau\to\infty} M(\tau,z) = \overline{M}_{46}(z)$. This indicates that aggregation by chemotaxis enhances the probability of individuals' survival. The details will be presented in Ei, Mimura and Takigawa [in prep.].

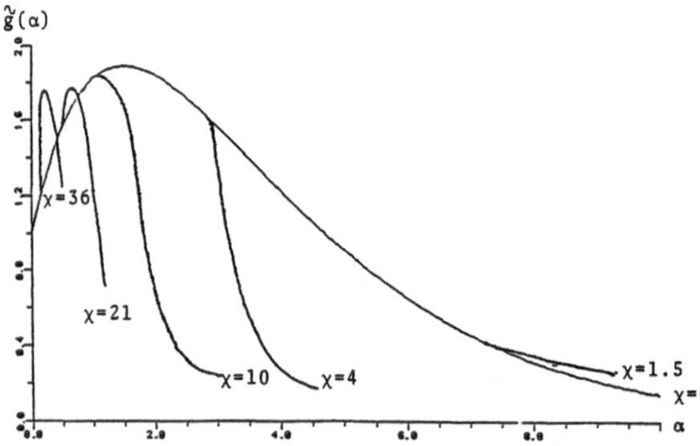

Figure 7. Functional forms of $\tilde{g}(\alpha)$ for various values of χ when $D = \ell = \beta = 1$ and $g(r) = (1+2r)\exp(-\frac{1}{2}r)$.

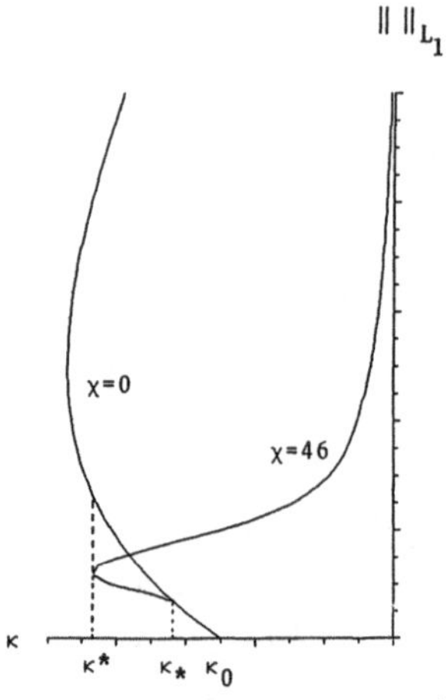

Figure 8. Bifurcation diagrams for (3.9) when $\chi = 0$ and $\chi = 46$, respectively.

References

S. Childress and J. K. Percus, Nonlinear aspects of chemotaxis, Math. Biosci. 56 (1981), 217-237.

O. Diekmann, H. J. A. M. Heijmans and H. R. Thieme, On the stability of the cell size distribution, J. Math. Biology. 19 (1984), 227-248.

S.-I. Ei and M. Mimura, Transient and large time behaviors of solutions to heterogeneous reaction-diffusion equations, Hiroshima Math. J. 14 (1984), 649-678.

S.-I. Ei, M. Mimura and S. Takigawa, (in preparation).

M. Gyllenberg, Nonlinear age-dependent population dynamics in continuously propagated bacterial cultures, Math. Biosci. 62 (1982), 45-74.

S. Ishii, Biologically active substances produced by insects, Nankodo, Tokyo, 1969 (in Japanese).

E. F. Keller and L. A. Segel, Initiation of slime mold aggregation viewed as an instability, J. Theor. Bio. 26 (1970), 399-415.

M. Mimura and S. Takigawa, A size-distribution model with density-dependent growth rates, (in preparation).

R. Schaaf, Global branches of one dimensional stationary solutions to chemotaxis systems and stability, Lecture Notes in Biomathematics, Springer-Verlag, Berlin, 55 (1984), 341-349.

G. F. Webb, Theory of nonlinear age-dependent population dynamics, Marcel Dekker, Inc., New York, 1985.

EVOLUTION OF THE NUMBER OF SEXES

Yoh Iwasa and Akira Sasaki

Department of Biology, Faculty of Science,
Kyushu University, Fukuoka 812, Japan

Most diploid organisms have two sexes, but some have three or more. For instance, among ciliates in which mating occurs by the contact of two diploid cells and the subsequent exchange of haploid genome, cells are grouped into several mating types (or sexes) so that mating occurs only between cells of different sexes (sonneborn, 1939). *Stylonychia* spp. have as many as 48 sexes (Ammermann, 1965). Other organisms with three or more sexes are fungi, in which mating occurs by the fusion of isogamous haploid gametes (Raper, 1966).

In this paper we study the evolutionary stability of the number of sexes in an isogamous species. We ask such questions as whether a third sex invading a bisexual population can increase, and under what condition will sexes in a population be lost until only two remain.

When three or more sexes exist, the genetic dynamics depend on the *mating kinetics*, the rule of heterosexual pair formation in the gamete pool, which determine the proportion of matings occuring between a particular pair of sexes given the proportion of cells of various sexes. Note that we don't need to consider this if only 2 sexes exist, because all the matings then occur between the two sexes irrespective of their relative abundance. To model this new element of genetic dynamics, we analyze four mating kinetics.

By nature of random collision processes, cells of rare sexes tend to find heterosexual partners quicker than those of common sexes. If the period suitable for mating is limited, rare sexes have an advantage over common sexes in contributing to the next generation.

Diploid Sex Determination

In ciliates, sexual reproduction begins with the conjugation of two diploid cells which then exchange haploid nuclei produced by meiosis. In many other organisms, however, haploid gametes produced by meiosis combine with gametes of different sexes to produce diploid zygotes. The genetic dynamics of diploid individuals for such gamete-fusion models are the same as those for diploid-cell-conjugation models provided that the sex of gametes is determined by the diploid genome of their "parent" cells. Later we will study models in which the sex of a gamete is determined by the haploid genome.

Sex 1: A_1A_1

Sex 2: A_1A_2, A_2A_2

Sex 3: A_1A_3, A_2A_3, A_3A_3

We assume that sex determination is at a single locus, with 3 alleles of pecking-order dominance ($A_1 < A_2 < A_3$), which is reported among ciliates (Heckmann, 1964; Nobili, 1966; Dini and Luporini, 1985). Prevalent XY-sex determination system is a special case of two sexes, i.e. A_1A_1 is sex 1, and A_1A_2 is sex 2.

We denote the fractions of diploid cells with genotype A_1A_1, A_1A_2, A_2A_2, A_1A_3 A_2A_3, and A_3A_3 by X_{11}, X_{12}, X_{22}, X_{13}, X_{23}, and X_{33}, respectively and the total fractions of diploid cells having sex 1, sex 2, and sex 3 by $S_1 = X_{11}$, $S_2 = X_{12} + X_{22}$, and $S_3 = X_{13} + X_{23} + X_{33}$, respectively. Since we assume that cells of different genotypes have the same fertility, S_i is the fraction of gametes having sex i.

Although we assume that mating occurs randomly between gametes (or diploid cells in ciliates) of different sexes, there are several equally plausible assumption about how cells of different sexes mate during a suitable mating period, and these assumptions yield different results on the number of matings occurring between each pair of sexes among all possible pairs when the relative abundances of sexes are given. Let M_{ij} be the proportion of matings between sexes i and j among all the matings. By definision, $M_{ij} = M_{ji}$. Mating kinetics gives M_{ij} as functions of S_1, S_2, and S_3: the fractions of gametes (or diploid cells) of three sexes.

The genetic dynamics can be calculated by counting the number of offspring from parents of various genotypes. For example in diploid cell conjugation models, the fraction of matings between cells of genotype A_1A_2 (sex type 2) and those of genotype A_1A_3 (sex type 3) is

$$M_{23}(X_{12}/S_2)(X_{13}/S_3), \tag{1}$$

where M_{23} is the fraction of matings between cells of sex 2 and sex 3. This conjugation will produce four types of "offspring" with genotypes, A_1A_1, A_1A_3, A_1A_2, and A_2A_3, in equal fraction, contributing each to the corresponding genotypes in the next generation. The abundance of offspring with genotype-ij is calculated as

$$\begin{aligned}
X'_{11} &= M_{12}p/2 + M_{13}q/2 + M_{23}pq/4,\\
X'_{12} &= M_{12}[p/2+(1-p)] + M_{13}(1-q)/2 + M_{23}[pq/4+p(1-q)/4+(1-p)q/2],\\
X'_{22} &= M_{23}[p(1-q)/4+(1-p)(1-q)/2],\\
X'_{13} &= M_{13}/2 + M_{23}p/4,\\
X'_{23} &= M_{23}[pq/4+p(1-q)/4+(1-p)q/2+(1-p)(1-q)/2],\\
X'_{33} &= 0,
\end{aligned} \tag{2}$$

where $p = X_{12}/S_2$, $1-p = X_{22}/S_2$, $q = X_{13}/S_3$, $1-q = X_{23}/S_3$, and $X_{33} = 0$. The same genetic dynamics can be derived for the models of gamete fusion mating by considering offspring for given parents' genotypes. Zygote formation is followed by a stage of growth and development in multicellular organisms (or of clonal growth in unicellular ones). We assume that the survivorship and the fertility of multicellular individuals (or the clonal growth rate of unicellular organisms) are the same for all genotypes.

To describe genetic dynamics, we need to specify mating kinetics. In the following we study four cases.

Mating kinetics 1

The simplest mating kinetics would incorporate the assumption that the number of matings between two sexes is proportional to the product of their abundances:

$$M_{ij} = \begin{cases} S_i S_j / C, & (i \neq j), \\ 0 & (i = j), \end{cases} \tag{3}$$

where $C = \sum_i \sum_{j \neq i} S_i S_j / 2$ is the normalization factor. This kinetics holds, for example, if gametes are paired at random irrespective of sex and then same-sex pairs are discarded because time available for mating is so short that gametes failing to find a heterosexual partner on the first attempt do not have a second chance.

The dynamical system (2) with mating kinetics (3) has four steady states, a central equilibrium:

$$\begin{aligned} \text{(I)} \quad & \hat{x}_{11} = 1/3,\ \hat{x}_{12} = (4\sqrt{3}-6)/3,\ \hat{x}_{22} = (7-4\sqrt{3})/3, \\ & \hat{x}_{13} = (\sqrt{3}-1)/3,\ \hat{x}_{23} = (2-\sqrt{3})/3,\ \hat{x}_{33} = 0. \end{aligned} \tag{4}$$

in which the abundance of each sex is the same ($S_1 = S_2 = S_3 = 1/3$), and three equilibria in which only two of the three sexes are present:

$$\text{(II)} \quad \hat{x}_{11} = \hat{x}_{12} = 1/2;\ \hat{x}_{22} = \hat{x}_{13} = \hat{x}_{23} = \hat{x}_{33} = 0. \tag{5}$$

$$\text{(III)} \quad \hat{x}_{11} = \hat{x}_{13} = 1/2;\ \hat{x}_{12} = \hat{x}_{22} = \hat{x}_{23} = \hat{x}_{33} = 0. \tag{6}$$

$$\text{(IV)} \quad \hat{x}_{22} = \hat{x}_{23} = 1/2;\ \hat{x}_{11} = \hat{x}_{12} = \hat{x}_{13} = \hat{x}_{33} = 0. \tag{7}$$

The central equilibrium (4) is stable with the dominant eigenvalue $\lambda = 0.69765$. The marginal equilibria (5) and (6) are unstable having the dominant eigenvalues $\lambda = 2$, and $\lambda = 1.781$ respectively, indicating that a rare allele (either A_2 or A_3) increases in a population with two other sexes. In contrast, equilibrium (7), in which the least dominant sex A_1 is absent, has the dominant eigenvalue 1. We can prove that the steady state is in fact unstable using analysis of higher order (center maniforld theorem).

Mating kinetics 1 gives a large advantage to a rare sex, because the chance for a gamete to find a suitable "mate", and thereby to contribute genetically to the next generation, is proportional to the fraction of gametes belonging to the differernt sexes. The eigenvalue $\lambda=2$ for (5) implies that the most dominant allele A_3 can produce twice as many copies as a resident gene when it invades the population with two other sexes. This is because half of the randomly formed pairs involving a cell of the resident sex occur with another cell of the same sex, and therefore doomed to failure. Most pairs formed by a cell of the rare invader sex are with the resident sexes and hence lead to successful matings. This model predicts the evolutionary increase of sexes.

Less dominant alleles do not enjoy as high a rate of increase as the most dominant one when they invade, as is indicated by smaller dominant eigenvalues for

(6) and (7) relative to (5). When rare, the least dominant allele (A_1) exists in heterozygotes (A_1A_2 or A_1A_3), which behave as common sexes, without enjoying the rare sex advantage. The invasion of the least dominant allele A_1 is not supported strongly by the dynamics. With some additional factors, such as a small differential viability or random drift, A_1 may possibly be lost from the population; in contrast both A_2 and A_3 are strongly protected.

Mating kinetics 2

Suppose that mating proceeds by two steps: a fraction of cells becomes mature first and each of these cells then randomly searches for a heterosexual partmer among nonmature gametes of a different sex, as in some fungi. Then the proportion of matings between the i-th and j-th sexes is:

$$M_{ij} = \begin{cases} S_i S_j/(1-S_i) + S_j S_i/(1-S_j), & \text{if } i \neq j, \quad (8a) \\ 0, & \text{if } i = j. \quad (8b) \end{cases}$$

The first term of (8a) is the fraction of matings in which a gamete of the i-th sex mature first and then captures a partner of the j-th sex chosen among sexes other than i. The second term corresponds to matings between a mature j-th sex and a nonmature i-th sex.

The genetic dynamics (2) with kinetics (8) have the same four equilibria as in mating kinetics 1: the central equilibrium (4) (with the dominant eigenvalue λ= 0.77874), and three marginal ones (5)(λ=1.5), (6)(λ=1.394), and (7)(λ=1).

The advantage for an invader sex when rare is smaller than in mating kinetics 1 but still considerable. The dominant sex (A_3) invading a population with two sexes (A_1 and A_2) is 50% more fertile (λ=1.5) than the resident. Again we are led to the evolutionary increase in the number of sexes.

Mating kinetics 3

Rare sexes have a higher per capita probability that the randomly chosen partners of pairing are suitable mates; hence a rare sex has an advantage in mating under kinetics 1. However, the proportion of rare sexes in the pool of unmated gametes decreases due to their higher rate of finding suitable mates, and the mating success of common sexes increases with time. This tendency may compensate for the initial high mating rate of rare sexes if the period available for mate-finding is sufficiently long.

In the following, we consider the dynamics of unmated gametes during a mating period, assuming that: (1) at the onset of the period, all the gametes are ready for mating; (2) the number of gametes decreases with time as zygotes are produced, and (3) the period suitable for mating is sufficiently long and mating continues until gametes remaining in the pool cannot find suitable mates.

Let $Y_i(t)$ be the number of gametes of sex type i at time t. If the total number of gametes is Y_o, then $Y_i(0) = Y_o S_i$ and $Y_i(t)$ decreases as follows:

$$dY_i/dt = -\sum_{j\neq i} b\, Y_i Y_j, \quad (i=1,2,3;\ t>0), \tag{9}$$

where the summation is calculated for all sexes other than i. The fraction of zygotes produced by sex i and sex j is:

$$M_{ij} = \begin{cases} \int_0^\infty b\, Y_i(t) Y_j(t)\, dt/C, & (i\neq j) \\ 0, & \end{cases} \tag{10}$$

where the normalization factor is $C = \sum_i \sum_{j\neq i} b \int_0^\infty Y_i(t) Y_j(t)\, dt/2$.

We assume $S_1 \geq S_2 \geq S_3$ without loss of generality. At the end of the mating period (expressed as $t=\infty$ in equation (9)), all the sexes are exhausted except for the sex with the largest abundance ($Y_1(\infty)$). Let $R(S_1,S_2,S_3)$ be the fraction of cells remaining from among the total initial amount Y_o ($R = Y_1(\infty)/Y_o$), given as a function of frequencies of n sexes at the beginning of the mating period. $R(S_1,S_2,S_3)$ is positive except when the two most abundant sexes are equally present ($S_1 = S_2$), in which case it is zero. Using $R(S_1,S_2,S_3)$, M_{ij} can be expressed as

$$\begin{aligned} M_{12} &= (1 - R - 2S_3)/(1-R), \\ M_{23} &= (1 + R - 2S_1)/(1-R), \\ M_{13} &= (1 - R - 2S_2)/(1-R). \end{aligned} \tag{11}$$

The cases with other rankings of the three sexes are obtained by permutation of suffixes.

The dynamical system (2) with this mating kinetics (11) exhibits curves of equilibria which connect the central equilibrium (4) to the 3 marginal ones, (5), (6), and (7). On these curves, the two most common sexes are equally abundant ($S_1 = S_2 \geq S_3$, etc.), hence $R = 0$. According to numerical analysis, the system converges to a point on these curves; i.e. these curves constitute an attracting invariant set. The particular equilibrium toward which the system converges depends on the initial condition.

Unlike previous models, the central equilibrium (4), with the three sexes in equal abundance, is neither asymptotically nor globally stable. Since rare invader sexes do not have any particular advantage over common resident sexes, the number of sexes may decrease if additional factors are operative, such as random drift or differential viability among sexes.

Mating kinetics 4

Mating kinetics 3 assumed simultaneous maturation of all gametes at the onset of the mating period. Alternatively gametes may mature over a prolonged period. Most matings would then occur during the steady state at which the supply and the loss of gametes balance. We consider the following kinetics:

$$dY_i/dt = c\, S_i - b\, Y_i \sum_{j\neq i} Y_j - e\, Y_i, \tag{12}$$

where c is the total rate of gamete supply. Gametes are consumed by the mating processes indicated by the second term. We also consider the loss of gametes due to a very small mortality with coefficient e (<<1). After a relatively short transient period, the system converges to a steady state where it stays for the rest of the mating period. The number of zygotes formed between sexes i and j is:

$$M_{ij} \propto b\, \hat{Y}_i \hat{Y}_j, \tag{13}$$

where $\hat{Y}_i$ is the steady state abundance of Y_i.

Depending upon whether or not the fraction of supply of one of the sexes exceeds 50%, we have very different solutions. If all the sexes are less than 50% (i.e. $S_1 < 1/2$, $S_2 < 1/2$, and $S_3 < 1/2$ -- we call it no-majority-sex region), the system has a steady state in which no Y_i is extraordinarily large. Hence we can neglect the third term of (12) for the mortality loss, because the loss rate e is assumed to be very small:

$$c\,\hat{S}_i = b\,\hat{Y}_i \sum_{j\neq i} \hat{Y}_j, \qquad (i=1,2,3). \tag{14}$$

Combining this with (13), and with the normalization $M_{12}+M_{23}+M_{13}=1$, we have

$$\begin{aligned} M_{12} &= 1 - 2S_3, \\ M_{23} &= 1 - 2S_1, \\ M_{13} &= 1 - 2S_2. \end{aligned} \tag{15}$$

In contrast, if the fraction of a sex in the total inflow of cells is greater than 50% (say $S_1 > 1/2$), some cells of this sex fail to find mates and unmated cells (Y_1) should accumulate in the system until the abundance $\hat{Y}_1$ becomes so large that the third term eY_1 in (12) is no longer small. The most common sex should then be regulated by mortality. The abundance of gametes of other sexes should be very small (order of e) because they are "captured" by the cells of the most common sex, which exists in a large concentration, and mating occurs almost exclusively between sex 1 and others. Therefore, if $S_1 > 1/2$ (then necessarily $S_2 < 1/2$, and $S_3 < 1/2$),

$$\begin{aligned} M_{12} &= S_2/(S_2 + S_3), \\ M_{23} &= 0, \\ M_{13} &= S_3/(S_2 + S_3). \end{aligned} \tag{16}$$

The cases with $S_2 > 1/2$ and with $S_3 > 1/2$ can be obtained by permutation of suffixes in (16).

Now we consider the following 2 functions:

$$\Phi_2 = X_{12} + 2X_{22} + X_{23}, \tag{17a}$$

$$\Phi_3 = X_{13} + X_{23} + 2X_{33}, \tag{17b}$$

indicating the abundance of alleles A_2 and A_3, respectively. Using Eq. (2) and Eq. (15), which holds in the no-majority-sex region, we can show that $\Phi_2' = \Phi_2$ and $\Phi_3' = \Phi_3$ -- these two are "constants of motion". The system has no globally stable equilibrium. Instead, there are infinitely many equilibria, and the state point converges to an equilibrium having the same Φ_2 and Φ_3 as in the initial state. These equilibria constitute an attracting surface, on which Φ_2 and Φ_3 are curve coordinates.

If the system starts from outside the no-majority-sex region, it eventually enters the region and converges to one of the equilibria there.

Haploid Sex Determination

Evolution of the number of sexes can also be analyzed in haploid sex determination models, in which 3 sexes are controlled by one locus with three alleles, A_1, A_2, and A_3, with the fractions X_1, X_2, and X_3 respectively.

After the zygote formation stage, haploid gametes are produced by meiosis. The proportion of haploid gametes of the i-th genotype in the next generation is

$$X'_i = \sum_j M_{ij}/2, \tag{18}$$

where M_{ij} is the fraction of matings between sexes i and j. The proportions of sexes are the same as those for alleles:

$$S_i = X_i, \qquad (i=1,2, \text{ and } 3). \tag{19}$$

The mating kinetics defines M_{ij} as functions of S_1, S_2, and S_3, giving the dynamics for allele frequencies X_i.

We examined the same four mating kinetics considered in the diploid sex determination models. In models with mating kinetics 1 and 2, the central equilibrium is globally stable. The models with kinetics 3 and 4 have three line segments and a region of equilibria, respectively, which are attracting -- the state points outside of these sets move toward them. In short, the models of haploid sex determination are similar in qualitative behavior to the diploid sex determination models with the same mating kinetics.

Autogamy

Many ciliates sometimes undergo "autogamy", in which gamete nuclei derived from the same meiotic product unite to form a zygote. As a consequence, the cells immediately become homozygous at all the loci (Grell, 1967). Since ciliates are also known to include species with multiple sexes, we suspect possible effects of autogamy on the evolution of the number of sexes.

The genetic dynamics are then similar to the diploid case except for an additional step in which a fraction of heterogametic sex genotypes turns homogametic. Obeying the genetic dynamics described by (2), autogamy with the fraction a causes the change from $X'_1,..,X'_3$ to $X''_1,..,X''_3$:

$$X''_{ii} = X'_{ii} + (a/2)\sum_{j\neq i} X'_{ij},$$
$$X''_{ij} = (1-a)X'_{ij}, \qquad (i,j=1,2,3;\ i\neq j). \tag{20}$$

Combining (20) and (2) with the 4 types of mating kinetics yields the genetic dynamics. Analysis shows that, with mating kinetics 1 and 2, the system has central equilibrium with three sexes of equal abundance and three bisexual equilibria where the genotype frequencies depend on autogamy fraction a. Bisexual equilibria are susceptible to invasion by the third sex, and all of the sex alleles are protected from loss. In contrast, with the mating kinetics 3, there are three arcs of equilibria, each connecting the central equilibrium and one of the marginal equilibria. With mating kinetics 4, there is a surface of equilibria.

In summary, autogamy per se does not change the quantitative conclusion concerning evolutionary stability of the number of sexes.

Discussion

The main conclusion of genetic models is that mating kinetics is most important factor in the evolution of the number of sexes maintained in a population. Whether the sex determining genome is haploid or diploid, models with mating kinetics 1 and 2 causes frequency dependent selection favoring rare sexes, and invader sexes are surely accepted and coexist with the resident; hence the number of sexes in the population will increase until it is so large that the advantage to a rare sex becomes very small. In contrast, in models with kinetics 3 and 4 there is an attracting set (curves or a region) of equilibria, along which the system dynamics are neutral. Hence it is very likely that a population having multiple sexes will lose its variety of sex determining alleles due either to random drift or to differential viability between alleles until only two remain.

The advantage for rare sexes is a natural outcome of delay during mate search because, under the random collision assumption, the cells of common sexes often hit others of the same sex. The cost of this delay is the reason for the difference between mating kinetics. If the period available for mate search is so short that an appreciable fraction of cells (or gametes) cannot find suitable mates, the rare sex advantage is significant, with mating kinetics 1 being an extreme case. In contrast, the rare sex advantage vanishes if there is no cost associated with a long mate search, as is the case in mating kinetics 3 and 4.

This dichotomous picture, however, may be too simple, because the same tendency, though much weaker, exists in the situation described by kinetics 3 and 4. The weak advantage to rare sexes would appear if, for example, the mortality e in Eq. (12) is small but positive, as the neutral region would then collapse and the central equilibrium (4) would become globally stable (though the movement toward the equilibrium is slow). If the advantage enjoyed by rare sexes is small, a slight differential mortality between sex alleles is enough to make increase in the number of sexes difficult.

The number of sexes maintained in a natural population should probably be explained as a subtle balance of several forces: (1) rare sex advantage, (2) differential fitness, (3) random drift due to finite population size, and (4) invasion of new sexes. The observed prevalence of bipolarity (two sex system) could be explained both by the facts that the production of new sexes is infrequent and the remark that the mating kinetics of many species is approximated by kinetics 3 or 4 rather than by kinetics 1 or 2.

A probable outcome of weak advantage for rare sexes is turnover of sexes, in which a new sex, either immigrant or mutant, will increase in the population and replace one of the resident sexes after a temporary coexistence. In a model in which many sex factors control only two sexes, Bull and Charnov (1977) discussed similar processes of turnover of sex factors. Asymmetry between sexes exists in the diploid models. In the models with mating kinetics 1 and 2, the marginal equilibrium with two dominant sexes is neutrally stable against invasion by the least dominant sex (A_1), while the other two marginal equilibria are unstable -- the dominant sexes are strongly protected from loss. This suggests a directional turnover of sexes -- a new sex can invade a population easily when it is dominant to the existing sex alleles, and a population with three sexes is then likely to lose the recessive one first due to random drift.

A potentially important process affecting sex turnover is the accumulation of recessive lethal mutations in the region of the heterogametic sex chromosome (e.g. Y-chromosome in XY-sex determination system) that is strongly linked to the sex determining alleles (Muller, 1914; Fisher, 1935; Nei, 1970; Charlesworth, 1978). In a population with two sexes (say A_1 and A_2), the dominant one (A_2) tends to be accompanied by recessive lethals because homozygotes A_2A_2 are absent. When a more dominant sex (A_3) invades, mating between A_1A_2 and A_2A_3 produces A_2A_2 zygotes, which are lethal. This will eventually cause loss of A_2 alleles if the rare sex advantage is not very strong. The A_2 allele will be lost, rather than the least dominant A_1, which is now strongly protected because A_2 and A_3 cannot constitute a stable population.

Now we consider the situation in which autogamy occurs with a positive fraction in each generation. Autogamy prevents the accumulation of recessive lethals in the population by producing homozygotes from heterozygous cells. The invasion of a new more dominant sex allele may then result in a net increase in the number of sexes instead of the simple replacement of a resident sex which has accumulated recessive deleterious mutations as discussed above. The same logic applies to a population with haploid sex determination in which recessive lethals do not accumulate on sex chromosomes. Given these arguments, we conjecture that the number of sexes tends to be larger in organisms with haploid sex determination (in freshwater green algae and some fungi) or in diploid organisms with occasional autogamy (in ciliates).

To conclude, the number of sexes in an isogamous population will, in the course of evolution, increase when, in order to mate, cells must find suitable partners within a relatively short period; otherwise bipolarity (two sex system) should evolve. Haploid sex determination or occasional autogamy in diploid sex determination will prevent the accumulation of recessive lethals on sex chromosomes and may thereby help to maintain a greater number of sexes in the population. Further investigation may reveal many more isogamous organisms with three or more sexes than has up to the present been reported. Experimental study will tell us about the relative importance of the various forces affecting the number of sexes in the light of evolutionary population biology.

Acknowledgements

We thank Professor H. Matsuda for his encouragement throughout the study. We also thank Drs. K. Aoki, D. Cohen, K. Hiwatashi, M. Iizuka, A. Miyake, T. Miyata, H. Toh, and N. Yamamura for their helpful comments. This work was supported in part by a Grant-in-Aid for Special Project Research of Optimal Strategy and Social Structure from the Ministry of Education, Science, and Culture of Japan.

Literature cited

Ammermann, D. 1965 Cytologische und genetische Untersuchungen an den Ciliaten Stylonychia mytilus Ehrenberg. Arch. Protistenk 108: 109-152.

Bull, J.J. and E.L. Charnov, 1977 Changes in the heterogameic mechanism of sex determination. Heredity 39: 1-14.

Charlesworth, B. 1978 Model for evolution of Y chromosomes and dosage compensation. Proc. Natl. Acad. Sci. US 75: 5618-5622.

Dini, F. and P. Luporini, 1985 Mating-type polymorphic variation in Euplotes minuta (Ciliophora: Hypotrichida) J. Protozool. 32: 111-117.

Fisher, R.A. 1935 The sheltering of lethals. Am. Nat. 69: 446-455.

Grell, K.G. 1967 Sexual reproduction in protozoa. pp147-213. In T.T. Chen (ed.) Research in protozoology. Vol. 2 Oxford: Pergamon Press.

Heckmann, von Klaus, 1964 Experimentelle Untersuchungen an Euplotes crass. I. Paarungssustem, Konjugation und Determination der Paarungs typen. Z. Verergunsl. (Mol. Gen. Genet.) 95: 114-124. (In German with English summary).

Iwasa, Y. and A. Sasaki, 1987 Evolution of the number of sexes. Evolution (in press)

Muller, H.J. 1914 A gene for the fourth chromosome of Drosophila. J. Exp. Zool. 17: 325-336.

Nei, M. 1970 Accumulation of nonfunctional genes on sheltered chromosomes. Am. Nat. 104: 311-322.

Nobili, R. 1966 Mating types and mating type inheritance in Euplotes minuta Yocom (Ciliata, Hypotrichida). J. Protozool. 13: 38-41.

Raper, J.R. 1966 Life cycles, basic patterns of sexuality, and sexual machanisms. pp 473-511. G.C. Ainsworth and A.S. Sussman (eds.), The fungi: an advanced treatise. vol. 2, the fungal organism. New York: Academic Press.

Sonneborn, T.M. 1939 Paramecium aurelia: mating types and groups; lethal interactions; determination and inheritance. Am. Nat. 73: 390-413.

A Lattice Model for Population Biology

H.Matsuda, N.Tamachi, A.Sasaki, and N.Ogita*

Department of Biology, Faculty of Science,
Kyushu University, Fukuoka 812 Japan
*The Institute of Physical and Chemical Research,
Wako 351-01 Japan

1. Introduction

In statistical physics we study macroscopic properties of matter on the basis of constituent particles, and in theoretical population biology we study features of populations on the basis of behaviors of individuals or, more basically and generally, on the basis of properties of self-replicating entities such as genes or chromosomes. Let us refer to any object that we broadly regard as a unit of replication as a 'replicon', thereby extending the original meaning used by molecular geneticists. Each replicon has a definite genetic state and undergoes birth and death. Therefore, in addition to 'attraction and repulsion', interactions between replicons typically includes 'attacking and helping', which affects the birth and death of recipients. The particular mode of interaction depends on a replicon's state. This state is inherited from its parent replicon, and we can therefore study what type of interaction is prevalent in a population by examining that population's dynamics. This is simply the evolution of behavior by natural selection.

In nature the migration distance per generation of a replicon is usually much smaller than the linear dimension of the whole region inhabited by a population. Therefore, in the modelling of the population dynamical aspects of evolutionary ecology, spatial heterogeneity or geographical structure of the population must be considered seriously. For instance, in his famous kinship theory, Hamilton (1964, 1972) gave the quantitative conditions for the evolution of altruism towards one's immediate family. However, he did not seriously study the evolution of helping or attacking behavior between neighbors that are not necessarily close kin. To study this we must take account of the fact that the states of neighboring replicons will be more similar than those of replicons well separated in space. We must also take account of the fact that a replicon cannot live without enough space. In conventional population dynamics models in continuous space, it is mathematically difficult to treat such correlation of states and mutual exclusion of replicons. In this report we introduce a lattice model in the same spirit as lattice gas models have been introduced in statistical physics to overcome difficulties associated with strong repulsive interactions between particles, and apply it to study conditions governing for the evolution of helping or attacking instinct towards neighbors.

2. Extinction-Invasion Lattice Model

In our model, the space consists of many sites which constitute a lattice. The distance between nearest neighbor sites corresponds in nature to a distance within which an organism or a replicon interacts with other organisms or replicons during its life. For simplicity we assume that the genetic state of a replicon is either + or -. Since we are concerned with inter-kin interactions rather than intra-kin interactions, we make the simplifying assumption that all the replicons in a site are of the same type because they are close kin. Thus, the state of the ith site can be represented by a variable σ_i whose three possible states $\{+, -, 0\}$ correspond to the sites being occupied by + type or by - type or empty. The time development of the overall state $\Omega(t)$ of a population is given by a Markov process with the following transition probabilities:

$$\text{Extinction:} \quad \Pr(\Omega(t+\delta t)=\Omega_i \,|\, \Omega(t)=\Omega) = e_i\delta t, \quad (\sigma_i \neq 0) \tag{2.1}$$

$$\text{Invasion:} \quad \Pr(\Omega(t+\delta t)=\Omega_{ij} \,|\, \Omega(t)=\Omega) = m_{ij}\delta t, \quad (\sigma_i = 0,\ \sigma_j \neq 0) \tag{2.2}$$

$$\text{Mutation:} \quad \Pr(\Omega(t+\delta t)=\overline{\Omega}_i \,|\, \Omega(t)=\Omega) = \mu\ \delta t, \quad (\sigma_i \neq 0) \tag{2.3}$$

$$(i,j = 1,2,\ldots,M;\ \delta t \to 0+)$$

Here, for given

$\Omega = (\sigma_1,\sigma_2,\ldots,\sigma_i,\ldots,\sigma_j,\ldots,\sigma_M)$, we define

$\Omega_i \equiv (\sigma_1,\sigma_2,\ldots,\sigma_{i-1},0,\sigma_{i+1},\ldots,\sigma_j,\ldots,\sigma_M)$,

$\Omega_{ij} \equiv (\sigma_1,\sigma_2,\ldots,\sigma_{i-1},\sigma_j,\sigma_{i+1},\ldots,\sigma_j,\ldots,\sigma_M)$, and

$\overline{\Omega}_i \equiv (\sigma_1,\sigma_2,\ldots,\sigma_{i-1},-\sigma_i,\sigma_{i+1},\ldots,\sigma_j,\ldots,\sigma_M)$.

Non-negative quantities e_i, $m_{i,j}$ and μ are the probability per unit time of the extinction at the ith site, of the duplicated migration from the jth to the ith site, and of mutation to another type, respectively.

Extinction probability at the ith site is given by

$$e_i = \alpha_{\sigma i} - \Sigma_{j(i)} \beta_{\sigma j}\sigma_j^2, \tag{2.4}$$

where $\Sigma_{j(i)}$ denotes the sum over those j which are nearest neighbors (n.n)of the ith site, and $\sigma_j^2 = 1$ for $\sigma_j \in \{+,-\}$. Constant parameters $\alpha_\pm$ and $\beta_\pm$ represent, respectively, the intrinsic mortality and the helping towards a neighbor. Negative value of $\beta_\pm$ represents attacking towards a neighbor.

In the nearest neighbor migration case (NNM), we assume that

$$m_{ij} = \begin{cases} m & (i,j\text{: nearest neighbors}) \\ 0 & (\text{otherwise}) \end{cases} \tag{2.5}$$

In the infinite range migration case (IRM), we assume that

$$m_{ij} = mz/(M-1) \qquad (i \neq j,\ M \to \infty) \tag{2.6}$$

Here, z is the number of nearest neighbors per site. A positive constant m is set such that, if all sites except one site are occupied, then the probabiligy of invasion of the empty site per unit time invaded per unit time is the same (= mz) for both NNM and IRM.

3. Excess Growth Rate of + Sites over − Sites

Let ρ_σ be the fraction of sites of state $\sigma \in \{+,-,0\}$, (σ site), where we have

$$\rho_+ + \rho_- + \rho_0 = 1. \tag{3.1}$$

The average of $d\rho_\sigma/dt$ of the above Markov process for the case NNM is:

$$\begin{aligned} \langle d\rho_\sigma/dt\rangle = {} & mz(1 - p_\sigma)\rho_\sigma \\ & - [\alpha_\sigma - z\{\beta_\sigma p_\sigma \phi_\sigma + \beta_{-\sigma} p_\sigma (1-\phi_\sigma)\}]\rho_\sigma \\ & + \mu(\rho_{-\sigma} - \rho_\sigma), \qquad (\sigma \in \{+,-\}) \end{aligned} \tag{3.2}$$

where $\rho \equiv \rho_+ + \rho_-$ is the degree of habitat saturation. Here, p_σ is the probability that a (randomly chosen) n.n. of σ site is occupied, and ϕ_σ is the probability that the occupied n.n. of σ site is also a σ site. In the IRM case, the first term on the right hand side of (3.2) is replaced by $mz(1 - \rho)\rho_\sigma$.

Writing $\alpha_{\pm} = \alpha \pm (1/2)\delta\alpha$, $\beta_{\pm} = \beta \pm (1/2)\delta\beta$, and $p_{\pm} = p \pm (1/2)\delta p$, we obtain the average excess growth rate of + sites over - sites for NNM in the limit $\delta\alpha \to 0$, $\delta\beta \to 0$, and $\delta p \to 0$:

$$\begin{aligned}\delta s &\equiv \langle(1/\rho_+)(d\rho_+/dt) - (1/\rho_-)(d\rho_-/dt)\rangle \\ &= -\delta\alpha + z\{(\beta - m)\delta p + p(\phi_+ + \phi_-)\delta\beta\} \\ &\quad + \mu(\rho_-/\rho_+ - \rho_+/\rho_-) \qquad (3.3)\end{aligned}$$

As is usually done in population genetics theory we assume that

$$\phi_\sigma = \phi + (1 - \phi)\rho_\sigma/\rho, \qquad (\sigma \in \{+,-\}) \qquad (3.4)$$

where ϕ is the probability that the replicons contained by a pair of n.n. occupied sites are 'identical by descent', which means that no mutation has occurred in their phylogenies since their most recent common ancestor. In (3.4) we assume that if a pair of replicons are not identical by descent, then the conditional probability of one of the two replicons being of a particular state can be approximated by the overall frequency of that state in the population.

Then, we have in the NNM case:

$$\begin{aligned}\delta s &= -\delta\alpha + z\{p\phi\delta\beta + (\beta - m)\delta p\} \\ &\quad + \mu(\rho_-/\rho_+ - \rho_+/\rho_-) \qquad (3.5)\end{aligned}$$

Similarly, we have in the IRM case:

$$\delta s = -\delta\alpha + z\beta\delta p + \mu(\rho_-/\rho_+ - \rho_+/\rho_-) \qquad (3.6)$$

In (3.6), we have used $\phi = 0$, because the range of migration is infinite.

4. Evolution of Helping or Attacking Instinct

Assuming without loss of generality that the replicon of the + state is kinder towards a neighbor than that of the - state, which means $\delta\beta > 0$, let us see which of the two replicon states, + or -, evolves; if $\delta s > 0$, then the + state will evolve, and if $\delta s < 0$, then the - state will evolve.

In (3.5) and (3.6), terms proportional to $\delta\alpha$ and $\delta\beta$ represent the direct effects of excess intrinsic mortality and excess helping, respectively, on the excess growth rate of the + state. The direct effect of excess helping is proportional to the probability of identity by descent ϕ, which is missing in the IRM case. This is the kinship effect between neighbors. In the weak mutation and strong habitat saturation limit: $\mu \to 0$, $\rho \to 1$, we find, following Kimura and Weiss (1964), that

$$\phi = \begin{cases} 1 - O(\sqrt{\mu/m}) & \text{(in one dimension)} \\ 1 - O(1/|\log \mu/m|) & \text{(in two dimensions)} \end{cases} \qquad (4.1)$$

Therefore, in a one-dimensional array of habitats a considerable kinship effect can operate which favours the evolution of helpers. However, in two dimensional habitats, because of the weakness of divergence of the logarithmic function, ϕ may not be sufficiently positive to make the kinship effect outweigh other effect in (3.5); only at a low degree of habitat saturation, can ϕ be sufficiently positive.

Finally, the term proportional to δp in (3.5) and (3.6) represents the indirect effect attributable to the fact that + sites with kinder replicons tend to have fewer vacant n.n. sites ($\delta p > 0$) in the stationary state. Therefore, in the IRM case generally the evolution of helping behavior ($\beta > 0$) in a population generally induces the evolution of helping intensity, while the prevalence of attacking behavior ($\beta < 0$) generally induces the evolution of attacking intensity. On the other hand, for NNM case, the coefficient of δp in (3.5) cannot be positive. Because the extinction probability in (2.4) is non-negative, we must have $\alpha \geqq z\beta$. Thus, if $\beta - m > 0$, we should have $\alpha > zm$, which would lead to extinction of the whole population: $\rho \to 0$. Therefore, in the NNM case, the indirect effect always suppresses the evolution of helping in the stationary state.

Introducing a decoupling approximation, as is usually done in statistical physics, we can estimate δp for the NNM case:

$$\delta p = [p^2/m\{1 + (z - 1)r\}]\delta\beta \quad , \qquad (4.2)$$

where r is the probability that the n.n. of empty site will be occupied; and in the IRM case:

$$\delta p = (p^2/mz\rho)\delta\beta. \qquad (4.3)$$

Substituting (4.2) and (4.3) into (3.5) and (3.6) we obtain for the NNM case:

$$\delta s = -\delta\alpha + zp[\phi + p(\beta/m - 1)/\{1 + (z - 1)r\}]\delta\beta, \qquad (4.4)$$

and in the IRM case:

$$\delta s = -\delta\alpha + (p^2\beta/m\rho)\delta\beta. \qquad (4.5)$$

Thus, our result suggests that, under natural conditions, when the migration range per generation is much larger than the range of social interactions such as attacking or helping between conspecifics which are not immediate kin, evolution of the intensity of the attacking or helping instinct depends primarily on which behavior is more prevalent in a population. It is weakly dependent of mutation rate or dimensionality affecting the probability of neighbors being idential by descent. On the other hand, if the migration range per generation is almost the same as the range of the above-mentioned social interactions, the helping instinct may evolve in a one-dimensional array of habitats, but, in two-dimensions, it may only do so for a sufficiently low degree of habitat saturation and sufficiently low mutation rate.

5. Discussion

Our study of a lattice model for population biology (Matsuda, 1981) was originally motivated by the work of Felsenstein (1975) on Malécot's model of isolation by distance. In this model it was tacitly assumed that (I) individuals are distributed randomly with equal expected density in continuous space and that (II) each individual reproduces independently and (III) migrates independently. Felsenstein showed that the above assumptions are not always compatible because the time-development based on (II) and (III) causes clumping of individuals, which violates (I). A corresponding model in discrete space, the stepping-stone model introduced by Kimura (1953), has no such difficulties because assumtion (II) is in effect lifted by invoking the Wright-Fisher model to keep the population size of each colony (stone) constant. On the other hand, the assumption of constant population sizes precludes study of the dependence of the state of a population upon the mode of reproduction of each individual from a more basic point of view. For instance, the model does not tell how the probability of identity by descent may depend on the degree of habitat saturation.

The effects of neighbors on the evolution of altruistic traits has already been studied by Eshel (1972). He concluded in his model that, if the effect of migration distance and the density-dependent mortality are neglected, low demographic mobility is the most crucial factor for the evolution of altruism. Recently, Boyd (1982) studied the latter effect, but he considered only social interactions between siblings. Since Eshel's work, the effects of group selection on altruism have been studied extensively (see, for instance, Wilson, 1980; Kimura, 1983). In these studies, the existence of groups as well as contributions to the group survival by altruists are assumed from the outset. However, groups may be formed or disintegrate depending on the mode of behaviors of each individual. Therefore, although group selection theory may be of use, we think that studying the conditions for the evolution of altruism may be intellectually more interesting if initial existence of groups is not assumed.

In any case, we hope that, in view of its versatility and its amenability to computer simulation, the lattice model, which proved useful in statistical physics (see, for instance, Lee and Yang, 1952; Matsubara and Matsuda, 1956), may also be effective in studies of population biology.

Acknowledgements

We thank the Japan Ministry of Education, Science and Culture for support by a Grant-in Aid for Special Project Research on Biological Aspects of Optimal Strategy and Social Structure. We wish to take the opportunity to dedicate this paper to Emeritus Professor T. Matsubara who inspired us with the idea of a unified understanding of nature through simple and comprehensive models.

References

Boyd,R. (1982) Density-dependent mortality and the evolution of social interactions. Anim. Behav. 30, 972-982.

Eshel,I. (1972) On the neighbor effect and the evolution of altruistic traits. Theor. Pop. Biol. 3, 258-277.

Felsenstein,J. (1975) A pain in the torus: some difficulties with models of isolation by distance. Amer. Nat. 967, 359-368.

Hamilton,W.D. (1964) The genetical evolution of social behavior. and II. J. Theor. Biol. 7, 1-16, 17-52.

Hamilton,W.D. (1972) Altruism and related Phenomena, mainly in social insects. Ann. Rev. Ecol. Syst. 3, 193-232.

Kimura,M. (1953) "Stepping ston" model of population. Ann. Rep. Natl. Inst. Genet. Japan 3, 63-65.

Kimura,M. (1983) Diffusion model of intergroup selection, with special reference to evolution of an altruistic character. PNAS 80, 6317-6321.

Kimura,M. and Weiss,G.H. (1964) The stepping stone model of population structure and the decrease of genetic correlation with distance. Genetics 49, 561-567.

Lee,T.D. and Yang,C.N. (1952) Statistical theory of equations of state and phase transitions. II. lattice gas and Ising model.

Matsubara,T. and Matsuda,H. (1956) A lattice model of liquid helium, I. Prog. Theor. Phys. 16, 569-582.
Matsuda,H. (1981) The Ising model for population biology. Prog. Theor. Phys. 66, 1078-1080.
Wilson,D.S. (1980) The natural selection of populations and communities. Menlo Park, California: Benjamin/Cummings Pub. Co.

POPULATION GENETICAL MECHANISM OF MOLECULAR EVOLUTION
-- Stochastic Selection as an Alternative to Random Drift --

K. Ishii, H. Inutsuka*, H. Matsuda* and N. Ogita+
College of General Education, Nagoya University
* Department of Biology, Faculty of Science, Kyushu University
+ The Institute of Physical and Chemical Research

The neutralist-selectionist controversy persists with respect to the population genetical mechanism of molecular evolution. In this paper, we first show how we can explain, from a selectionist perspective, the fact that molecular evolution rates are smaller than total mutation rates. Next, using a statistical analysis of electrophoretic data on local differentiation, we show that, in most cases, stochastic selection is more responsible for local differentiation than is random drift.

Introduction

During the past two decades, the development of techniques of sequence analysis and electrophoresis has enabled us to study extensively evolution at the molecular level. It has been found that molecular evolution rates are roughly constant per year and that the electrophoretically revealed polymorphism is rather insensitive to the species of organisms, although they are sensitive to the molecular functional constraint. However, the neutralist-selectionist controversy continues with respect to population genetical mechanisms. The range of applicability of the neo-Darwinian view of evolution is still obscure (for an extensive review of the controversy from a neutralist side, see Kimura (1983)).

Zuckerkandl and Pauling (1965) interpreted the nearly constant rate of evolution as a kind of shuttling motion between functionally similar amino acid residues. They supposed that the spread of a mutant through the population will be caused by its slight advantage in natural selection, although the evolutionarily effective changes in amino acid sequence will be limited to the functionally nearly neutral changes.

In contrast to this neo-Darwinian view of molecular evolution, Kimura (1968) proposed the neutral theory and subsequently developed it such that both evolution rates and polymorphism are derived in a unified manner. The neutral theory insists that positive selection is rarely responsible for mutational substitutions in molecular evolution and that random drift due to finite population size is responsible for the spread of a mutant in a population. According to this theory, the rate of evolution equals that of neutral mutation. It is therefore unclear as to why the constancy of actual evolution rate is per year rather than per generation. The neutral theory admits the effects of negative selection to explain the upper bound of polymorphism. This brings up the question as to why positive selection should be excluded as a force in molecular evolution.

Gillespie and Langley (1974) proposed a model in which fluctuating environments can give rise to frequency dependent selection capable of sustaining polymorphism. However, they made no quantitative comparison with actual data on polymorphism. Takahata, Ishii and Matsuda (1975) pointed out that if the autocorrelation of temporal fluctuations in environments is taken into account, their stochastic selection model, which includes the effects of mutation, can explain data on polymorphism more adequately than does the neutral theory. They did not, however, develop a model for deriving the evolution rate.

Based on these previous studies, we have tried with simple models to elucidate comprehensively the way in which evolution and polymorphism depend on parameters such as selection intensity s, duration time τ of a stochastic selection, mutation rate μ, and the effective population size N. We have also tried to analyze a given data set in order to see which parameter region it corresponds to and which view of molecular evolution it supports. In this paper, however, we will concentrate on two of the main results we have so far obtained.

Firstly, we consider an empirical rule that molecular evolution rates are smaller than total mutation rates. This is often cited by neutralists as crucial supporting evidence for their theory since its explanation is trivial if the evolution rate is equal to the neutral mutation rate (Kimura (1982)). However, irrespective of the existence or strength of positive selection, we found that this rule can be explained by an assumption of adaptively-determined total mutation rate. Our explanation is based on the general relationship $\partial\bar{m}/\partial\mu=\upsilon/\mu-1$ among the average Malthusian parameter $\bar{m}$ of a population, the evolution rate υ, and the total mutation rate μ, which is proved for an infinitely large asexual population under arbitrarily fluctuating selection with the condition that Malthusian parameters are independent of μ.

Secondly, we show the results of a statistical test developed for the purpose of determining whether stochastic selection or random drift is more likely to be responsible for the local differentiation of a species gene frequencies. Our test is based on the relation between the average $\bar{x}$ and the difference d of frequencies of an allele in two subpopulations in which migration has kept d small. The results indicate that in most cases it is stochastic selection rather than random drift which is mainly responsible for local differentiation.

Molecular evolution rate and the total mutation rate

Comparing several base sequences of a homologous gene from different species, we can measure the molecular evolution rate as the base substitution rate per nucleotide site along a phylogenic line. It is approximately constant per year along any phylogenic line for each homologous gene (see, however, a critical study by Gillespie (1986)). It is faster for a gene with less functional constraint on its sequence and is greatest for "pseudo" genes or "dead" genes (Miyata (1982)). Since selection pressure on dead genes must be negligibly weak, their evolution rates are believed to be nearly equal to their total mutation rates. This is what we mean when we say that the molecular evolution rate υ is smaller than the total mutation rate μ.

On the other hand, according to population genetics theory, the evolution rate υ varies according to evolutionary conditions. For example, it may be equal to (i) the neutral mutation rate μ_n (Kimura (1968)) or (ii) the product $2Ns\mu_a$ of twice the population size N, selective advantage s, and the adaptive mutation rate μ_a (Kimura and Ohta (1971)) or (iii) the inverse $1/\tau$ of the average duration time τ of the selective environment (Ishii, Matsuda and Ogita (1982)), and so on.

Now, if the evolution rate υ is equal to the neutral mutation rate μ_n, it is obvious that υ never exceeds the total mutation rate μ. On the other hand, the evolution rate υ under natural selection, as we see, for example, from (ii) and (iii) above, will not, in general, necessarily be smaller than the total mutation rate μ. Thus, it is often claimed by neutralists that the fact that v is smaller than μ is crucial supporting evidence for the neutral theory (Kimura (1982)). So our problem lies in explaining how v can be smaller than μ under conditions of natural selection.

(*i*)*Replicon model of evolution*

In order to study comprehensively evolution rate under natural selection, we introduce a replicon model of evolution. Here, "replicon" is a general term we use to denote an element of an evolving population. In our context of molecular evolution, it can, for example, be a segment of DNA, a gene, a chromosome, or a gamete, etc.

The key concepts in this model are the step number n and the phylogenic state φ of each replicon. The step number n of a replicon is the total number of mutations it has experienced during its descent from an ancestor at a given time, say t=0. Then the evolution rate v is given as the rate $d\bar{n}/dt$ of increase of the average step number $\bar{n}$; that is the population average of the step number of each replicon.

In order to study the change of $\bar{n}$ over time, we classify replicons according to their phylogenic states (φ's), and we study the change over time of the frequency x_φ of replicons in phylogenic state φ. Here, the phylogenic state φ of a replicon will in general differ from its genetic state σ, which, for example, may be specified by its DNA sequence. The phylogenic state φ of a replicon is defined as the sequence of genetic state σ's it has experienced through mutation during its deescent from an ancestor at time t=0. Thus, for a replicon with step number n, its phylogenic state φ is described as $\varphi = (\sigma_0, \sigma_1, \sigma_2, \ldots, \sigma_n)$, where σ_0 is the genetic state of the first ancestor at time t=0, and σ_i is the genetic state produced by the ith mutation which occurred to a replicon in the genetic state σ_{i-1} (i=1,2,...,n).

With this definition of phylogenic state, each phylogenic state φ has a unique parental phylogenic state φ' from which the offspring in phylogenic state φ was produced by mutation. (The step number $n_{\varphi'}$ is smaller than n_φ by 1.) Therefore, the frequencies x_φ satisfy

$$dx_\varphi/dt = (m_\varphi(t) - \bar{m}(t))x_\varphi + \mu(f_{\varphi,\varphi'}x_{\varphi'} - x_\varphi), \quad (1)$$

Here, $m_\varphi(t)$ is the Malthusian parameter at time t of a replicon in phylogenic state φ, and $\bar{m}(t) = \sum_\varphi m_\varphi(t)x_\varphi(t)$ is the average Malthusian parameter of the population. The total mutation rate μ per replicon and the mutation rate $\mu f_{\varphi,\varphi'}$ from φ' to φ are assumed to be constants independent of time t.

(ii)Extended Haldane-Muller principle of mutation load

For this replicon model of evolution, we can derive a general formula relating the evolution rate v and the average Malthusian parameter $\bar{m}$ (Ishii and Matsuda (1985)). Firstly, when each Malthusian parameter $m_\varphi(t)$ is independent of mutation rate μ, we find that

$$\partial\bar{m}/\partial\mu = v/\mu - 1. \tag{2}$$

Next, remembering that real organisms incur a cost for reduction of mutation rate, we consider a case in which the cost $\partial m_\varphi(t)/\partial\mu$ of preventing mutation is the same for all replicons and is equal to $c(t)$. Then, formula (2) is generalized as

$$\partial\bar{m}/\partial\mu = v/\mu - 1 + c. \tag{3}$$

We emphasize here that formulas (2) and (3) hold quite generally at any time t for any genic selection scheme $\{ m_\varphi(t) \}$ regardless of time fluctuation, strength of selection, and mutation structure $\{ f_{\varphi,\varphi'} \}$. As a special case, we may consider a strong constant selection scheme which gives $v/\mu=0$ in the limit of weak mutation. Then, the right hand side of (2) reduces to -1, and this is simply the continuum version of the well known Haldane-Muller principle of mutation load (Haldane (1937), Muller (1950)).

Now, building upon this general formula (3) of mutation load, we explain why the molecular evolution rate v should be smaller than the total mutation rate μ. To do this, we need only assume that μ is adaptively determined so as to maximize $\bar{m}$. Then, from (3), we find that the relative evolution rate v/μ is equal to $1-c$, and hence v must be smaller than μ.

Up to this point, we have neglected the effects of recombination, the finite population size, and non-genic selection. However, we expect our results to apply beyond these restrictions.

(iii)Size effect on the extended Haldane-Muller principle

As a first step of study in this direction, we consider the size effect. We can study this by computer simulation of the Wright-Fisher model version of our replicon model.

Fig.1 shows the mutation rate μ dependence of the evolution rate v and the average Malthusian parameter $\bar{m}$ of a replicon model under a regime of simple stochastic selection with strength s and average duration time τ. More specifically, the Malthusian parameter $m_\varphi(t)$ was assumed to depend on φ only through step number n_φ in that

$$m_\varphi(t) = \begin{cases} s(t) & \text{for} \quad n_\varphi = \text{even} \\ -s(t) & \text{for} \quad n_\varphi = \text{odd}, \end{cases} \tag{4}$$

where $s(t)$ is a Markov process with the two possible values $+s$ and $-s$ and an average relaxation time τ.

In Fig.1, $s\tau$ is fixed at $\sqrt{10}$. Solid lines and broken lines represent v/μ and $\overline{m}/s$ respectively. Thick lines denote the case $N=\infty$, and thin lines are interpolations of simulation results for the case $Ns=1$. Simulation values of v/μ and $\overline{m}/s$ were obtained for ten different values of μ/s and are marked by symbols + and ×, respectively.

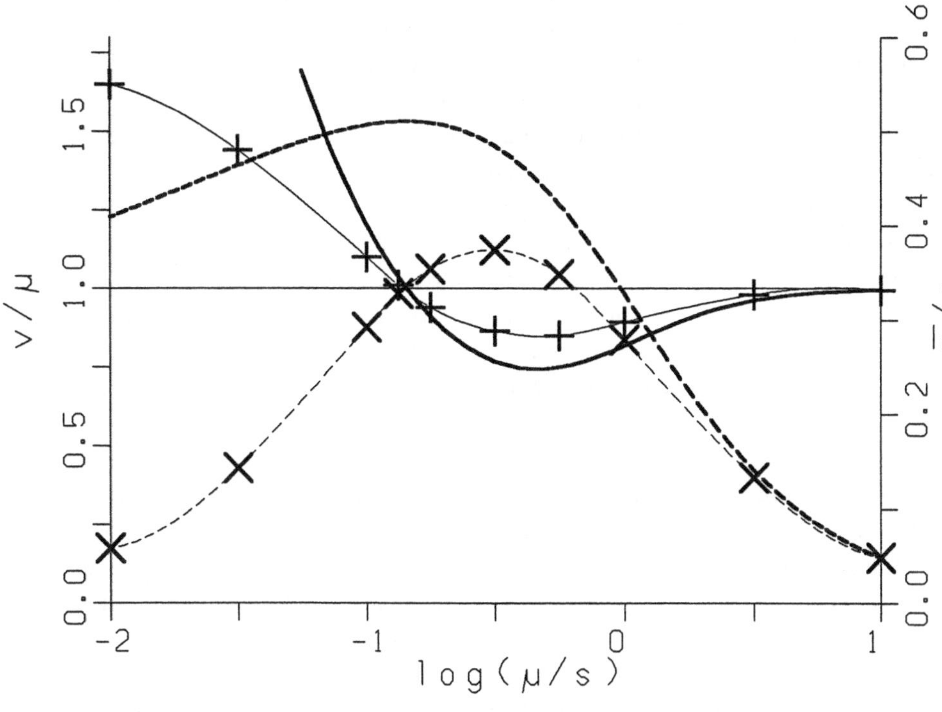

The result for $N=\infty$ was obtained analytically as follows. According to (1) and (4), the frequency x of even step number replicons satisfies

$$dx/dt = 2s(t)x(1-x) + \mu(1-2x),$$

and the stationary joint distribution density for $(x(t), s(t))$ has been explicitly obtained (Matsuda and Ishii (1980)). Hence, we calculate $\overline{m}$ as the average of $s(t)(2x(t)-1)$ with this distribution. We then obtain the evolution rate v via general formula (2) by numerically differentiating $\overline{m}$ with respect to μ.

First, we consider the result for the $N=\infty$ case. For weak mutation the relative evolution rate v/μ is larger than 1, but, for strong mutation, becomes smaller than 1 as μ increases and approaches 1 from below. The average Malthusian parameter $\bar{m}$ increases with μ if v/μ is larger than 1, attains its maximum at the point where v/μ is equal to 1, and decreases thereafter.

We next consider the computer simulation results for the $Ns=1$ case. Because of the size effect, relative evolution rate v/μ is taken to be closer to 1, and the average Malthusian parameter $\bar{m}$ is reduced substantially. However, it also happens in this case that, if v/μ is larger than 1, $\bar{m}$ increases monotonically when we increase μ. However, the point at which $\bar{m}$ attains its maximum is shifted in the direction of greater μ, where we find v/μ to be smaller than 1.

This suggests that when the size effect is taken into account, formula (2) should be modified as follows:

$$\partial\bar{m}/\partial\mu > 0 \quad \text{for } v/\mu > 1. \tag{5}$$

When we consider an identical cost c of preventing mutation for all replicons, v/μ remains the same but $\partial\bar{m}/\partial\mu$ increases by c. Thus, in this case, inequality (5) holds all the more strongly. Then, according to inequality (5), if the total mutation rate μ is adaptively determined, we conclude that the evolution rate v should be smaller than μ.

Given these results, we can say that neutralists make an overstatement when they claim that v being smaller than μ is crucial supporting evidence for the neutral theory. This observation can also be explained from a selectionist standpoint by an assumption of adaptively-determined total mutation rate.

Electrophoretic data of local differentiation

Detailed studies of local differentiation have been done for a number of species, and we have at our disposal electromorph frequency data for several loci in a number of subpopulations. Our problem here is to determine whether random drift or stochastic selection is more likely to be responsible for observed local differentiation.

In order to devise a means of testing this statistically, we first consider two subpopulations with migration between them. Let x_1 and x_2 be the frequencies of an electromorph in each subpopulation. Frequencies x_1 and x_2 are assumed to vary with time according to

$$dx_1/dt = f_1(t) + m_1(x_2-x_1), \qquad dx_2/dt = f_2(t) + m_2(x_1-x_2),$$

where m_i is the migration rate from the ith subpopulation, and $f_i(t)$

is stochastic variation responsible for local differentiation due to random drift and/or stochastic selection. We assume that migration is frequent enough to keep the difference $d = |x_1-x_2|$ of the two frequencies small.

If local differentiation is due to random drift, we have

$\langle f_i(t)\rangle = 0, \qquad \langle f_i(t)f_j(t')\rangle = \delta_{ij}\delta(t-t')x_i(t)(1-x_i(t))/N_i,$

for the averages of the $f_i(t)$'s. Here, N_i is the size of the ith subpopulation. Then, denoting by $\bar{x}$ the average $(x_1+x_2)/2$ of the two frequencies, it can be shown that the average of d^2 conditional on $\bar{x}=x$ is

$\langle d^2\rangle_{\bar{x}=x} = x(1-x)/Nm,$

with $N = 2N_1N_2/(N_1+N_2)$ and $m = m_1+m_2$.

On the other hand, if differentiation is due to stochastic genic selection $f_i(t) = s_i(t)x_i(t)(1-x_i(t))$, with $s_i(t)$ having averages

$\langle s_i(t)\rangle = 0, \qquad \langle s_i(t)s_j(t')\rangle = \delta_{ij}s^2\exp(-|t-t'|/\tau),$

then, it can be shown that the conditional average of d^2 is

$\langle d^2\rangle_{\bar{x}=x} = 2s^2x^2(1-x)^2/m(m+1/\tau).$

So, if we define X, Y and Z as $X = \bar{x}(1-\bar{x})$, $Y = d^2/X$, $Z = d^2/X^2$, then, if local differentiation is due to random drift, this leads approximately to the *drift hypothesis* that <u>Y is independent of X</u>, while differentiation due to stochastic selection leads approximately to the *selection hypothesis* that <u>Z is independent of X</u>.

We tested the drift hypothesis using actual data on electromorph frequencies as follows. For each subpopulation pair, we let n be the total number of electromorph frequency pairs (x_1,x_2) collected from available polymorphic loci. We transform each (x_1,x_2) to (X,Y) and calculate the averages $\bar{X}$ and $\bar{Y}$ of X and Y. We then test the independence of the event $X < \bar{X}$ from the event $Y < \bar{Y}$ by means of a χ^2 test, based on the incidence numbers n_1, n_2, n_3, n_4 of n pairs (X,Y) for ($X\leq\bar{X}$ and $Y\leq\bar{Y}$), ($X\leq\bar{X}$ and $Y>\bar{Y}$), ($X>\bar{X}$ and $Y\leq\bar{Y}$), and ($X>\bar{X}$ and $Y>\bar{Y}$). Finally, we calculate the fraction f_D of subpopulation pairs for which the drift hypothesis is rejected at the 5% level.

Similar analysis can also be performed for the selection hypothesis, and we obtain its rejection fraction f_S. If a species has a smaller rejection fraction for the selection hypothesis than for the drift hypothesis, then we may judge its local differentiation to be mainly due to stochastic selection.

Table 1 is the result of our statistical analysis of some actual data. In this table we also give the average number $\bar{n}$ of the total number n of frequency pairs (x_1,x_2) used in the analysis of a subpopulation pair. The results suggest that, in most cases, stochastic selection is more responsible for local differentiation than is random drift.

Table 1 Rejection fractions f_D, f_S

Species		f_D	f_S	$\bar{n}$	Source
Drosophila	*equinoxialis*	0.67	0	105.3	1
D.	*paulistorum*	1.00	0.50	95.5	1
D.	*robusta*	0.40	0.20	49.8	2
D.	*silvestris*	1.00	0.33	59.3	3
D.	*tropicalis*	1.00	0	78.5	1
D.	*willistoni*	0.60	0.60	109.4	1
Philaenus	*spumarius*	0.90	0.60	32.0	4
Astyanax	*mexicanus*	0.47	0.33	39.7	5
Menidia	*beryllina*	0.43	0.43	19.0	6
M.	*menidia*	0.75	0.83	22.9	6
Bufo	*viridis*	0.46	0.38	36.1	7
Anolis	*brevirostris*	0.33	0.20	9.4	8
A.	*carolinensis*	1.00	1.00	23.7	9
Peromyscus	*boylli*	0.43	0.14	10.4	10
Thomomys	*talpoides*	0.98	0.89	29.7	11

†Source: 1. Ayala et al. (1974), 2. Prakash (1973), 3. Sene and Carson (1977), 4. Saura et al. (1973), 5. Avise and Selander (1972), 6. Jhonson (1975), 7. Nevo et al. (1975), 8.Webster and Burns, (1973), 9. Webster et al. (1975), 10. Avise et al. (1974), 11. Nevo et al. (1974).

Conclusion

In view of these findings we feel that at the protein level there are ample grounds for accepting the neo-Darwinian view of evolution. At the DNA level the neutral theory might apply to evolution due, for example, to synonymous mutations. However, this is by no means obvious in the absence of further careful study.

Acknowledgements

This work was supported in part by a Grant-in-Aid for Special Project Research on Biological Aspects of Optimal Strategy and Social Structure from the Japan Ministry of Education, Science and Culture. Numerical computations were carried out on computers at the Institute of Physical and Chemical Research, the Computer Center of Kyushu University, and the Nagoya University Computation Center.

References

Haldane, J.B.S. (1937). The effect of variation on fitness. *Amer. Nat.* 71, 337-349.

Gillespie, J.H. (1986). Natural selection and the molecular clock. In *Proceedings of the International Symposium on Mathematical Biology, Nov.10-15, 1985, Kyoto, Japan* (ed. E.Teramoto) New York: Springer Verlag.

Gillespie, J.H. & Langley, C.H. (1974). A general model to account for enzyme variation in natural populations. *Genetics* 76, 837-848.

Ishii, K. & Matsuda, H. (1985). Extension of the Haldane-Muller principle of mutation load with application for estimating a possible range of relative evolution rate. *Genet. Res., Camb.* 46, 75-84.

Ishii, K., Matsuda, H. & Ogita, N. (1982). A mathematical model of biological evolution. *J. Math. Biology* 14, 327-353.

Kimura, M. (1968). Evolutionary rate at the molecular level. *Nature* 217, 624-626.

Kimura, M. (1982). The neutral theory as a basis for understanding the mechanism of evolution and variation at the molecular level. In *Molecular Evolution, Protein Polymorphism and the Neutral Theory* (ed. M.Kimura), pp.3-56. Tokyo: Japan Scientific Societies Press.

Kimura, M. (1983). *The Neutral Theory of Molecular Evolution.* Cambridge: Cambridge University Press.

Kimura, M. & Ohta, T. (1971). On the rate of molecular evolution. *Nature* 217, 624-626.

Matsuda, H. & Ishii, K. (1980). Stationary gene frequency distribution in the environment fluctuating between two distinct states. *J. Math. Biology* 11, 119-141.

Miyata, T. (1982). Evolutionary changes and functional constraint in DNA sequences. In *Molecular Evolution, Protein Polymorphism and the Neutral Theory* (ed. M.Kimura), pp.233-266. Tokyo: Japan Scientific Societies Press.

Muller, H.J. (1950). Our load of mutations. *Amer. J. Hum. Genet.* 2, 111-176.

Takahata, N, Ishii, K. & Matsuda, H. (1975). Effect of temporal fluctuation of selection coefficient on gene frequency in a population. *Proc. Nat. Acad. Sci. USA* 72, 4541-4545.

Zuckerkandl, E. & Pauling, L. (1965). Evolutionary divergence and convergence in proteins. In *Evolving Genes and Proteins* (ed. V.Bryson and H.J.Vogel), pp.97-166. New York: Academic Press.

Evolutionary and Ecological Stability of Prey-Predator Systems with Predatory Switching

Hiroyuki MATSUDA[1], Kohkichi KAWASAKI[2], Nanako SHIGESADA[3]
Ei TERAMOTO[3] and Luigi M. RICCIARDI[4]

[1] Information Processing Center of Medical Sciences, Nippon Medical School, Sendagi, Bunkyo-ku, 113, JAPAN
[2] Science and Engineering Research Institute, Dohshisha University, Kyoto, 602, JAPAN
[3] Department of Biophysics, Faculty of Science, Kyoto University, Kyoto, 606, JAPAN
[4] Dipartimento di Matematica, Università di Napoli, ITALY

1. Introduction

Most predators utilize more than one prey species, and it is known that their diets do *not directly* reflect population densities of available prey species. We consider two prey species which live in two different patches and one predator which allocates its foraging (including searching and handling) activity between the two prey species according to their relative densities. Murdoch(1969) introduced the term "switching" to refer to the case in which the relative amount of prey in the predator's diet increases *more than* proportionally to the relative prey density. Predatory switching has been postulated for predators ranging from protozoa to birds (reviewed by Murdoch & Oaten,1975), and it may be a simple consequence of the predator's searching behavior (May,1977).

Comins & Hassell(1976), Tansky(1978), and Teramoto, Kawasaki & Shigesada(1979) analyzed dynamical behaviors of prey-predator systems with predatory switching properties (reviewed by Holt,1983). These studies show that predatory switching can stabilize the system as a whole and permit the coexistence of competing prey species. Matsuda(1985) considered the possibility that switching properties might be determined by predators' behavioral strategies. He found the evolutionarily stable strategies (ESS, termed by Maynard Smith & Price,1973) for predatory switching, by analyzing the conditions under which a mutant predator with a switching property different from that of the dominant (wild type) can invade in a two-prey and one-predator system. Further, Matsuda et al.(1986) verified that the equilibrium state of a system with an evolutionarily stable switching strategy is ecologically stable in the absence of invasion of mutant predators. They used the following switching functions given by Elton & Greenwood(1970):

$$f_1 = \frac{u^n x_1{}^n}{u^n x_1{}^n + (1-u)^n x_2{}^n}, \quad f_2 = \frac{(1-u)^n x_2{}^n}{u^n x_1^n + (1-u)^n x_2^n}, \quad (n>0,\ 0<u<1) \tag{1}$$

which represent the fractional allocation of searching efforts to prey 1 and 2, respectively. Here x_i is the density of prey i, u is the relative selectivity for the first prey species, and n is the intensity of switching.

However, these studies ignored the effects of handling time and intraspecific competition. These two factors are known to have major effects on the stability of many ecological systems. Since there is no theoretical reason to expect an evolutionary change in some behavior to yield ecological stability of the system, it may be useful to investigate the relationship between evolutionary changes in switching properties and ecological stability of more general prey-predator systems which incorporate effects of handling time and intraspecific competition.

2. A General Model for Switching Properties

We assume the number of individuals of prey species i eaten by a predator per unit time (denoting by N_i, i=1,2) to be

$$N_i = f_i r_i x_i, \quad r_i = a_i/(1+a_i h_i x_i), \qquad (i=1,2) \tag{2}$$

where f_i is the fraction of time allocated by the predator to prey i, h_i is the handling time per individual of prey i, and a_i is the encounter rate between prey i and the predator. Here we assume that

a) $f_1+f_2 \equiv 1, \quad 0 < f_i < 1,$ (3a)

b) f_i is a function of the relative prey density

$p=x_1/(x_1+x_2),$ (3b)

and

c) $\partial f_1/\partial p > 0.$ (3c)

We first analyze the simple case in which $h_1=h_2=0$ (r_i becomes a_i) a restriction which we will subsequently remove. From assumption (3a), the per capita total energy intake rate of the predator (denoting by E) is written as

$$\begin{aligned} E &= G_1N_1+G_2N_2 = f_1G_1a_1x_1+f_2G_2a_2x_2 \\ &= f_1\cdot(G_1a_1x_1-G_2a_2x_2)+G_2a_2x_2, \end{aligned} \tag{4}$$

where G_i represents the energy content of the ith prey individual. For given x_1 and x_2, it is clear that the following relation gives an optimal foraging strategy which maximizes the rate of total energy intake:

$$f_1=1,\ f_2=0 \text{ if } G_1a_1x_1 > G_2a_2x_2,$$
$$f_1=0,\ f_2=1 \text{ if } G_1a_1x_1 < G_2a_2x_2. \tag{5}$$

Fig.1 shows the predator's per capita total energy intake rate as a function of the relative density of the first prey (p) type for given values of n and u of Eq.(1) (see Matsuda,1985). When the relative prey density is $p_s=1-u$, the predator allocates the same effort to the two prey, namely $f_1=f_2=\frac{1}{2}$. We call this state the *switching point*. From assumption (3c),

$$f_1 > \tfrac{1}{2} > f_2 \tag{6}$$

if and only if $p>p_s$.

When $p=p_e=G_2a_2/(G_1a_1+G_2a_2)$, i.e. $G_1a_1x_1=G_2a_2x_2$, the total energy intake is independent of f_1 ($\equiv 1-f_2$). Since $G_ia_ix_i$ represents the energy intake rate when the predator exclusively searches for prey species i, we call this state p_e the *equivalence point*. In Fig.1(a), the switching point coincides with the equivalence point.

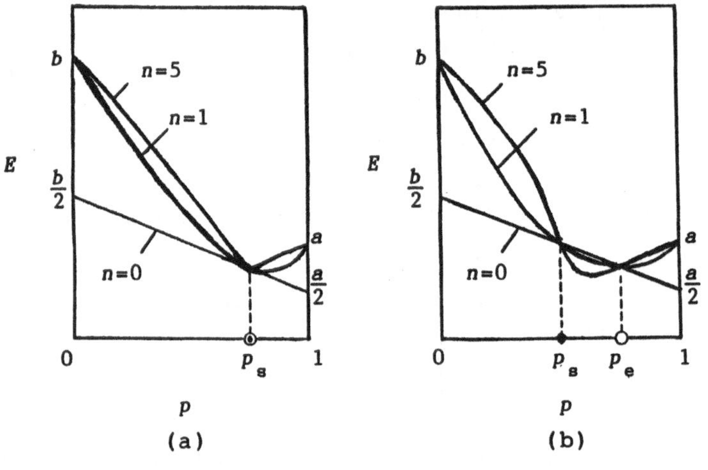

Fig.1. The per capita total energy intake rate of a predator(E) as a function of the first prey relative density(p) when $x_1+x_2=1$ (constant), $a_1=1$, $a_2=3$, $G_1=G_2=1$. Here $p=1-u$ (closed circle) and $p_e=a_1/(a_1+a_2)$ (open circle). In (a) $p_s=p_e$, and in (b) $p_s\neq p_e$

In Fig.1(b), there is a gap between them. Further, when $p=x_1/(x_1+x_2)$ takes an intermediate value between p_s and p_e, the predator allocates a larger amount of its searching effort to the less efficient prey type ($f_1>\frac{1}{2}$ and $G_1a_1x_1<G_2a_2x_2$) so that the total intake E of a switching predator becomes smaller than that of a non-switching predator with $n=0$ ($f_i\equiv\frac{1}{2}$).

To investigate the evolution of switching, we consider a system which consists of two prey species, a resident predator, and an invading (mutant) predator. For simplicity, we assume that all species reproduce asexually. Then, this system is described by

$$\begin{aligned} dx_1/dt &= [\varepsilon_1 - f_1a_1y_0 - g_1a_1y_1]x_1 = F_1x_1, \\ dx_2/dt &= [\varepsilon_2 - f_2a_2y_0 - g_2a_2y_1]x_2 = F_2x_2, \\ dy_0/dt &= [-\delta + f_1G_1a_1x_1 + f_2G_2a_2x_2]y_0 = F_3y_0, \\ dy_1/dt &= [-\delta + g_1G_1a_1x_1 + g_2G_2a_2x_2]y_1 = F_4y_1, \end{aligned} \qquad (7)$$

where y_0 and y_1 are the densities of the dominant and mutant predators, respectively, ε_i is the intrinsic growth rate of prey i, δ is the death rate of the predator, and g_i is the fractional allocation of foraging effort of the mutant predator to prey i satisfying assumption (3a) to (3c). We assume that the invading predator differs from the resident only in its switching property. Suppose that a system composed of two prey species and a wild type predator (which we call "the pre-existing system") is invaded by a mutant predator after the system has established a steady state. We consider that such an invasion occurs randomly, but the interval between successive invasions is long enough for the system to reach a new steady state before a new mutant invades. Predatory switching will be said to evolve if the wild type predator becomes extinct and the mutant predator dominates the system.

The steady state of the pre-existing system corresponds to the boundary equilibrium point $X^*=(x_1^*,x_2^*,y_0^*,0)$ in system (7) which satisfies $F_1=F_2=F_3=0$ and $y_1=0$. We have,

$$\begin{aligned} x_1^* &= p^*D^*, \\ x_2^* &= (1-p^*)D^*, \\ y_0^* &= (a_1\varepsilon_2+a_2\varepsilon_1)/a_1a_2, \\ p^* &= x_1^*/(x_1^*+x_2^*) = f_1^{-1}(f_1^*), \\ f_1^* &= a_2\varepsilon_1/(a_2\varepsilon_1+a_1\varepsilon_2), \\ D^* &= x_1^*+x_2^* = \delta(a_1\varepsilon_2+a_2\varepsilon_1)/a_1a_2[G_1\varepsilon_1p^*+G_2\varepsilon_2(1-p^*)]. \end{aligned} \qquad (8)$$

Matsuda et al.(1986) showed that the equilibrium point X^* is ecologically stable in the pre-existing system for a wide range of values of n and u in Eq.(1). Thus, it is clear that the conditions for invasion by a small number of mutant predators of the pre-existing system at equilibrium reduce to the condition that $F_4(\boldsymbol{X}^*)$ of Eq.(7) is positive.

Assume that a mutant predator having a different switching property appears in the system. If $G_i a_i x_i^* > G_j a_j x_j^*$ $(j \neq i)$ and if the mutant allocates a larger amount of its effort to prey i than does the resident predator $(f_i^* < g_i^*)$, then the mutant will have get a higher per capita total energy intake. Under such circumstances, the stable steady state of the pre-existing system is not considered to be an evolutionarily stable state, and the resident predator will be replaced by the invader. Matsuda(1985) analyzed evolution of the switching property described by Eq.(1) and concluded that the evolutionarily stable steady state with respect to the switching property corresponds to the equivalence point.

However, it is obvious that the equilibrium state in the system with the evolutionarily stable switching strategy will be realized only if it is ecologically stable. Matsuda et al.(1986) showed that an equilibrium state that has attained the equivalence point is *always* ecologically stable when the strategy set of switching functions is given by Eq.(1). They also analyzed the effect of predatory switching on the ecological stability of a three-trophic-level system in which species 2 (x_2) feeds on prey species 1 (x_1) and the top omnivorous predator feeds on prey 1 and 2. In the appendix, we will show that the evolutionarily stable equilibrium state of system (7) is realizable from the dynamical point of view for any case satisfying assumptions (3a), (3b) and (3c).

3. Effects of the Handling Time and Intraspecific Competition

In this section, we consider a more general model described by

$$
\begin{aligned}
dx_1/dt &= [\varepsilon_1 - k_1 x_1 - f_1 r_1 y_0 - g_1 r_1 y_1] x_1 = F_1 x_1, \\
dx_2/dt &= [\varepsilon_2 - k_2 x_2 - f_2 r_2 y_0 - g_2 r_2 y_1] x_2 = F_2 x_2, \\
dy_0/dt &= [-\delta + f_1 G_1 r_1 x_1 + f_2 G_2 r_2 x_2] y_0 = F_3 y_0, \\
dy_1/dt &= [-\delta + g_1 G_1 r_1 x_1 + g_2 G_2 r_2 x_2] y_1 = F_4 y_1,
\end{aligned}
\tag{9}
$$

$$
\begin{aligned}
r_i &= a_i/(1+a_i h_i x_i), \qquad (i=1,2) \\
f_1 &= u^n x_1^n/[u^n x_1^n + (1-u)^n x_2^n], \qquad f_2 \equiv 1-f_1, \\
g_1 &= v^m x_1^m/[v^m x_1^m + (1-v)^m x_2^m], \qquad g_2 \equiv 1-g_1,
\end{aligned}
$$

where k_i is the intraspecific competition coefficient of prey i and h_i is the handling time of prey i (i=1,2).

Unfortunately, we cannot generally find a solution of the equilibrium point X^* satisfying $F_1=F_2=F_3=y_1=0$ in system (9). If, however, the equilibium point attains the equivalance point under natural selection, we can find by a procedure similar to that used in system (7), the ESS for the case that k_i and/or h_i are positive. At the equivalence point, the energy intake rate for predators which search for prey 1 is the same as that for predators searching only for prey 2. Therefore,

$$G_1 r_1 x_1^* = G_2 r_2 x_2^* = \delta. \tag{10}$$

From Eq.(10), we can easily find the value of the equilibrium point as follows:

$$\begin{aligned} x_i^* &= \delta / a_i(G_i - a_i h_i), \qquad (i=1,2) \\ y_0^* &= \sum_{i=1}^{2} \left(\frac{\varepsilon_i}{a_i(G_i - h_i\delta)} - \frac{k_i\delta}{a_i^2(G_i - h_i\delta)^2} \right). \end{aligned} \tag{11}$$

The argument that at (ecologically stable) ESS, the relative prey density coincides with the equivalence point corresponds to the Bishop-Cannings' theorem (Bishop & Cannings,1978, see also Maynard Smith,1982) which can be applied to mixed ESS for a wide range of problems in behavioral ecology (with a few exceptions, *e.g.* local mate competition). In this sense, we would expect the assumption of Darwinian evolution to simplify the analysis of population dynamics. The parameter values of n and u which realize an ESS satisfy the following relation:

$$\frac{u^n}{(1-u)^n} = \frac{G_1 a_1^{n-2}(G_1 - h_1\delta)^{n-2}[a_1\varepsilon_1(G_1 - h_1\delta) - k_1\delta]}{G_2 a_2^{n-2}(G_2 - h_2\delta)^{n-2}[a_2\varepsilon_2(G_2 - h_2\delta) - k_2\delta]}. \tag{12}$$

Thus, the parameter set (n,u) satisfying Eq.(12) is not unique.

If equilibrium (11) is an ecologically stable steady state of the pre-existing system, the parameter set (n,u) satisfying Eq.(12) is considered to be an evolutionarily stable strategy as shown in Fig.2(a).

However, when this equilibrium is ecologically unstable, the population densities x_1, x_2 and y_0 in the pre-existing system do not attain an equilibrium, and there appears a stable limit cycle of the type shown in Fig.2(b). In this case, we must search for an ESS by

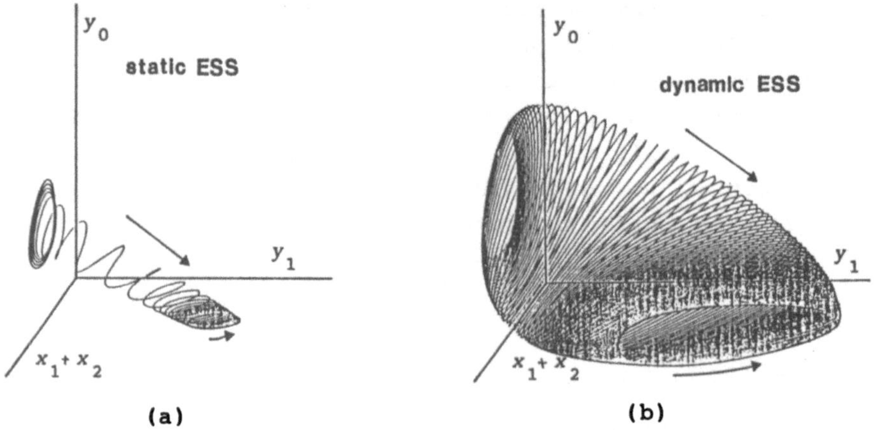

Fig.2. Trajectories of the dynamical system (9) obtained by computer simulation. a_1=1.2, a_2=0.8, ε_1=1.1, ε_2=0.9, δ=1, k_1=k_2=0.07, h_1=0.08, h_2=0.12, G_1=G_2=1. (a)(u,v,n) =(0.45,0.59,2.7). A trajectory moving from a state in which y_1=0 to one in which a state y_0=0 (i.e. the mutant has an ecologically stable ESS). (b)(u,v,n)=(0.5,0.6,5). The pre-existing system exhibits a stable limit cycle, and, after the mutant predator drives the pre-dominant predator to extinction, the system exhibits an evolutionarily stable limit cycle (dynamic ESS).

Table 1. ESS for switching parameters in system (9)

n	u(static)	u(dynamic)	n	u(static)	u(dynamic)
0.6	(0.5326)	0.5301±0.0025	4.0	(0.5991)	0.6021±0.0025
0.8	0.5524	(stable)	4.2	(0.5997)	0.6047±0.0025
1.0	0.5642	(stable)	4.4	(0.6002)	0.6042±0.0025
1.2	0.5721	(stable)	4.6	(0.6007)	0.6057±0.0025
1.4	0.5776	(stable)	4.8	(0.6011)	0.6061±0.0025
	:		5.0	(0.6015)	0.6065±0.0025
2.7	0.5936	(stable)	5.2	(0.6018)	0.6068±0.0025
	:		5.4	(0.6022)	0.6072±0.0025
3.8	(0.5986)	0.5986±0.0025	5.6	(0.6025)	0.6075±0.0025

Other parameters are chosen as Fig.2. u(static) is the value satisfying Eq.(12). u(dynamic) is the value of ESS calculated by means of the Runge-Kutta-Gill method using S-3300 digital computer at the Nippon Medical School Information Processing Center of Medical Sciences. The word "(stable)" means an ecologically stable ESS.

which mutant predators cannot invade a pre-existing system exhibiting a stable limit cycle (such an ESS is called a "dynamic ESS" introduced by Auslander et al.,1978). Using computer simulation, we have found that the set of parameters (n,u) for an ESS *approximately* satisfies Eq.(12), as is shown in Table 1.

4. Conclusion

This study is based on four main assumptions. First, $f_1+f_2\equiv 1$. Second, f_1 is a monotonically increasing function of the relative density of prey 1 (p). Third, f_i varies with p instantaneously but the function f_i, including the parameters n and u, does not change from generation to generation unless mutations occur. Finally, the prey density (especially at the equilibrium state) influences the switching parameters.

From these assumptions, we obtained four major results. First, if the set of strategies available to a switching predator is given by Eq.(1), the ESS for predatory switching, which is not unique is specified by Eq.(12). Second, the optimal energy intake strategy given in (5) is an ESS. Third, if h_i=0, the equilibrium realized at an ESS is always ecologically stable. Finally, in our model, evolutionary changes in parameters n and u will proceed until the relative prey density attains the equivalence point.

Holt(1983) discussed optimal foraging strategies for various forms of switching functions. He concluded that predatory switching is not adaptive when either prey 1 or 2 is scarce, and that such a switching model depicts non-optimal predators. In his arguments concerning optimal foraging strategy, he assumes *a priori* that prey densities are *fixed* at constant values.

However, interactions between prey and predator should affect their population sizes, and different switching properties may lead to different steady states. Since predatory switching prevents either prey from becoming too rare, a non-optimal switching strategy may be an evolutionarily stable strategy.

Acknowledgements: The authors would like to express their sincere thanks to Professor M. Yamaguti and Dr. Y. Suzuki of Kyoto University for valuable comments. We also thank to Dr. T.Namba of Senshu University for critical reading the manuscript and Professor Y.Shinagawa of Nippon Medical School for his encouragement throughout the study. This work was supported in part by a Grant-in-Aid for special Project Research on Biological Aspects of Optimal Strategy and Social Structure from the Japan Ministry of Education, Science and Culture (to E.Teramoto, No.59115006). The numerical computations were performed on the S-3300 digital computers at the Nippon Medical School Information Processing Center of Medical Sciences.

References

Auslander,D., Guckenheimer,J. & Oster,G.(1978). Random evolutionarily stable strategies. *Theor.Popul.Biol.*13,276-293.

Bishop,D.T. & Cannings,C. (1978). A generalized war of attrition. *J.theor.Biol.*70,85-124.

Comins,H.N. & Hassell,M.P.(1976). Predation in multi-prey communities. *J.theor.Biol.*62,93-114.

Greenwood,J.J.D. & Elton,R.A.(1979). Analysing experiments on frequency-dependent selection by predators. *J.Anim.Ecol.* 48,721-737.

Holt,R.D.(1983). Optimal foraging and the form of the predator isocline. *Amer.Natur.* 122,521-541.

Matsuda,H.(1985). Evolutionarily stable strategies for predator switching. *J.theor.Biol.* 115,351-366.

Matsuda,H., Kawasaki,K., Shigesada,N., Teramoto,E. & Ricciardi,L.M. (1986). Switching effect on the stability of the prey-predator system with three trophic levels. *J.theor.Biol.* 122,251-262.

May,R.M.(1977). Predators that switch. *Nature* 269,103-104.

Maynard Smith,J.(1982)."*Evolution and the Theory of Games*". Cambridge: Cambridge University Press.

Maynard Smith,J. & Price,G.R.(1973). The logic of animal conflict. *Nature* 246,15-18.

Murdoch,W.W.(1969). Switching in general predators:experiments on predator specificity and stability of prey populations. *Ecol.Monogr.* 39,335-354.

Murdoch,W.W. & Oaten,A.(1975). Predation and population stability. *Adv.Ecol.Res.* 9,1-131.

Tansky,M.(1978). Switching effect in prey-predator system. *J.theor.Biol.* 70,263-271.

Teramoto,E.,Kawasaki,K. & Shigesada,N.(1979). Switching effect of predation on competitive prey species. *J.theor.Biol.* 79,303-315.

Appendix. Ecological Stability at an ESS

To investigate the ecological stability of the boundary equilibrium point $X^*=(x_1^*,x_2^*,y_0^*,0)$ of system (7), we examine the characteristic equation for the linearization of system (7) at equilibrium X^* which is written as

$$-\begin{vmatrix} -a_1f_{11}^*x_1^*y_0^*-\lambda & -a_1f_{12}^*x_1^*y_0^* & -a_1f_1^*x_1^* \\ -a_2f_{21}^*x_2^*y_0^* & -a_2f_{22}^*x_2^*y_0^*-\lambda & -a_2f_2^*x_2^* \\ (G_1a_1f_1^*+\sum_i G_ia_if_{i1}^*x_i^*)y_0^* & (G_2a_2f_2^*+\sum_i G_ia_if_{i2}^*x_i^*)y_0^* & -\lambda \end{vmatrix}$$

$$= \lambda^3 + A\lambda^2 + B\lambda + C = 0, \qquad (13)$$

where $f_{ij}^*=\partial f_i/\partial x_j$ at the equlibrium. From assumption (3a), $f_{1i}\equiv-f_{2i}$

(i=1,2). Further, from assumption (3c), $f_{ii}>0$ (i=1,2). Thus, A, B, and C of Eq. (13) are written as

$$\begin{aligned} A &= a_1 f_{11}^* x_1^* y_0^* + a_2 f_{22}^* x_2^* y_0^* > 0 \\ B &= (G_1 a_1{}^2 f_1^{*2} x_1^* + G_2 a_2{}^2 f_2^{*2} x_2^*) y_0^* + \\ &\quad (G_1 a_1 x_1^* - G_2 a_2 x_2^*)(a_1 f_1^* x_1^* y_0^* f_{11}^* - a_2 f_2^* x_2^* y_0^* f_{22}^*) \\ C &= a_1 a_2 x_1^* x_2^* y_0^{*2} (G_1 a_1 f_1^* f_{22}^* + G_2 a_2 f_2^* f_{11}^*) > 0. \end{aligned} \tag{14}$$

Since A and C are always positive, the stability condition reduces to $AB-C>0$. From assumptions (3a), (3b), and the fact that $G_1 a_1 x_1^* = G_2 a_2 x_2^*$ at an ESS, we obtain $x_1^* f_{11}^* = x_2^* f_{22}^*$. Thus, $AB-C$ reduces to

$$AB-C = G_1 a_1 x_1^{*2} y_0^{*2} f_{11}^* (a_1 f_1^* - a_2 f_2^*)^2 \geq 0. \tag{15}$$

Here the case in which $AB-C=0$ occurs only if $\varepsilon_1=\varepsilon_2$. Thus the Equilibrium state of system (7) with an evolutionarily stable switching property is always ecologically stable unless $\varepsilon_1 \neq \varepsilon_2$.

II. MATHEMATICAL THEORIES OF PATTERN AND MORPHOGENESIS

Morphogenesis and Pattern Formation

Pattern Formation in *Dictyostelium discoideum*

Pattern formation by coupled oscillations: The pigmentation patterns on the shells of molluscs

Hans Meinhardt and Martin Klingler
Max-Planck-Institut für Entwicklungsbiologie
Spemannstr. 35, 7400 Tübingen, F. R. G.

The diversity and beauty of the pigmentation patterns on the shells of snails and bivalved molluscs invites us to construct models to understand their formation. The similarity of patterns in unrelated species on the one hand and the diversity in closely related species on the other encourage the assumption that most of them are generated by a common mechanism.

Pattern formation on the shells of molluscs proceeds in most species in a strictly linear manner since new pattern elements are added only along the growing edge of the shell. The second dimension is a protocol of what has happened at the growing edge as function of time. The shell pattern represents, so to say, a space-time plot. Frequent are pigmentation lines or ridge-like structures which are oriented perpendicular, parallel or oblique to the growing edge.

Keeping in mind the space-time character of the shell pattern, lines perpendicular to the growing edge indicate the formation of a spatial periodic pattern of pigment production along the edge which is stable in time. Other patterns indicate that pigment deposition oscillates. A particular cell produces pigment only during a certain time interval and enters then into an inactive (refractory) period until the next pigment production takes place. A synchronous oscillation in pigment production leads to lines which are parallel to the shell margin. Oblique lines originate from travelling waves of pigment production. Such waves arise if pigment-producing cells trigger their neighboring cells so that - after a certain delay - these cells also produce pigment.

In an earlier paper, the basic idea of using reaction-diffusion systems to model patterns of mollusc shells has been published (Meinhardt, 1984). A model, based on similar principles but focussed on the involvement of the nervous system has been proposed by Ermentrout, Campbell and Oster (1986).

Oscillations and travelling waves can be generated with reaction-diffusion systems containing an autocatalytic and an antagonistic component. Similar ingredients are required to produce spatial patterns which are stable in time (Gierer and Meinhardt, 1972; Meinhardt, 1982). Thus, the pattern of mollusc shells seems to be a special application of general pattern forming mechanisms. The diversity of patterns found in nature provides an inroad for the investigation of the range of possible patterns which can be generated by modifications of a basic mechanism.

The nonlinear partial differential equations 1a and 1b describe an interaction, which is able to generate the basic shell patterns:

$$\frac{\partial a}{\partial t} = \rho s a^{2*} - \mu a + D_a \frac{\partial^2 a}{\partial x^2} \qquad \text{(Eq. 1a)}$$

$$\text{with } a^{2*} = \frac{a^2}{1+\kappa a^2} + \rho_0$$

$$\frac{\partial s}{\partial t} = \sigma - \rho s a^{2*} - \nu s + D_s \frac{\partial^2 s}{\partial x^2} \qquad \text{(Eq. 1b)}$$

According to these equations, the autocatalytic activator <u>a</u> is a molecule which controls pigment production; <u>a</u> has a non-linear feedback on its own production which saturates at high <u>a</u> concentrations. The autocatalysis is limited due to the depletion of a substrate <u>s</u>, which is consumed in the course of autocatalysis; σ describes the constant production rate of <u>s</u>; μ and ν are the decay rates; D_a and D_s are the diffusion constants and ρ_0 is a small basic activator production. Alternatively, the reaction antagonizing the autocatalysis may be accomplished by an inhibitor. According to Eq. 2, an inhibitor <u>h</u> is produced under the control of the activator which inhibits the activator autocatalysis.

$$\frac{\partial a}{\partial t} = \frac{\rho(a^{2*}+\rho_0)}{h} - \mu a + D_a \frac{\partial^2 a}{\partial x^2} \qquad \text{(Eq. 2a)}$$

$$\text{with } a^{2*} = \frac{a^2}{1+\kappa a^2}.$$

$$\frac{\partial h}{\partial t} = \rho a^{2*} - \nu h + D_h \frac{\partial^2 h}{\partial x^2} + \rho_1 \qquad \text{(Eq. 2b)}$$

Depending on the parameters, the basic types of shell patterns mentioned above can be simulated with both equations. Simulations of more complicated patterns, for instance, those containing intersections or bifurcations of pigment lines, require systems in which at least three substances interact. By computer simulations we will show that the overall patterns as well as many fine details of shell patterning can be understood on the basis of these mechanisms. The details of the simulations can be derived from the computer program which has been used for these simulations. It is given in the appendix together with the list of parameters.

Lines perpendicular to the growing edge

The formation of stripes perpendicular to the growing edge (Fig. 1a) requires, as mentioned, that groups of cells permanently produce pigment, and that the pigment-producing regions are separated from each other by regions in which pigment production never occurs. Such patterns can be generated by local autocatalysis and long ranging inhibition. Required is that the diffusion of the antagonistic substance is much higher than that of the activator, i.e. $D_s \gg D_a$ or $D_h \gg D_a$. The resulting patterns will be stable in time if the antagonistic reaction has a shorter time constant, i.e. $\sigma > \mu$ or $\nu > \mu$, otherwise oscillations are likely to occur (see below). Small random fluctuations are sufficient to initiate the pattern formation from a homogeneous initial state. Fig. 1 shows a corresponding simulation with Eq. 2.

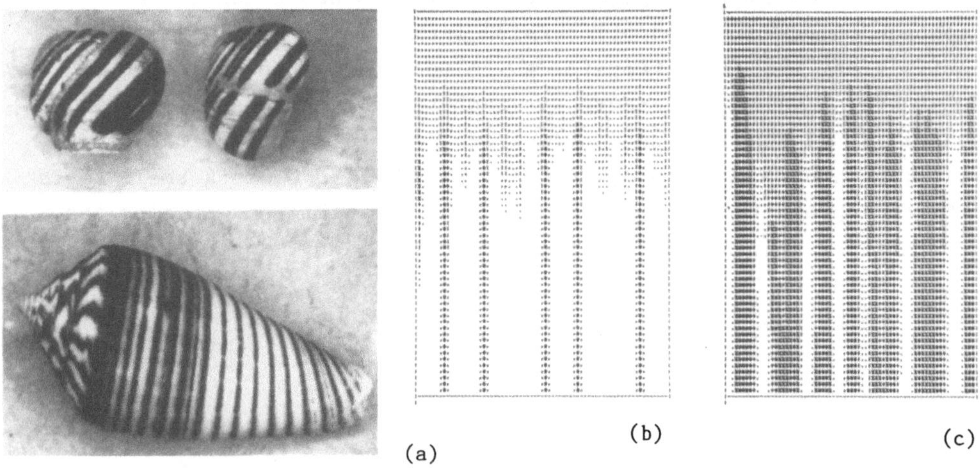

Fig.1 Stripes perpendicular to the growing edge. (a) Pattern on the shells of Cepaea nemoralis L. and Conus macronatus REEVE (b, c) Simulation with eq. 2; the parameters of these and the following simulations are given in the appendix. A saturation of the autocatalysis (c) leads to broader stripes with different width.

Lines parallel to the growing edge

This pattern results if the cells underneath the growing edge deposite pigment more or less simultaneously for certain time intervals and if these pigment producing phases are separated by phases in which no pigment production takes place. In a reaction scheme containing an (autocatalytic) activator and an antagonistic substance, oscillations can occur if the antagonistic reaction follows a change in the concentration of the activator too slowly, i.e. if $\sigma < \mu$ (Eq. 1) or $\nu < \mu$ (Eq. 2) (Meinhardt and Gierer, 1974). Stripes parallel to the growing edge appear if the oscillation of the cells occurs in synchrony. Such a synchronization results if the activator shows a substantial diffusion (i.e. if $D_a \gg D_s$ or $D_a \gg D_h$) since a faster oscillating cell can advance the phase of a delayed neighbor due to the activator exchange. Usually, the pigment lines are not precisely parallel to the growing edge (and the growth lines), indicating that - despite the fact that neighboring cells are in a similar phase - a substantial phase difference can accumulate over the total extension of the growing edge. Figure 2 shows a natural pattern together with a simulation.

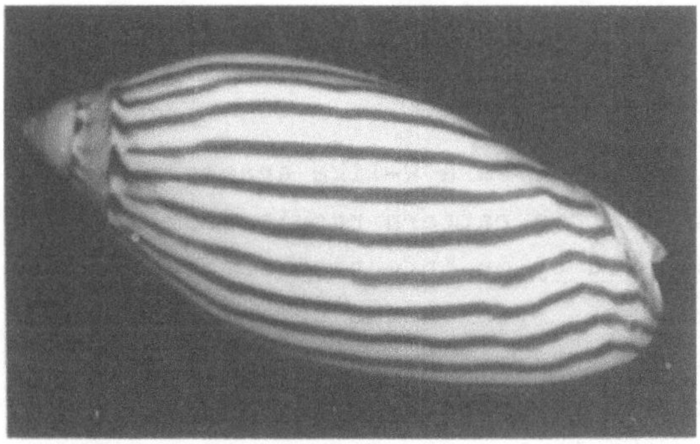
(a) (b)
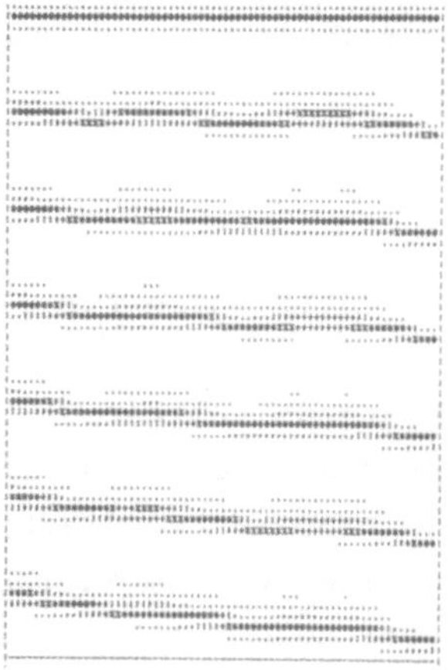

Fig. 2 Stripes parallel to the growing edge. They result from a (more or less) synchronous oscillation in the pigment production. (a) Amoria ellitoi. (b) Simulation with Eq. 2.

Oblique lines resulting from travelling waves

If the activator diffusion is smaller than that required for the formation of parallel pigment lines, travelling waves can occur. A just activated cell can "infect" its neighbor which becomes, with some delay, also activated and so on. The resulting pigment pattern on the shells consists of oblique lines. The larger the speed of the wave (in relation to the speed of shell growth) the smaller the angle between the line and the growing edge.

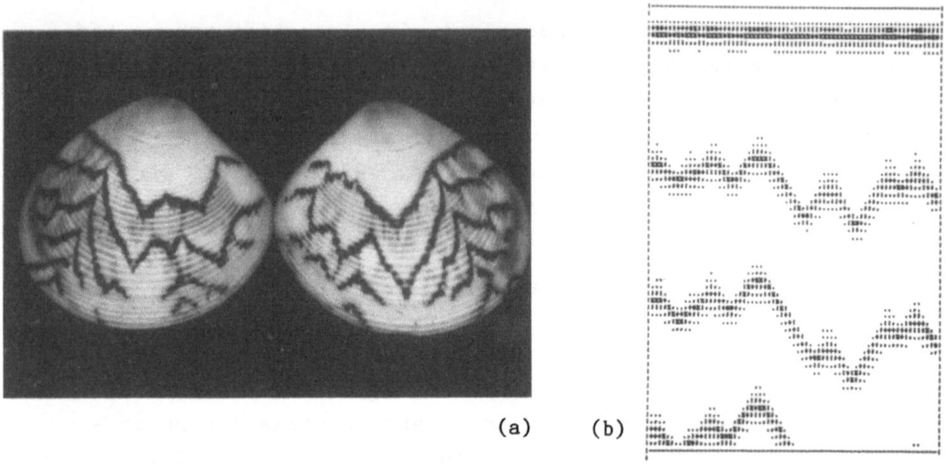

(a) (b)

Fig. 3 Oblique pigmentation lines. They result from travelling waves of pigment production. (a) A natural pattern. (b) Simulation: Faster oscillating cells form the initiation points of two diverging lines. At the point where two waves collide, both waves extinguish each other.

Characteristic for many shell patterns is a W-like arrangement of the oblique lines. The Λ-element of this pattern result from a spontaneous trigger of a cell followed by an infection of both neighbors, and so on. In other words, a spontaneous trigger of a cell gives rise to two diverging oblique lines. If two such lines merge, both lines become extinct because all neighboring cells are still in the refractory period, leading to the V-element of the pattern (Fig. 3). According to this view, the formation of the oblique lines on the shells of many molluscs has much in common with the formation of travelling waves in aggregating slime molds (Gerisch, 1968). Only that in molluscs the pattern occurs along a line and the second dimension is a record of the temporary pattern.

Figure 3 shows a simulation with Eq. 2. In adddition to a basic activator production a basic inhibitor production has been assumed. With such selection of parameters, small random fluctuations are sufficient that some cells oscillate much faster than others. The faster oscillating cells become the initiation points of two divergent lines. Other examples of oblique line patterns will be given in Fig. 4, 6 and 8.

Formation of branches originating from oblique lines

The formation of a branch indicates the sudden formation of a backwards wave. This requires two features (Fig. 4b). (i) The refractory period must be short in relation to the time interval between one activation and the next spontaneously occurring activation. (ii) Occasionally, the activation of a small group of cells lasts longer than the refractory period of a neighboring cell such that a backwards infection can take place. In the simulation of the tent-like pattern of the shell of Olivia porphyria L. (Fig. 4), the interaction has been designed in such a way that branching occurs if the number of travelling waves becomes too few. (The number of travelling waves becomes reduced due to the mutual annihilation of two travelling waves by collision.) To generate a branch, it has been assumed that a hormone-like, homogeneously distributed substance is produced by the activated cells which has an inhibiting influence on the inhibitor decay. Thus, whenever the number of activated cells becomes too small, the inhibitor life time becomes shortened such that the cells are shifted from the oscillating mode into a steady state. After passing the refractory period of a neighboring oscillating cell, a backwards wave is initiated. This causes an increase of the number of travelling waves and thus an increase in the inhibitor lifetime. All cells return to the oscillating mode and

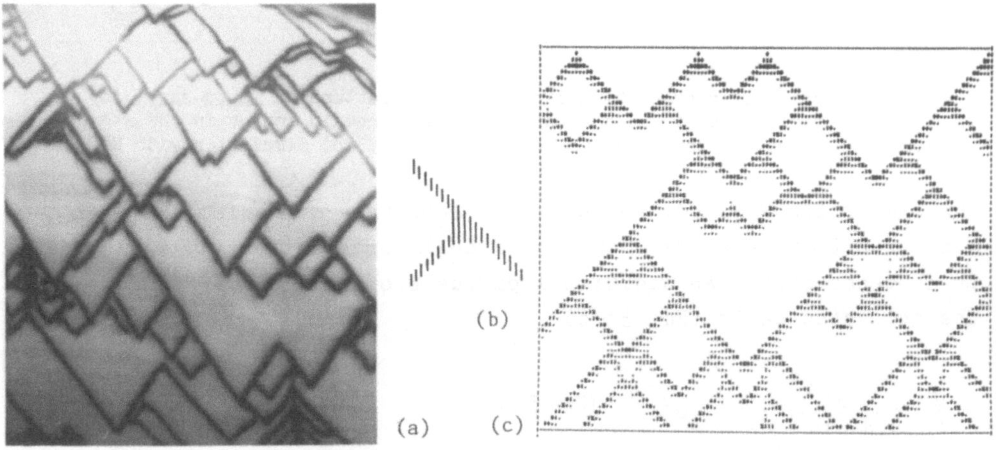

Fig. 4 Oblique lines with branches. (a) Pattern on Olivia porphyria. (b) The principle of branch formation: a particular cell remains long enough in a steady state to trigger a backwards wave. (c) Simulation: branch formation is initiated whenever the number of travelling waves is lower than a threshold value. The controlling hormone-like agent is produced by each cell which actually secretes pigment (activated cell) and is rapidly distributed in the organism.

the formation of the branch is completed (Fig. 4c). This global control of branching leads to the formation of several branches at a particular time, in agreement with the natural pattern.

Crossings of oblique lines

Pigmentation lines which cross each other can be found on shells of Tapes litteratus L. (Fig. 5). Crossings of oblique lines can be regarded as the formation of two backwards waves at the position where two waves collide. Similar as in the case of branch formation, the backwards waves can result from cells which remain activated for a prolonged period due to a shift into the steady state (Fig. 5b). In

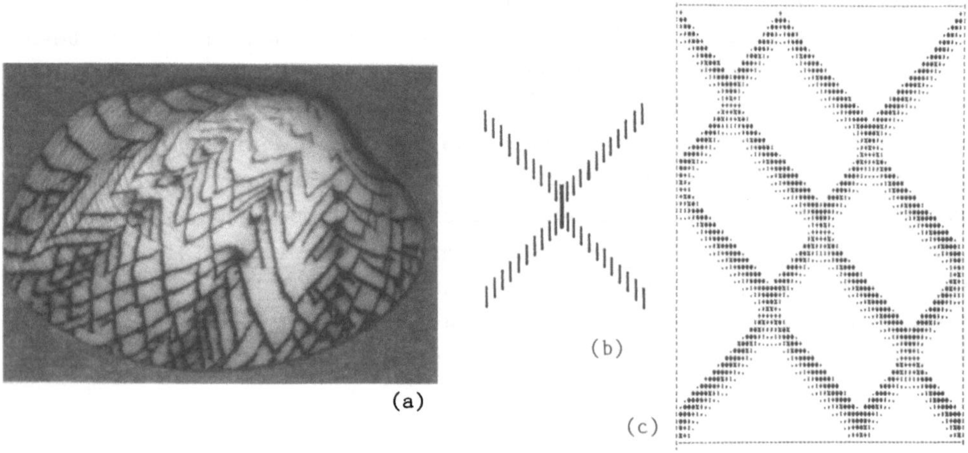

Fig. 5 Crossing of pigmentation lines. (a) The pattern on Tapes spec. (b) The principle: cells at the confrontation point remain longer in the steady state and induce in this way two backwards waves. (c) Simulation with Eq. 3.

the simulation Fig. 5c, an activator-substrate mechanism has been assumed with parameters such that a cell, once activated, would remain in a steady state. An additional diffusible inhibitor is produced by the activated cells (Eq. 3a-c):

$$\frac{\partial a}{\partial t} = \frac{\rho s a^{2*}}{1+\gamma h} - \mu a + D_a \frac{\partial^2 a}{\partial x^2} \qquad \text{(Eq. 3a)}$$

$$\frac{\partial h}{\partial t} = \nu(a-h) + D_h \frac{\partial^2 h}{\partial x^2} \qquad \text{(Eq. 3b)} \qquad \text{with } a^{2*} = \frac{a^2+\rho_0}{1+\kappa a^2}$$

$$\frac{\partial s}{\partial t} = \sigma - \frac{\rho s a^{2*}}{1+\gamma h} - \varepsilon s \qquad \text{(Eq. 3c)}$$

A travelling wave results since each newly activated cell extinguishes the activation of the preceding cell. If two waves collide, no newly activated cell is available to extinguish the activation of the cells at the point of collision. These cells would remain in the steady state until one or two backwards waves are triggered. The newly activated cells of the backwards waves extinguish the prolonged activation of the cells at the point of collision.

Formation of checkerboard-like patterns

The formation of a checkerboard-like pattern requires a spatial periodic and a time-wise periodic pattern of about the same wave length. In addition, the pigmented and non-pigmented regions must have the same extension in both dimensions, space and time.

The mechanism outlined for the generation of crossings was based on the superposition of two inhibitory reactions. One inhibitory substance diffuses rapidly, generating in this way a periodic spatial pattern (see Fig. 1). The second inhibitory substance does not diffuse and has a long time constant. It causes therefore a periodic pattern in time. In the example of Eq. 3, this inhibition results from the depletion of a substrate. The superposition of both inhibitory effects leads to patterns which are periodic along the

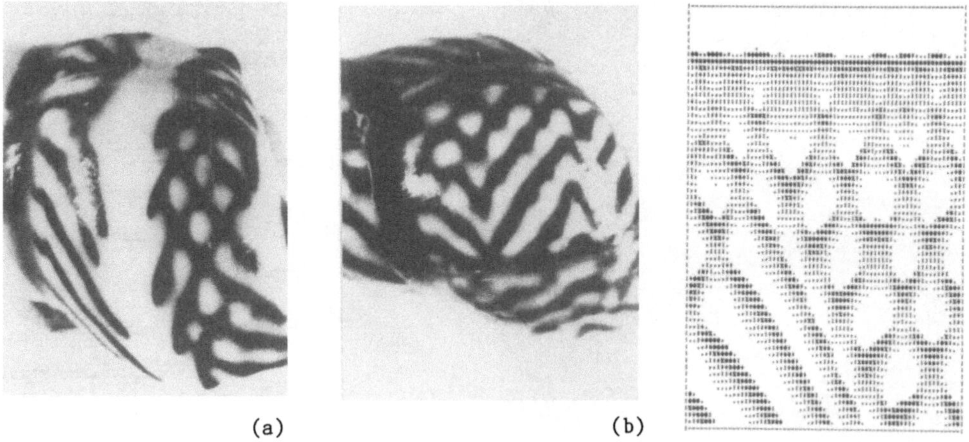

(a) (b) (c)

Fig. 6 Checkerboard-like pattern and oblique lines. (a, b) Patterns on *Bankivia fasciata* (after photographs kindly supplied by J. Campbell, see Ermentrout, Campbell and Oster, 1986. (c) Simulation: Autocatalysis is antagonized by two inhibitory substances, one is diffusible, the other is not (Eq. 3).

space as well as along the time coordinate. A series of oblique lines or of a checkerboard-like pattern has this feature. For this reason, minor changes of parameters can lead to the transition from one of these patterns to the other, as it is frequently observed in the natural pattern of Bankivia fasciata Menke.

Superposition of a stable spatial periodic and a temporal periodic pattern

Many shell patterns can be explained under the assumption that they are generated by the superposition of two patterns, one which is stable in time and a second one which is based on oscillations. The stable pattern controls the parameters of the oscillations. Two examples will be given. In the two simulations, Fig. 7 and 8, it has been assumed that the production rate of the substrate (σ in Eq. 1) is a function of the position: $\sigma(x)$. In the simulation of the shell pattern of Amorina undulata L. (Fig. 7a) the cells oscillate faster in the region of high σ compared with those in a region of low σ . Instead of lines parallel to the growing edge (Fig. 2), the lines obtain a wavy shape (Fig. 7b). Since the cells oscillate with different frequencies, an increasing phase difference between the

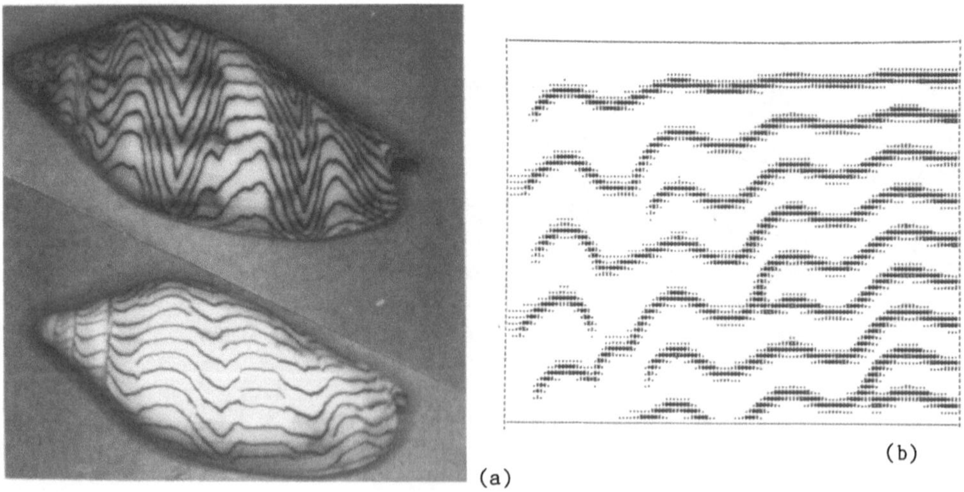

(a) (b)

Fig. 7 Wavy lines. (a) The shell pattern on Amorina undulata. (b) Simulation: a stable periodic pattern controls the parameter of the oscillation. The assumed stable pattern consists of the superposition of a sinoidal and a gradient-like pattern.

oscillators would accumulate with time. However, diffusion of the activator leads to a partial synchronization of the oscillating cells as long as the phase difference does not become too large. Otherwise, the synchronization may fail with the consequence that a line of pigmentation terminates (Fig. 7c).

In the simulation of the shell pattern of Cyprea ziczac L. it has been assumed that the cells in the regions of high substrate production (high σ) remain in a steady state. Thus, in that region pigment is deposited permanently. Stripes perpendicular to the growing edge appear. These permanently activated cells trigger the cells between these stripes which are in the oscillating mode. Thus, a series of travelling waves (oblique lines) are initiated at regular intervals at both sides of the steady state line. Since two oppositely moving waves extinguish each other, the oblique lines have the shape of a V.

(a)

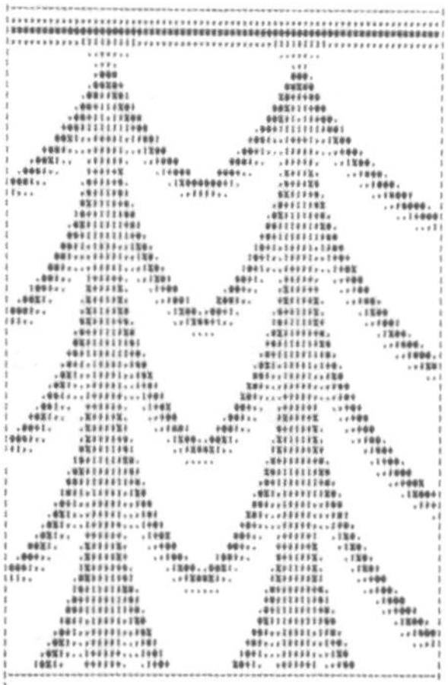
(b)

Fig. 8 Oscillating and non-oscillating cells (a) Shell of Cyprea ziczac L. (b) Simulation: The spatial stable pattern causes some cells to remain permanently in a steady state of pigment production. Travelling waves (V-pattern) spread out from these stripes.

Conclusion

The overwhelming richness of pattern on shells of molluscs can be explained by relative simple reaction-diffusion mechanisms. Most of the patterns are explicable by nearest neighbor interactions, in some cases they have to be supplemented by an evenly distributed hormon-like substance. By the computer simulations, one can only uncover the logic behind the pattern formation, not how these principles are biologically realized. A paper describing a more detailed comparison between the simulations and the natural patterns as well as between naturally occurring pattern regulation after injuries and those expected from the model is in preparation.

References

Ermentrout, B., Campbell, J. and Oster, G. (1985): A model for shell patterns based on neural activity. The Veliger (in press)

Gerisch, G. (1968). Cell aggregation and differentiation in Dictyostelium. Curr.Top-Dev.Biol. 3, 157-232.

Gierer, A. and Meinhardt, H. (1972): A theory of biological pattern formation. Kybernetik 12: 30-39

Meinhardt, H. (1982): Models of biological pattern formation. London: Academic Press

Meinhardt, H. (1984): Models for positional signalling, the threefold subdivision of segments and the pigmentation pattern of molluscs. J. Embryol.exp. Morph.83, 289-311 (Supplement)

Meinhardt, H. and Gierer, A. (1974): Application of a theory of biological pattern formation based on lateral inhibition. J. Cell Sci. 15, 321-346.

Appendix: Computer program and parameter lists

A computer program written in FORTRAN is provided to show the details of the simulations. Each simulation begins with the settings and a printout of parameters in the subroutine PARAME. The first 10 integers control the total number of iterations (IT), the number of iterations between the printouts (IP), the field size (KX to KY), the scaling of the printout (KZ), the selection of initial conditions (IA), the type of equation (IB) ect. The next five floating point constants (AA-AE) are used to specify further the initial concentrations. Each following set of five floating point numbers determines the parameters of one substance of the reaction-diffusion system, e.g. diffusion (DA = D_a), the decay rate (TA = μ), or the basic production (QA = ρ_o) of the activator. Each iteration starts with setting the boundary condition (impermeable). The activator concentrations are plotted with a line printer in a schematic way with the subroutine PLOT. The darkness of the characters provides a measure for the local activator concentration at a particular time. For each simulation shown in Fig. 1 - 8, the parameter lists are given. The program has many comment statements (C--- or after !-sign) which should allow a direct understanding of the program. (For more details of programming, see Meinhardt, 1982).

```
C----PROGRAM SHELL PATTERNS
      DIMENSION A(131),B(131),C(131),D(131),Z(130),Y(130)
      COMMON/D/ IT,IP,KX,KY,KZ,IA,IB,IC,ID,IZ,IFIM,
     1AA,AB,AC,AD,AE,
     2DA,TA,QA,RA,SA,  DB,TB,QB,RB,SB,
     3DC,TC,QC,RC,SC,  DD,TD,QD,RD,SD
      IFIM=3           ! storage file for datas
  150 CALL PARAME      ! input of constants
   50 TYPE 920
      READ(5,915) IZK
      IF(IZK.EQ.0) GOTO 150    !  IZK=1: start of a new calculation
      GOTO (1,2),IZK           !  IZK=2: continuation
C---------- Initial conditions ------------------------
    1 AM=KZ   ! Scaling of the plot
      DO 141 I=KX,KY
      Z(I)=1.
      Y(I)=TA*(1.+SA*(RAN(0,0)-.5))! source density with fluctuations
      A(I)=AA
      B(I)=AB
  141 C(I)=AC
      AHO=AD ! Initial hormon concentration (Eq.7)
      GOTO  (131,132,133,134,139),IA
  131 A(IC)=AE
      GOTO 139
  132 A(IC)=AE
      A(1)=AE
      GOTO 139
  133 DO 143 I=1,KY
  143 Z(I)=1.+AD*SIN(I*AE)**IC+I*RB ! Arteficial spatial stable pattern
      TYPE 900
      CALL PLOT (Z,1.+AD+RB*I,KX,KY,1)   ! printout
      TYPE 915
      GOTO 139
  134 TYPE 917
      READ (5,915) I
      IF (I.EQ.0) GOTO 139
      A(I)=AE
      GOTO 134
  139 CALL PLOT (A,AM,KX,KY,2) ! printout initial activator
      CALL PLOT (A,AM,KX,KY,1) !  distribution
      IF (IZK.EQ.5) GOTO 50
    2 ITOT=0
    3 ICC=0
      DAC=1.-TA-2.*DA
      DBC=1.-TB-2.*DB
      DBCC=DBC
      DCC=1.-TC-2.*DC
      DDC=1.-TD-2.*DD
```

```
  500 DO 160 IPRINT=1,IP !     begin of the iterations
      A1=A(KX)    !  --- boundary conditions: impermeable
      B1=B(KX)
      C1=C(KX)
      A(KY+1)=A(KY)
      B(KY+1)=B(KY)
      C(KY+1)=C(KY)
      BSA=0.
C---------- Reactions---------------
      DO 158 I=KX,KY   ! DO-loop in the linear array of cells
      AF=A(I)
      BF=B(I)
      CF=C(I)
      GOTO (201,202,203,204),IB     !Selection of the interaction
  201 AQ=Y(I)*BF*(AF**2/(1.+RA*AF**2)+QA)!activator-depletion mechanism
      A(I)=AF*DAC+DA*(A1+A(I+1))+AQ
      B(I)=BF*DBC+DB*(B1+B(I+1))-AQ+QB*Z(I)
      GOTO 159
  202 AQF=AF**2  !        activator - inhibitor mechanism
      AQ=AQF/(1.+RA*AQF)
      A(I)=AF*DAC+DA*(A1+A(I+1))+Y(I)*(AQ+QA)/BF
      B(I)=BF*DBCC+DB*(B1+B(I+1))+TB*AQ+QB
      GOTO 159
  203 AQ=AF**2     !   Crossings
      AQ=Y(I)*CF*(AQ+QA)/(1.+RA*AQ)/(1.+RB*BF)
      A(I)=AF*DAC+DA*(A1+A(I+1))+AQ
      B(I)=BF*DBC+DB*(B1+B(I+1))+TB*AF
      C(I)=CF*DCC+DC*(C1+C(I+1))-AQ+QC
      GOTO 159
  204 AFQ=AF**2        ! Branches controlled by a hormon (AHO)
      AQ=Y(I)*AFQ/(1.+RA*AFQ)
      A(I)=AF*DAC+DA*(A1+A(I+1))+AQ/(RB+BF)
      B(I)=BF*DBCC+DB*(B1+B(I+1))+AQ
      BSA=BSA+SB*AF
       IF (I.LT.KY) GOTO 159
      AHO=AHO*(1.-SB)+BSA/KY
      TBB=TB/AHO        !    !effective inhibitor decay rate
      DBCC=1.-2.*DB-TBB
  159 A1=AF
      B1=BF
      C1=CF
  158 CONTINUE     ! end of another iteration
  160 CONTINUE     ! end of IP iterations
      ITOT=ITOT+1
      ICC=ICC+1
C ----Plott of the activator distribution after IP iterations
      CALL PLOT (A,AM,KX,KY,1)
      IF (ICC.LT.IT)     GOTO 500
      CALL PLOT (A,AM,KX,KY,2) ! end of the plot
      GOTO 50
  900 FORMAT (' POSITIONAL INFORMATION:')
  915 FORMAT (I3,F8.4,I3)
  917 FORMAT ('$ cell # to be set =   AE ')
  920 FORMAT (' !')
        END
      SUBROUTINE PARAME
C----allows entry and individual change of 35 constants
C----constant 1-10 (IP-IZ) integer; 11-35 (AA-DS) floating point
      COMMON/D/ IT,IP,KX,KY,KZ,IA,IB,IC,ID,IZ,IFIM,
     1AA,AB,AC,AD,AE,DA,TA,QA,RA,SA,DB,TB,QB,RB,SB,
     2DC,TC,QC,RC,SC,  DD,TD,QD,RD,SD
      DIMENSION IDD(10),D(28),DT(35)
```

```
      EQUIVALENCE (IDD(1),IT),(D(1),AA)
      DATA DT/ 'IT','IP','KX','KY','KZ','IA','IB','IC','ID','IZ',
     1'AA','AB','AC','AD','AE','DA','TA','QA','RA','SA','DB','TB'
     2,'QB','RB','SB','DC','TC','QC','RC','SC','DD','TD','QD','RD','SD'/
    5 WRITE(5,905)          !  input constant #
      READ(5,900) IS
   10 IF (IS.EQ.0)  GOTO 500       !Return after data printout
      IF (IS.EQ.41) GOTO 41        !write present data on file
      IF (IS.EQ.44) GOTO 44        !read data from file
      IF (IS.EQ.45) GOTO 45        !input new data
      IF (IS.EQ.46) READ (5,900) IFIM    ! change file number
        IF (IS.GT.38) GOTO 5
      IF (IS.GT.10) GOTO 100
      WRITE (5,901) IS,IDD(IS),DT(IS)! write existend integer constant
      READ (5,900) IDD(IS)     !input new integer constant
      WRITE (5,901) IS,IDD(IS),DT(IS)
      GOTO 5
  100 WRITE (5,903) IS,D(IS-10),DT(IS)!write old floating point const.
      READ (5,902)  D(IS-10)  ! input new floating point constant
      WRITE (5,903) IS,D(IS-10),DT(IS)
      GOTO 5
   45 ID=10
      DO 125 IS=1,ID      ! all constants new
      WRITE (5,904) IS,DT(IS)
      IF (IS.GT.10)  GOTO 126
      READ(5,900)ISD
      WRITE (5,901) IS,ISD,DT(IS)
      IDD(IS)=ISD
      GOTO 125
  126 READ (5,902) D(IS-10)
      WRITE (5,903) IS,D(IS-10),DT(IS)
  125 CONTINUE
      GOTO 5
   41 WRITE (IFIM,931) (IDD(I),I=1,10),(D(K),K=1,ID-10)
      TYPE 932,IFIM ! write constants on disc
      GOTO 501
   44 READ (IFIM,931) (IDD(I),I=1,10),(D(K),K=1,ID-10) ! read disc
  501 CALL CLOSE (IFIM)    ! end of file
      WRITE (5,933) (IDD(I),DT(I),I=1,10),(D(K),DT(K+10),K+10,K=1,ID-10)
      GOTO 5
  500 WRITE (5,933) (IDD(I),DT(I),I=1,10),(D(K),DT(K+10),K+10,K=1,ID-10)
        RETURN
  900 FORMAT (I6)
  901 FORMAT (I3,I6,1X,A5)
  902 FORMAT (F8.4)
  903 FORMAT (I3,F10.4,1X,A5)
  904 FORMAT (/,/,' ?',I5,X,A3)
  905 FORMAT (' NEW CONSTANT #')
  931 FORMAT (1X,2I6,8I3,5(/,1X,5F8.4))
  932 FORMAT (' AUF FILE ',I4)
  933 FORMAT (1X,10(I3,'=',A3),6(/,1X,5(F8.4,'=',A3,I2)))
      END
      SUBROUTINE PLOT (A,AM,KA,KS,IQ)
C-----schematic plot of a linear array (A) with a line printer
C-----KA first cell, KS last cell, AM level of maximum density
      DIMENSION A(1),TY(16),AP(131),M(16)
      DATA TY/' ','.',',',':',';','+','%','$','*','#','A','-',
     1'!','Y',' ','N'/  !Symbols to be plottet
      GOTO (1,2),IQ
    1 AMX=10./AM
      DO 320 IX=KA,KS
      N=A(IX)*AMX+1.0001
```

```
      IF (N.LE.0) N=16
      IF (N.GT.10) N=10
  320 AP(IX)=TY(N)
      AP(KS+1)=TY(13)
  350 WRITE(5, 988) (AP(IX),IX=1,KS+1)
      RETURN
    2 DO 420 IX=KA,KS  !  |--------| at the begin or end of the plot
  420 AP(IX)=TY(12)
      AP(KS+1)=TY(13)
      WRITE(5, 988) (AP(IX),IX=1,KS+1)
      RETURN
  988 FORMAT (1X,'|',,135A1)
      END
```

Simulation Fig. 1c

```
60=IT   30=IP    1=KX   69=KY    4=KZ    5=IA    2=IB  25=IC   25=ID    0=IZ
 1.0000=AA 11   1.0000=AB 12   0.0000=AC 13   0.0000=AD 14   0.0000=AE 15
 0.0020=DA 16   0.0100=TA 17   0.0010=QA 18   0.0000=RA 19   0.0500=SA 20
 0.4000=DB 21   0.0200=TB 22   0.0000=QB 23   0.0000=RB 24   0.0000=SB 25
```

Simulation Fig. 1d

```
60=IT  100=IP    1=KX   69=KY    2=KZ    5=IA    2=IB  25=IC   25=ID    0=IZ
 1.0000=AA 11   1.0000=AB 12   0.0000=AC 13   0.0000=AD 14   0.0000=AE 15
 0.0020=DA 16   0.0100=TA 17   0.0010=QA 18   0.5000=RA 19   0.0500=SA 20
 0.4000=DB 21   0.0200=TB 22   0.0000=QB 23   0.0000=RB 24   0.0000=SB 25
```

Simulation Fig. 2b

```
60=IT   60=IP    1=KX   69=KY    2=KZ    5=IA    2=IB   0=IC   25=ID    0=IZ
 0.3000=AA 11   0.2000=AB 12   0.0000=AC 13   0.0000=AD 14   0.0000=AE 15
 0.1000=DA 16   0.0500=TA 17   0.0200=QA 18   0.0004=RA 19   0.1000=SA 20
 0.0000=DB 21   0.0300=TB 22   0.0075=QB 23   0.0000=RB 24   0.0000=SB 25
```

Simulation Fig. 3b

```
60=IT   25=IP    1=KX   69=KY    3=KZ    5=IA    2=IB   0=IC   25=ID    0=IZ
 0.3000=AA 11   0.2000=AB 12   0.0000=AC 13   0.0000=AD 14   0.0000=AE 15
 0.0100=DA 16   0.0500=TA 17   0.0200=QA 18   0.0004=RA 19   0.1000=SA 20
 0.0000=DB 21   0.0300=TB 22   0.0075=QB 23   0.0000=RB 24   0.0000=SB 25
```

Simulation Fig. 4c

```
60=IT   25=IP    1=KX  125=KY    2=KZ    4=IA    4=IB  10=IC   25=ID    0=IZ
 0.0100=AA 11   0.1000=AB 12   0.0000=AC 13   0.5000=AD 14   1.0000=AE 15
 0.0150=DA 16   0.1000=TA 17   0.1000=QA 18   0.2500=RA 19   0.1000=SA 20
 0.0000=DB 21   0.0140=TB 22   0.1000=QB 23   0.1000=RB 24   0.1000=SB 25
```

Simulation Fig. 5c

```
60=IT   30=IP    1=KX   69=KY    2=KZ    4=IA    3=IB  20=IC   30=ID    0=IZ
 0.0000=AA 11   0.0100=AB 12   8.0000=AC 13   0.0000=AD 14   1.0000=AE 15
 0.0100=DA 16   0.0800=TA 17   0.0000=QA 18   0.5000=RA 19   0.0000=SA 20
 0.4000=DB 21   0.0200=TB 22   0.0000=QB 23   3.0000=RB 24   0.0000=SB 25
 0.0000=DC 26   0.0050=TC 27   0.1000=QC 28   0.0000=RC 29   0.0000=SC 30
```

Simulation Fig. 6c

```
60=IT   40=IP    1=KX   69=KY    2=KZ    5=IA    3=IB  20=IC   30=ID    0=IZ
 0.1000=AA 11   0.0100=AB 12   5.0000=AC 13   0.0000=AD 14   0.0000=AE 15
 0.0100=DA 16   0.0500=TA 17   0.0050=QA 18   1.0000=RA 19   0.1500=SA 20
 0.4000=DB 21   0.0300=TB 22   0.0000=QB 23   2.0000=RB 24   0.0000=SB 25
 0.0000=DC 26   0.0000=TC 27   0.0300=QC 28   0.0000=RC 29   0.0000=SC 30
```

Simulation Fig. 7b

```
60=IT   30=IP    1=KX   69=KY    1=KZ    3=IA    1=IB   4=IC   25=ID    0=IZ
 0.0000=AA 11   2.5000=AB 12   0.0000=AC 13   1.4000=AD 14   0.1000=AE 15
 0.0150=DA 16   0.1000=TA 17   0.0100=QA 18   0.5000=RA 19   0.0040=SA 20
 0.0500=DB 21   0.0000=TB 22   0.0100=QB 23   0.0300=RB 24   0.0000=SB 25
```

Simulation Fig. 8b

```
60=IT   25=IP    1=KX   69=KY    2=KZ    3=IA    1=IB   8=IC   25=ID    0=IZ
 0.4000=AA 11   4.0000=AB 12   0.0040=AC 13   0.8000=AD 14   0.1000=AE 15
 0.0100=DA 16   0.1000=TA 17   0.0020=QA 18   0.5000=RA 19   0.0000=SA 20
 0.0500=DB 21   0.0020=TB 22   0.0500=QB 23   0.0000=RB 24   0.0000=SB 25
```

FROM MAP SYSTEMS TO PLANT MORPHOGENESIS

Hermann B. Lück and Jacqueline Lück

Laboratoire de Botanique analytique et Structuralisme végétal
Faculté des Sciences et Techniques de St-Jérôme
C.N.R.S. U.A. 563
Rue Henri Poincaré, 13397 MARSEILLE cedex 13, France

In a plant, growth is localized in some special, often distal areas, called meristems. Cell divisions and cell expansion contribute to the elongation of the axes, and also produce infinitely appendages such as leaves, stipules, and branches which are arranged in regular patterns. We want to give here a possible explanation for pattern inception in meristems on the basis of organized tissue growth.

Plant tissues can be described with the help of the rigid cell wall skeleton. This skeleton appears on a cell layer as a network of polygons. Cell divisions introduce into this network new cell walls which divide the polygons. Each of the insertion sites of a new cell wall subdivides each old wall on which it is anchored into two parts. We suppose that the wall insertion follows defined positioning rules which work at each cell division and which are specific for a given tissue structure. Biological observations on meristematic tissues emphasize that the positioning of division walls is determined by the regulated accumulation and orientation of microtubules [1] or cellulose fibrils [2].

DESCRIPTION OF A TISSUE BY DOUBLE-WALL MAPS

The consequences of deterministic rules for the positioning of division walls in meristematic cells have been investigated by means of *developmental models* called *Double-Wall Map 0L systems* [3,4,5,6,7,8]. The generated maps are topological representations of the cell wall skeleton of 2-dimensional cell layers.

A map is a finite set of non-intersecting regions which lie on a plane. Frontiers delimit bounded regions (r) [9] which represent cells. They are surrounded by an unbounded region (e) which stands for the environment. Frontiers stay for cell walls. Our maps are characterized by double-walls in opposition to BPM0L systems, e.g. [10], in which walls are simple (further biblio-

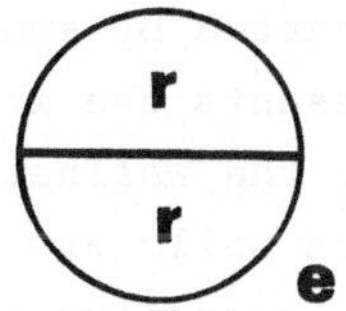

graphy in [11,12]). In a double-wall map each region has its own frontier. The frontiers of adjacent regions are closely touching [5]. The resulting double-edges represent *cell walls* while each of the complementary edges is called a *half-wall*. The polygonal envelop of a cell can thus be described by a sequence of half-wall edges. In a half-wall we distinguish basic units called *segments*; a half-wall may be composed of several segments. We accept the possibility that all cell boundaries have either the same number of segments or, between two sister cells, the boundaries may be composed of different numbers of segments. In any case, these different numbers reiterate in all pairs of sister cells.

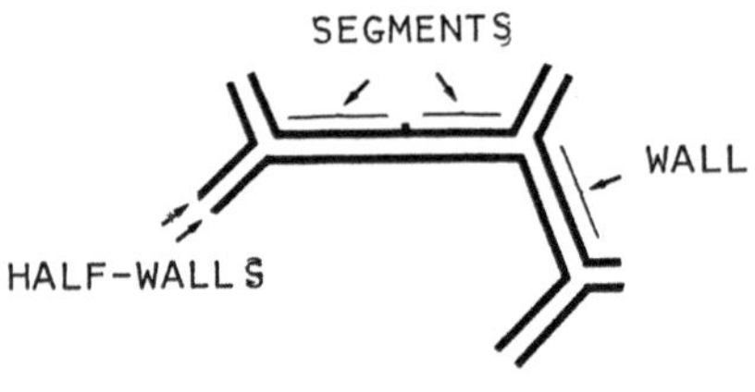

When a cell divides, the division wall will be anchored on two of all these segments. In order to specify the insertion sites, we label the segments clockwise viewed from inside the cell. The serial numbers begin always with the segment which was the division wall at the last division of the cell, i.e. the insertion sites are defined in reference to the youngest wall of a cell. Consequently, a complete cell wall is represented by two opposite running half-wall subsequences of labels. The length of such complementary subsequences are assumed to be equal, i.e. there is parity between complementary segments (systems without parity are not considered).

THE GENERATION OF MAPS

We assume that all cells divide synchronously, i.e. the successive map representations describe successive tissue states in discrete time steps, the time unit being the cell generation. The progressive regionalisation of maps is obtained by parallel substitution of labels and binary division of bounded regions. This procedure is formalized by a system $G = \{\Sigma, P, "/", M_0\}$. Σ is the alphabet over the segment labels, P the set of segment production rules. The slash symbol "/" is not rewritten and has no production. It just indicates the corners where the new walls are inserted by spanning a line between the two slashes in a cell. M_0 represents the initial map. A map derivation needs three interventions : (1) the wallnet is rewritten by parallel label substitution; (2) division walls are inserted according to the position of the slashes; (3) the outcoming topological maps are redrawn under the constraint of an additional geometrical device. The topological map becomes a topographical one.

We construct the production rules in such a way that a division of a mother cell with a boundary sequence of segments $w_1 \ldots w_m$ gives rise to two daughter cells with the sequential labelling starting by the division wall (cf. Fig.1).

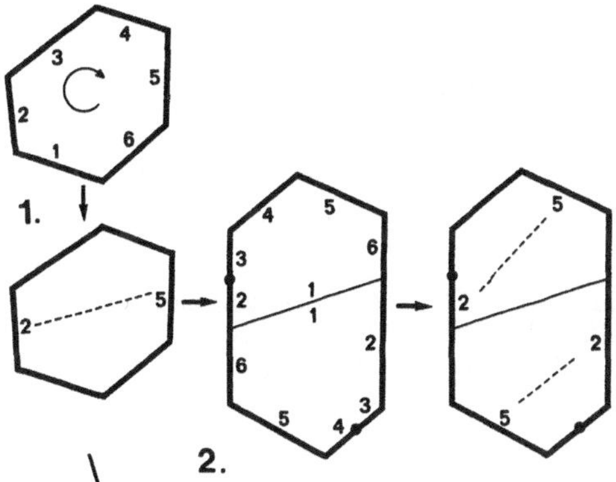

Fig.1 : A map system has the production rules

$$P = \begin{Bmatrix} 1 \to 5 & & 4 \to 5 \\ 2 \to 6/2.3 & ; & 5 \to 6/2 \\ 3 \to 4 & & 6 \to 3.4 \end{Bmatrix}$$

The segment labeled 1 is at the next step, i.e. after cell division, relabeled as segment 5. Segment 6 is divided into the two segments 3.4 (the point, a corner, is added for better reading). Segments 5 and 2 subdivide also in resp. two and three segments. Slashes instead of points indicate the division wall insertion sites. After division, the new wall is spanned between segments 2 and 5 of the mother cell.

The labels on the l.h.side of the productions (the alphabet Σ) show that the mother cell has a boundary 1.2.3.4.5.6. The r.h.side terms of the productions show that the mother cell boundary produces the sequence 5.6/2.3.4.5.6/2.3.4 surrounding a pair of daughter cells. This sequence is composed of twice the sequence of the mother cell. The labels 1 are lacking. They are replaced by slashes. These two labels are given to the halves of the division wall.

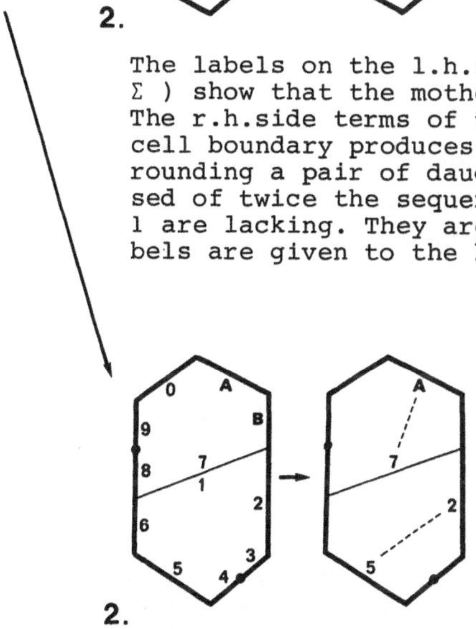

The case of a map system with differentiated wall insertions between sister cells

$$P = \begin{Bmatrix} 1 \to 5 & & 7 \to 6/8 \\ 2 \to 6/8.9 & & 8 \to 9 \\ 3 \to 0 & & 9 \to 0 \\ 4 \to A & ; & 0 \to A \\ 5 \to B/2 & & A \to B/2.3 \\ 6 \to 3.4 & & B \to 4.5 \end{Bmatrix}$$

The division wall in the cell with the boundary 1.2.3.4.5.6 is anchored on the segments 2 and 5, and in the sister cell with boundary 7.8.9.0. A.B the division wall is spanned between the segments 7 and A.

From step to step, i.e. from cell generation to cell generation, the number of wall segments is conserved but the polygonal degree can be modified if more than one segment build up a wall. Thus wall growth, i.e. multiplication of segments/half-wall, is a consequence of the distance between the insertion sites of a division wall, and this because a constant number of segments is maintained in an entire cell boundary.

DEVELOPMENTAL ARCHETYPES

The developmental sequence of maps obtained under a system G has to be transformed into topographical representations. This is achieved by giving to all segments a standard length. Under this condition, Fig.2 reveals to the botanist that the map represents a branched filament. If

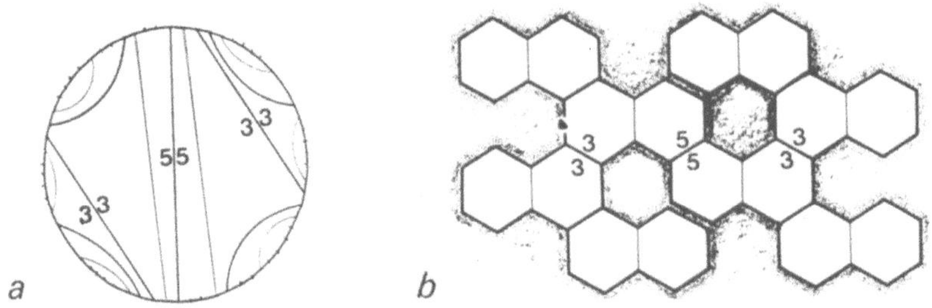

Fig.2 : (a) topological map and (b) corresponding topographical representation at step 5 derived by a system with the production rules P = {1 → 5, 2 → 6/2.3.4, 3 → 5, 4 → 6/2, 5 → 3, 6 → 4}, and an initial map given by a cell boundary 1.2.3.4.5.6.

the number of segments in a wall cannot be represented on a plane, the map is transformed into a 3-dimensional sheet where the curvature of the lines occurs always towards the same side. Fig.3 gives such a sample in which cells of a filament become coalescent and grow out like a two-rowed tube. The further developmental steps of this map derivation (Fig.4) show a plant body with branched biseriate fronds.

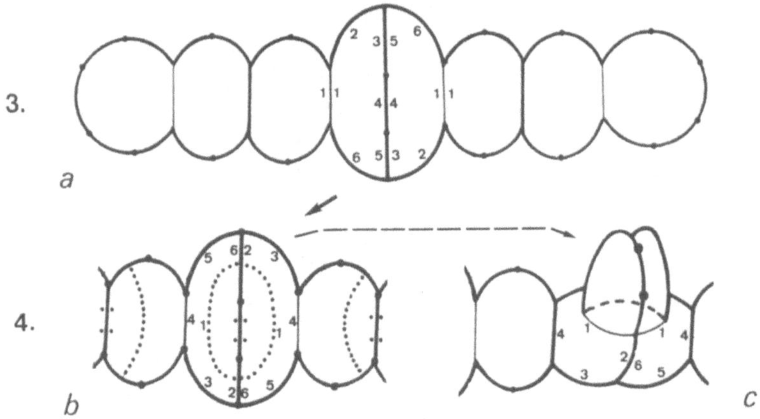

Fig.3 : Steps 3 and 4 in a map sequence obtained under following production rules : 1 → 4, 2 → 5, 3 → 6/2, 4 → 3.4.5, 5 → 6/2, 6 → 3, and an initial map with the cell border 1.2.3.4.5.6. (a) 3rd step, the two middle cells have a common wall of length 3. (b) The map is rewritten. The dotted lines indicate the new walls. (c) The map of (b) is redrawn topographically. The two emerging apical cells have a common wall, 5 segments long.

Till now, eleven archetypal plant developments have been recorded; they are cited in table I. Many variations around each archetypal theme may occur.

EQUIVALENT DOUBLE WALL MAP SYSTEMS

The different aspects of the cellular organization in the theoretical tissues as well as their morphology were investigated. We proposed for this purpose four equivalent systems which lead to possibly identic maps with the same cell wall insertions. They are based on progressively higher integrated alphabets. The structure of each of these alphabets and of the corresponding production rules gives specific insights into the generated maps and their properties.

A basic *system* G_I has an alphabet Σ_I of symbols representing segments of half-walls as described before. Systems G_I open the way to define exhaustively the possibilities of wall positioning for a given number of segments [3].

A *system* G_{II} has an alphabet Σ_{II} over entire cell walls. An element of this alphabet is built up by a double string composed of parallel and opposite running sequences of segment labels. Whereas in expression G_I all systems have the same alphabet, in expression G_{II} the size and com-

P_{II}, external walls :

$$\frac{1.2.3.4.5.6}{e} \rightarrow \frac{4.5.6/2.3.4.5.6/2.3}{e}$$

$$\frac{2.3.4.5.6}{e} \rightarrow \frac{5.6/2.3.4.5.6/2.3}{e}$$

$$\frac{2.3}{e} \rightarrow \frac{5.6/2}{e}$$

$$\frac{5.6}{e} \rightarrow \frac{6/2.3}{e}$$

$$\frac{6}{e} \rightarrow \frac{3}{e} \; ; \; \frac{3}{e} \rightarrow \frac{6/2}{e}$$

$$\frac{2}{e} \rightarrow \frac{5}{e} \; ; \; \frac{5}{e} \rightarrow \frac{6/2}{e}$$

P_{II}, internal walls :

$$\frac{1}{1} \rightarrow \frac{4}{4} \; ; \; \frac{2}{6} \rightarrow \frac{5}{3}$$

$$\frac{4}{4} \rightarrow \frac{3.4.5}{5.4.3} \; ; \; \frac{5}{3} \rightarrow \frac{6/2}{2/6}$$

$$\frac{2.3}{6.5} \rightarrow \frac{5.6/2}{3.2/6}$$

$$\frac{3.4.5}{5.4.3} \rightarrow \frac{6/2.3.4.5.6/2}{2/6.5.4.3.2/6}$$

$$\frac{2.3.4.5.6}{6.5.4.3.2} \rightarrow \frac{5.6/2.3.4.5.6/2.3}{3.2/6.5.4.3.2/6.5}$$

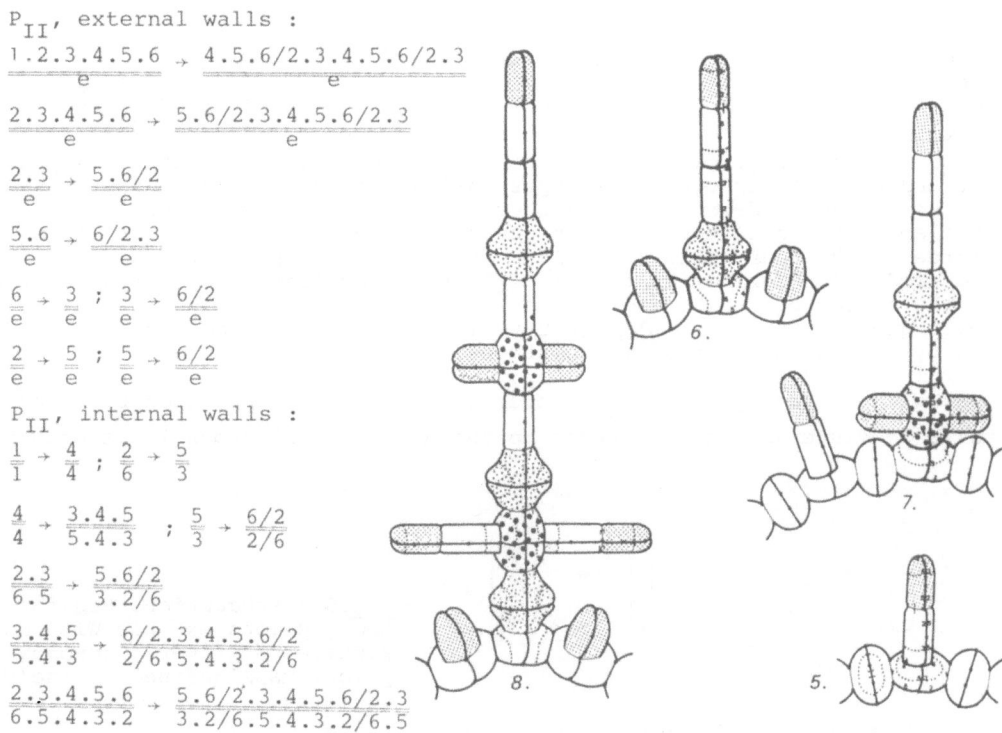

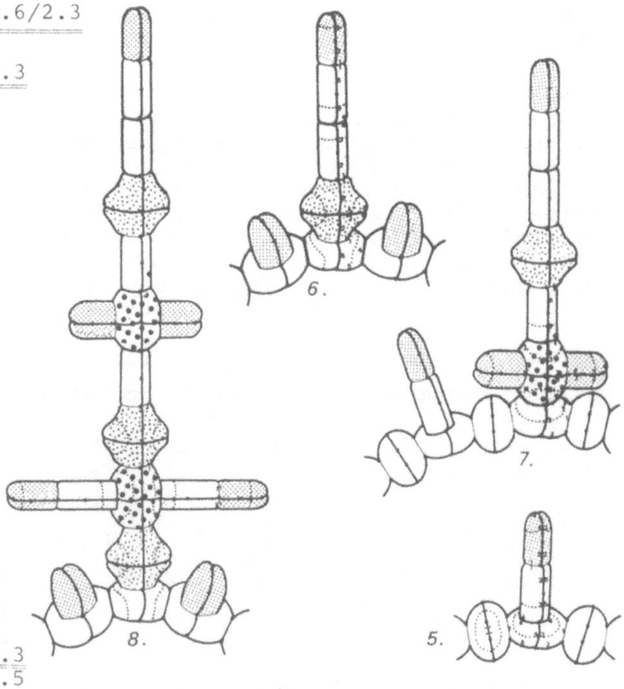

Fig.4 : Topographical maps of 6 further developmental steps of the same system than in Fig.3. The alphabet Σ_I indicated in Fig.3. The generated plant body is built up by 8 double-wall labels which are used in the procumbent part, and 7 wall labels used in the erect part. These entire wall labels of Σ_{II} are enumerated at the margin of the figure.

position of the alphabets is variable. The number and length of the entire wall labels furnish a measure of the complexity of the generated
tissue [4]. These complexities permit the construction of a filiation
tree of morphological archetypes of plants [6]. Furthermore, subgraphs
of the production derivation graph in systems G_{II} characterize the differentiation of the tissue.

A further step in the alphabet integration leads to composite labels
which describe complete cell borders including their complementary half-
walls in neighbor cells (*systems* G_{III} in [8]). It results an alphabet
Σ_{III} over *cell types* with different polygonal degree and different wall
composition. We distinguish trivial cells and apical cells which reproduce themselves in one of their daughter cells.

$P_{III} =$

$$\begin{Bmatrix} c_2 \rightarrow c_2 \ c_4^{(1)} \\ c_4^{(1)} \rightarrow c_4^{(1)} \ c_4^{(2)} \\ c_4^{(2)} \rightarrow c_2 \ c_5 \\ c_5 \rightarrow c_5 \ c_4^{(2)} \end{Bmatrix}$$

derivation graph :

$$c_2 \circlearrowright \rightarrow c_4^{(1)} \circlearrowleft \rightarrow c_4^{(2)} \rightleftarrows c_5 \circlearrowright$$

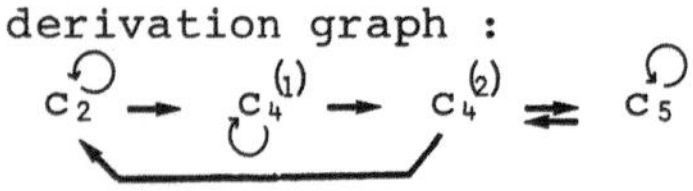

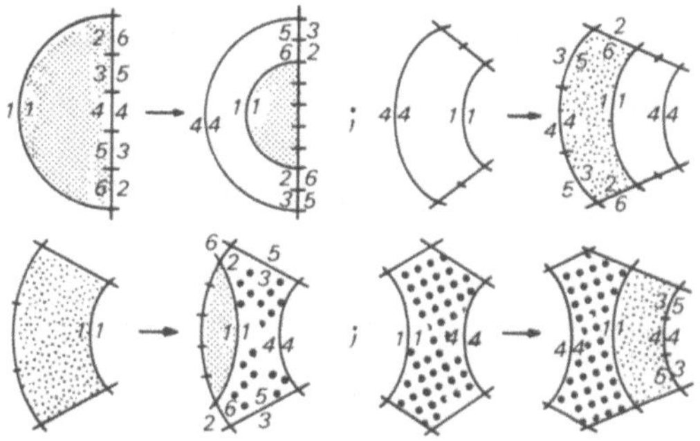

Fig.5 : System G_{III} generating also the example of Figs 3 and 4. Alphabet and productions over cell types $c_y^{(x)}$ which appear inside the tissue
(y the polygonal degree and x a specific variation for a given degree).
The labels for cell types in contact to the environment are not given.
 Three of the four cells are self-reproducing (cf. productions of P_{III}
and production derivation graph). ▨ self-replicating c_2 cells located
at the apices of stalks and branches; ▢ self-replicating internodal
cells grouped in pairs; ▨ self-replicating 5-sided nodal cells grouped by four at the basis of the stalks. ▨ trivial cells.

Systems G_{IV} specify an alphabet over *topographical maps* which represent
pairs of sister cells in respect to the defined cell types (An alphabet

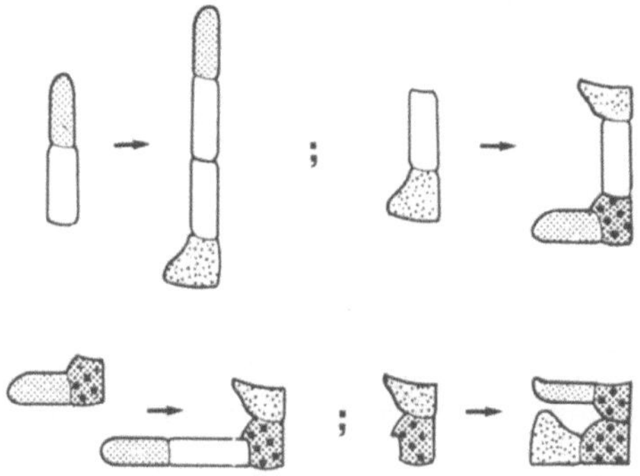

Fig.6 : Productions P_{IV} for
the previous example with an
alphabet Σ_{IV} over 4 topographical maps defined by their
geometry.

over unique cells would also work, but needs a reference point). Thus the production rules over geometrical figures open the way to a direct construction of the generated plant bodies as shown in Fig.6 which avoids the cumbersome map derivation procedure.

PATTERNS IN MERISTEMS

A direct consequence of the topographical constructions is the search for some mechanism of meristematic behavior. We are looking for cell plates composed of an aggregation of self-reproducing cell types and which conserve their configuration during reproduction. Different modalities of the apical organization of plants are found among our systems. They are spread over from the division of simple apical cells in cellular filaments [13] to the complex configuration in which leaves are initiated following a precise phyllotaxy. Decussate tetramerous phyllotaxy as well as 2-5 usuel phyllotaxy can be generated [7]. In the first case 4-celled apical plates are arranged like leaves on four spiral parasti-

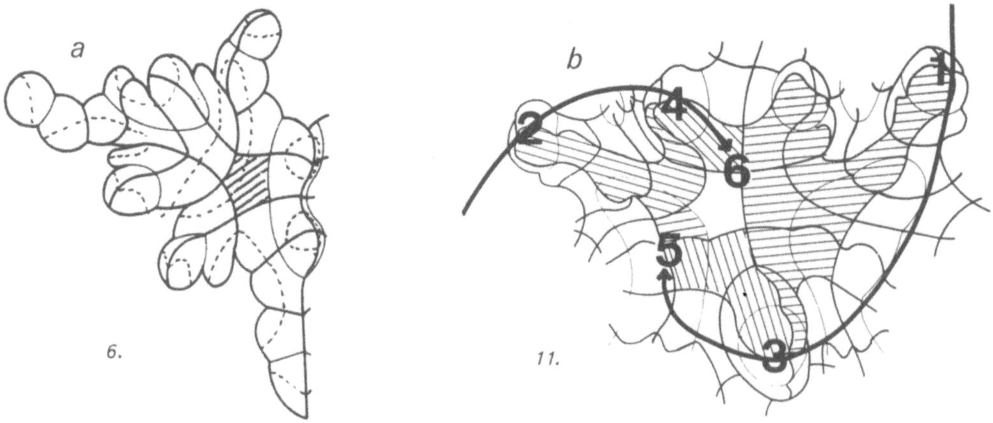

Fig.7 : Simulation of the 2-5 phyllotaxy with a map system based on following wall segment productions : 1 → 3 , 2 → 4, 3 → 5/7.8, 4 → 9, 5 → 0/2, 6 → 9, 7 → 0/2, 8 → 3, 9 → 4.5/7, 0 → 8. (a): 6th step in a map derivation initiated by a cell with the boundary 1.2.3.4.5 . The hatched cell embedded in a cellular environment is the initial cell of a map derivation which leads 5 steps later to the Fig7b. Lateral foliar-like appendages are disposed on spirals.

chies. In the second case apical cells produce at each division a subsidiary trivial cell which is at the origin of an area giving rise to a leaflike emergence. These leaves are arranged as in the usual phyllotaxy (Fig.7b). Each leaf itself bears at its top a 4-celled apical plate which is self-replicating and produces lateral outgrowths arranged in opposite decussate manner. That means that there are two types of apical structures on a same plant body.

If we look, in the 2-5 phyllotaxy figures, to larger cell plates composed of apical cells (Fig.8) we find a group of eleven such cells which reproduce themselves with their specific neighborhood. The 11-celled apical plate (Fig.8 state 6) then splitts into two identical 11-celled plates (Fig.8, state 7), one produced by a group of six cells which continues the main apical meristem, the other issued from the remaining five cells and which can be assimilated to the apex located at the axil of the first derivated leaf. The entire apical structure can thus be under-

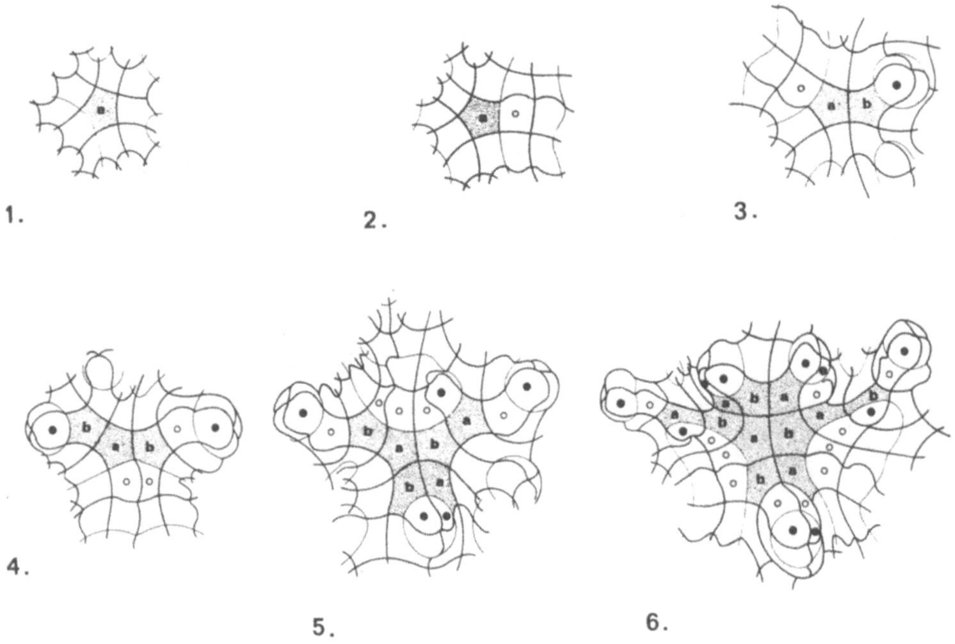

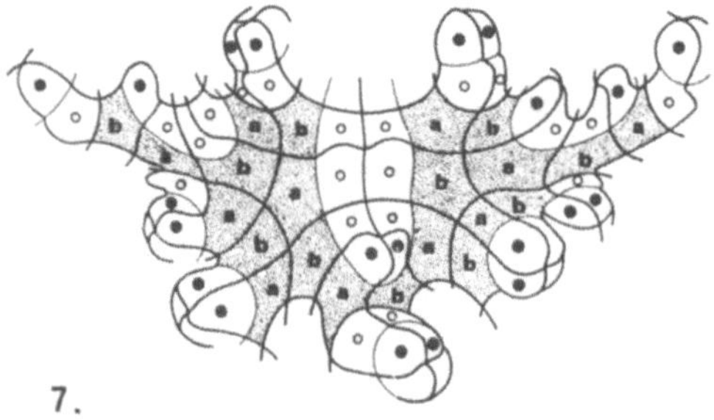

Fig.8 : Map derivation with the system of Fig.7. Initial map : part of Fig. 7a. (a) and (b) : self-reproducing 5-sided apical cells; (o) : 4-sided trivial cell; (•) : 3-sided self-reproducing apical cell on appendages.

stood by the activity of a group of eleven cells which are not permanent but which are reproduced in one of their derivatives. It becomes obvious how a complex intricated behavior of differently organized meristems can act together in a co-ordinate way.

CELL DIFFERENTIATION AND FORM

The theoretical constructs of meristems [14] depending on numbers and form of cells represent a precursory layout of a general theory of meristematic differentiation such as it is proposed here in an entire rational way. Our hypothesis upon an ordered positioning of new cell walls during cell divisions furnishes a general coherent framework which relates all developmental meristematic patterns obtainable on this basis.

Table I : Cell differentiation in meristematic structures

(1)	(2)	(3)	(4)	(5)
I		0 0 2 0	$\circlearrowleft c_2 \rightarrow \circlearrowleft c_4 \circlearrowright$	all cells
II	[3] Fig.5	0 0 2 1	$\circlearrowleft c_2 \rightleftarrows c_4 \rightleftarrows \circlearrowleft c_6$	1 apical cell 2 nodal cells
III		1 0 2 0	$\circlearrowleft c_3 \rightarrow \circlearrowleft c_4^{(1)} \rightarrow \circlearrowleft c_4^{(2)} \circlearrowleft$	all cells
IV	[3] Fig.6	1 0 1 8		1 apical cell all inner cells
V	[3] Fig.3b	1 0 0 8		all inner cells
VI	[4] Fig.2 [5] " 8 [6] " 7	3 1 . .	$\circlearrowright c_2 \rightarrow \circlearrowleft c_4^{(1)} \rightarrow c_4^{(2)} \rightleftarrows c_5 \circlearrowright$ ($c_4^{(2)} \rightarrow c_2$)	2 apical cells 2 internodal cells 4 nodal cells
VII	[5] Fig.7			
VIII	[8] Figs. 1,4,5,6			
IX	[3] Fig.7 [6] " 1	2 3 . .	$\circlearrowleft c_5 \rightleftarrows c_4^{(1)} \rightarrow \circlearrowleft c_3 \rightarrow c_4^{(2)} \rightleftarrows c_4^{(3)}$ ($c_4^{(2)} \rightarrow c_4^{(1)}$, $c_4^{(3)} \rightarrow c_4^{(1)}$)	1 inner apical cell 4 cells on appendages
X	[7] Figs. 2,3	2 0 . .	$\circlearrowleft c_5^{(1)} \rightleftarrows c_5^{(2)} \circlearrowright$	4 inner apical cells
XI	[7] Figs. 4,5	2 1 . .	$\circlearrowleft c_5 \rightleftarrows c_4 \rightleftarrows c_3 \circlearrowright$	11 inner apical cells 4 cells on appendages

(1): Morphogenetic archetypes : I. filaments; II. branched filaments; III. cellular grids; IV. fringed cell layers; V. layers with staggered cells; VI. procumbent filaments bearing fronds; VII. growth centers on fasciation ribs; VIII. leaves; IX. 3-sided cells with spiraled segmentation; X. decussate tetramerous verticillation; XI. 2-5 phyllotaxy.
(2): Bibliographical references for archetypal figures. (3): 2x2 table:

	s	t
i		
e		

(i) # of inner cell types,
(e) # of cell types in contact with the environment,
(s) # of self-replicating cell types,
(t) # of trivial cell types.

For archetypes VI to IX, only inner cell types are registered. (4): Derivation graph of cell types. (5): # of self-reproducing cells participating in a cell plate.

The main botanical result is that the apical division behavior differs from ordinary cellular behavior only by its autoreproductiveness which is nothing else than a special property of the cellular neighborhood relationship. This is the first attempt to describe the apical activity whithout assumptions of inhibitory or spacefilling mechanisms. The matching of segments belonging to the half-walls of two adjacent cells, determined by given wall insertions, is sufficient to generate a huge variety of morphogenetic patterns. These considerations may lead to a periodic table of morphogenetic behavior based on the number of self-replicating and trivial cell types. A first sketch of such a periodical table of morphogenetics is given in Table I. What is needed now is a regulation device for finite growth of lateral appendages based on cellular lifespan differentiation. The way to do this has been proposed previously [15].

REFERENCES

[1] Gunning, B.E.S. : Microtubules and cytomorphogenesis in a developing organ : The root primordium of *Azolla pinnata* . In "Cytomotphogenesis in Plants", O. Kiermayer, ed., Cell Biology Monographs 8: 301- 325 , Springer-Vlg (1981).

[2] Green, P.B. : Shifts in plant cell axiality: Histogenetic influences on cellulose orientation in the succulent, *Graptopetalum*.Dev. Biology 103 : 18-27 (1984).

[3] Lück, J. and Lück, H.B. : Proposition d'une typologie de l'organisation cellulaire des tissus végétaux. Act.1er Sém.de l'Ecole de Biol.théor., C.N.R.S.-E.N.S., H.Le Guyader et Th. Moulin, eds., E.N.S.T.A., Paris: 335-371 (1981).

[4] Lück, J. and Lück, H.B. : Sur la structure de l'organisation tissulaire et son incidence sur la morphogenèse. Act.2me Sém.de l'Ecole de Biol.théor.,C.N.R.S., H. Le Guyader,ed., Publ.Univ.Rouen: 385-397 (1982).

[5] Lück, J. and Lück, H.B. : Generation of 3-dimensional plant bodies by double walled map and stereomap systems. Lect.Notes in Comp. Sci. 153: 219-231, Berlin, Springer-Vlg (1983).

[6] Lück, J. and Lück, H.B. : Comparative plant morphogenesis founded on map and stereomap generating systems. In "Dynamical systems and cellular automata", J. Demongeot, E. Goles, and M. Tchuente, eds., Academic Press, London : 111-121 (1985a).

[7] Lück, J. and Lück, H.B. : Un mécanisme générateur d'hélices phyllotaxiques. Act.4me Sém.de l'Ecole de Biol.théor.,G. Benchetrit and J. Demongeot, eds. :317-330, Ed.du C.N.R.S. (1985b).

[8] Lück, H.B. and Lück, J. : Unconventional leaves (an application of map 0L systems to biology). In"The book of L", G. Rozenberg and A. Salomaa,eds., Springer-Vlg Berlin : 275-289 (1986).

[9] Rosenfeld, A. and Strong, J.P. : A grammar for maps. In"Software Engineering", J. Tou,ed.,Academic Press, NY, 2:227-239 (1971).

[10] Lindenmayer, A. and Rozenberg, G. : Parallel generation of maps: Developmental systems for cell layers. Lect.Notes in Comp.Sci. 73 : 301-316, Springer-Vlg, Berlin (1979).

[11] Ehrig, H.,Nagl, M. and Rozenberg, G. : Graph grammars and their application to computer science.Lect.Notes in Comp.Sci. 153, Springer-Vlg, Berlin (1983).

[12] Rozenberg, G. and Salomaa, A. : The book of L. Springer-Vlg(1986).

[13] Lück, H.B. and Lück, J. : Abstr.ISMB, Kyoto Japan: 149-150 (1985).

[14] Newman, I.V. : Pattern in the meristems of vascular plants. Proc. Lin.Soc.of New South Wales: 86 : 9-59 (1961).

[15] Lück, H.B. and Lück, J. : Automata theoretical explanation of tissue growth. Proc. ISMTB, Res.Inst.Math.,Kyoto:174-185 (1978).

THE CORTICAL TRACTOR:
A New Model For Epithelial Morphogenesis

Louis Y. Cheng[1], James. D. Murray[2], Garrett. M. Odell[3], George F. Oster[4]

[1] Department of Civil Engineering, University of California, Berkeley, CA 94720
[2]Centre for Mathematical Biology, University of Oxford, Oxford, England, OX1 3LB
[3]Department of Mathematical Sciences, Rensselaer Polytechnic Institute, Troy, NY 12181
[4]Departments of Biophysics and Entomology, University of California, Berkeley, CA 94720

1 INTRODUCTION

We present here a new theory of epithelial morphogenesis based on the assumption that the cytoplasmic machinery responsible for epithelial cell motility is fundamentally similar to that of freely migrating mesenchymal cells. The additional constraint imposed on epithelial cells is that their apical borders remain attached. This model is able to mimic all of the foldings and invaginations of the earlier apical constriction model (Odell, et al., 1981). Moreover, it provides a mechanism for epithelial cells to actively change neighbors while maintaining the apical seal. This is an essential feature of epithelial morphogenesis, and the application of the model to neural plate formation is discussed in detail in Jacobson, Odell & Oster (1985)and Jacobson, Odell, Oster & Cheng (1986).

A sheet of epithelial cells is viewed as a dynamic structure: each cell cortex cycles its cytoplasm in a flow pattern whose time average is shown in Fig. 1. Membrane and adhesive structures are inserted in the basal and lateral surfaces, flow apicalward, and are recycled to the cell interior in the apical region. Junction structures pile up at the apical circumference, and maintain the seal which isolates the embryo from the external chemical environment. The flow of the cortical tractor (CT) is driven by the actomyosin contractile machinery of the cell cortex (Oster and Perelson, 1985; Oster, 1984). Here we shall not be concerned with mechanism which drives the cortical tractor motion, but with the mechanical consequences of the motion itself. We emphasize that the cortical flow shown in Fig. 1 represents a time-averaged velocity field; on a shorter time scale, cytoplasmic motions appear much more disordered.

2 MODEL EQUATIONS FOR PLACODE FORMATION

The development of epithelial deformations generally proceeds in two steps. First, the cells form a 'placode': a thickened region where the cells have changed from a cuboidal to a columnar shape. Second, the epithelial sheet commences to buckle under the generation of bending moments. In the apical contraction model, both of these deformations were generated by the contraction of the circumferential band of microfilaments that encircle the apical ends

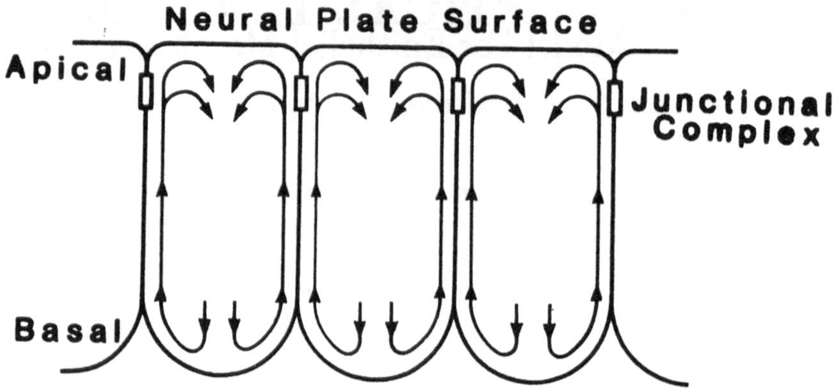

Figure 1: The flow of cortical cytoplasm according to the cortical tractor model.

of the cells. In the cortical tractor model these deformations arise in a quite different fashion, although apical contratction may still play some role in generating bending moments. In this paper we will treat only the first stage: the development of the placodes; the bending moment model is discussed in Jacobson, et al., 1985.

2.1 Finite Difference Equations for Normal Displacements

The simplest model for the cortical tractor can be derived by considering the free body diagram in Fig. 2. Here we represent two adjacent cells by trapezoidal elements, whose vertices are denoted by $(i-1, i, i+1)$, and refer to the cell with boundaries $i, i+1$ as cell i. We shall employ the following notation:

- $H_i =$ the height at node i (i.e. the boundary between cell i and cell i-1).
- $H_o =$ the initial (unstressed) height of each cell.
- $\lambda_i = H_i/H_o =$ the stretch ratio at node i.
- $W_i =$ the width of cell i.
- $W_o =$ the initial (unstressed) width of each cell.
- $v_i =$ the velocity of the cortical flow in cell i.
- $k =$ the elastic modulus of each cell.
- $G =$ the passive shear modulus of each cell.

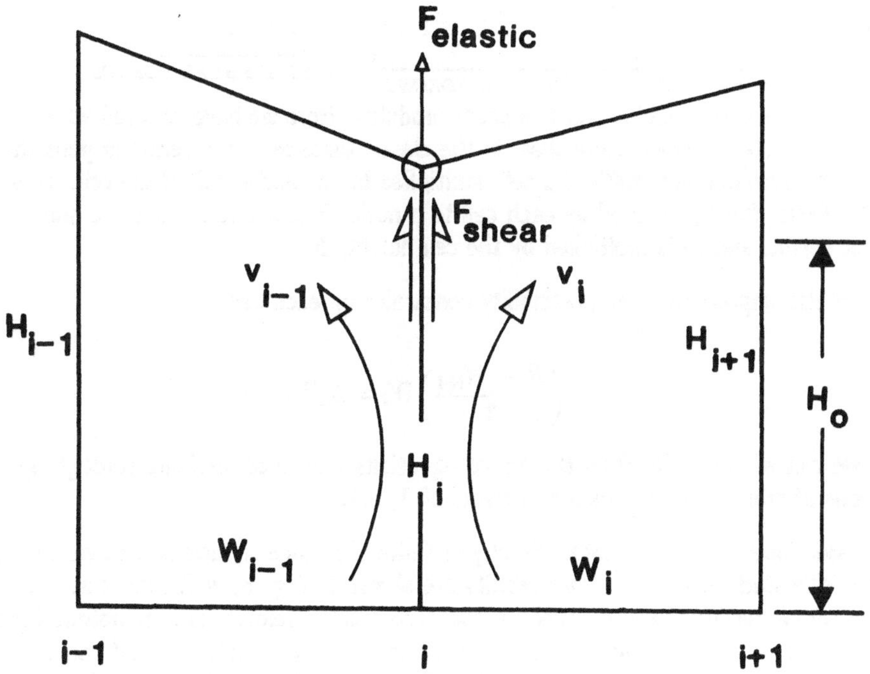

Figure 2: A free body diagram of two adjacent cells showing the elastic, shear and tractor forces.

- α = the active shear modulus for each cell due to tractor motion.
- μ = viscosity of the cortical cytoplasm.

The equations of motion for cell i are derived by writing down balance equations on node i for the vertical force components.

The equations of motion for node i are written down by equating the sum of the forces to the inertial force, md^2H_i/dt^2, where m is the mass of a cell. However, for this system the inertial forces are negligible (Odell, et al., 1981), so that we need not consider the acceleration term on node i. Thus the force balance equation for node i is

$$F_{ELASTIC} + F_{SHEAR} + F_{ACTIVE} + F_{VISCOUS} = 0 \tag{1}$$

which can be approximated by

$$\underbrace{\mu\frac{dH_i}{dt}}_{VISCOUS\ FORCES} = \underbrace{-k\left[\left(\frac{H_i + H_{i+1}}{2} - H_o\right)\frac{W_i}{2} + \left(\frac{H_{i-1} + H_i}{2} - H_o\right)\frac{W_{i-1}}{2}\right]}_{PASSIVE\ ELASTIC\ FORCES} \tag{2}$$

$$\underbrace{-G\left[\left(\frac{H_i - H_{i-1}}{2}\right) + \left(\frac{H_i - H_{i+1}}{2}\right)\right]}_{PASSIVE\ SHEAR\ ELASTIC\ FORCES} + \underbrace{\alpha\left|(v_i - v_{i-1})\, H_i\right|}_{ACTIVE\ SHEAR\ FORCES}$$

where k is an elastic modulus and G a shear modulus. Here we have counted as minus forces tending to shorten the cell. Note that, in the elastic force term the terms in parentheses are the elastic forces per unit width of a cell, multiplied by the half-width of the cell. This assigns half the elastic force generated by each cell to a node. The active shear force modulus, α, is per unit height, and so is multiplied by the cell height, H_i.

We shall also impose an incompressibility constraint on each cell:

$$\left(\frac{H_i + H_{i+1}}{2}\right) W_i = H_o W_o \tag{3}$$

Thus we can eliminate W_i from the above equations and, after dividing through by H_o, the equations of motion can be written in terms of λ_i only.

These equations may be simulated directly as finite difference equations, and we shall present an extensive study elsewhere. Two results are shown in Fig. 3, which illustrate some counterintuitive properties of the cortical tractor. The tractor velocity is a monotone function of local chemical conditions, and so the variation in the tractor velocity field can be taken as reflecting the spatial chemical concentration distribution. In Fig. 3(a) A uniform gradient in tractor velocity, v, produces a placode of constant height: i.e. a constant displacement field, u(x), where x measures distance along the cell sheet and u measures the height perturbation from the uniform, unperturbed state. (The taper at the edges is due to the imposition of zero displacement boundary conditions.) In 3(b), a periodic tractor velocity field, v(x), produces a displacement field, u(x), which is shifted with respect to the velocity field. Thus the morphogenetic movements generated by the cortical tractor do not correspond simply to the pattern of chemical concentrations, but to the *gradients* in concentration.

2.2 Field Equations

Some insight into the behavior of the model can be gleaned by converting the above equations into a continuum model. There is no unique correspondence between the above finite difference equations and a set of partial differential equations. However, we can proceed in the most elementary fashion by the following identifications:

$$\lambda_i \longrightarrow \lambda(x,t), \quad x_{i+1} - x_i \longrightarrow \delta$$

so that we can expand λ in a Taylor's series to second order, e.g.

$$\lambda_{i+1} \longrightarrow \lambda(x) + \delta\lambda_x + \frac{\delta^2}{2}\lambda_{xx} + \cdots$$

From (2), this yields the following nonlinear partial differential equation for $\lambda(x,t)$:

$$\mu\frac{\partial\lambda}{\partial t} = -K_o\left(1 - \frac{1}{\lambda}\right) - \frac{K_1}{\lambda^3}\left[\frac{\partial^2\lambda}{\partial x^2} - \left(\frac{\partial\lambda}{\partial x}\right)^2\right] + G_1\frac{\partial^2\lambda}{\partial x^2} + \lambda\left|\alpha_o\frac{\partial v}{\partial x} - \alpha_1\frac{\partial^2 v}{\partial x^2}\right| \tag{4}$$

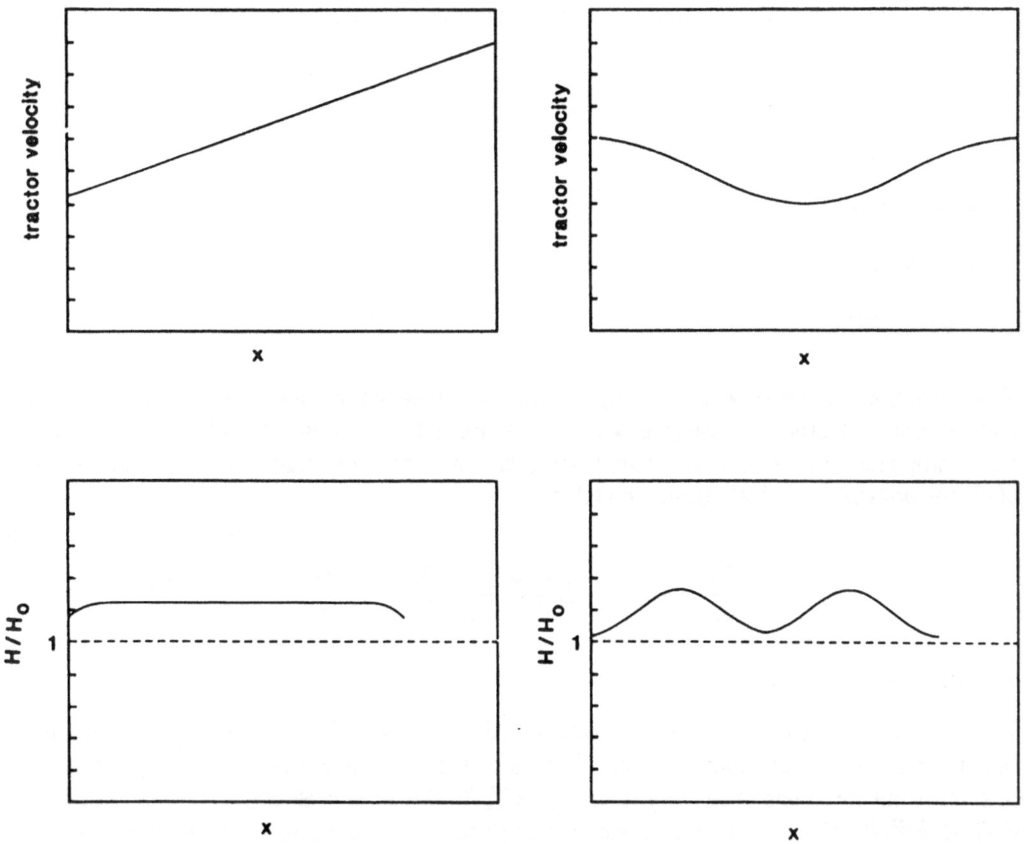

Figure 3: The deformation fields corresponding to (a) a linear tractor velocity field, and (b) a periodic velocity field.

where we have employed the constant volume constraint and defined the following quantities:

- $K_o = kW_o$
- $K_1 = \frac{1}{4}kW_o\delta^2$
- $G_1 = \frac{1}{2}G\delta^2$
- $\alpha_o = \alpha\delta$
- $\alpha_1 = \frac{1}{2}\alpha\delta^2$

The linearized version of equation (4) is easier to understand, and captures the essential pattern forming features. Therefore, we can substitute $\lambda = 1 + u$ into the above equation and retain only linear terms in u, the height perturbation. Thus for small velocity perturbations, $v(x)$, the linearized cortical tractor model is:

$$\mu\frac{\partial u}{\partial t} = -K_o u + D\frac{\partial^2 u}{\partial x^2} + \left|\alpha_o\frac{\partial v}{\partial x} - \alpha_1\frac{\partial^2 v}{\partial x^2}\right|. \tag{5}$$

where $D = G_1 - K_1$.

The first term on the right hand side of this equation is the elastic force tending to restore the cells to their unstrained cuboidal shape. The second term is the passive elastic forces tending to smooth out variations in cell height. If $G > kW_o/2$, this term is always stabilizing; however, if $G < kW_o/2$, the system may become numerically unstable under perturbations whose wavelength is shorter than the width, W_o, of the finite element. This is simply a limitation of the continuum limit; physically, the passive elastic forces cannot generate deformations. The third term on the right is a 'convective' force which tends to deform the sheet wherever there is a gradient in tractor velocity, v. The fourth term is the active shearing forces which tend to amplify height variations.

Note that because of the incompressiblility constraint, the taller a cell grows, the narrower it becomes. Therefore, the strain λ is proportional to the cell density. Therefore, we can substitute for u the normalized cell density deviation $(N - N_o)/N_o$, where N_o is the undeformed cell density (i.e. when $\lambda = 1$, $u = 0$), and obtain an equation in cell density, N, which has the same form as equation (5).

If a cortical tractor velocity, v, is prescribed by some means (e.g. a boundary condition, or a chemical gradient) then equation (4) predicts a variation in cell density—and therefore in cell shape—wherever variations in v arise. This is because the active shear forces which tend to deform the cells can only arise when gradients in v are present. The consequences of this simple mechanical effect are quite subtle, as discussed in detail by Odell and Bonner (1985) where a similar model was applied to the morphogenesis slime molds. In the present setting, it is worth noticing that a constant linear gradient, V, in the tractor velocity, v, can produce a uniform placode ($u > 0$). An asymmetric velocity profile (e.g. $v \propto sin(x)$) can produce a deformation of similar form, but displaced, as shown in Fig. 3. Since local chemical

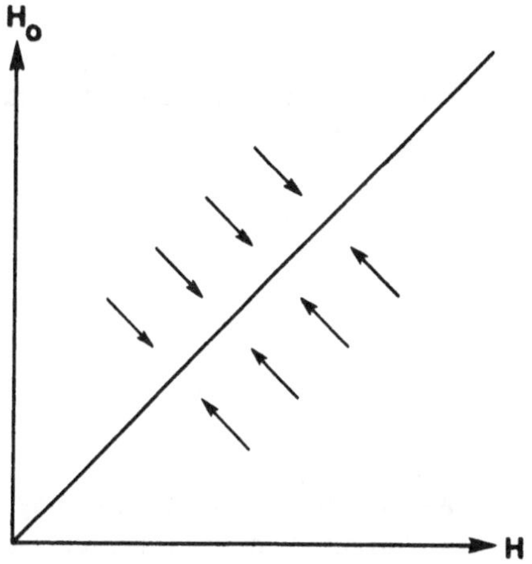

Figure 4: The (H, H_o) phase plane showing the line equilibrium, $H = H_o$

conditions regulate the tractor velocity, v, this implies that the cortical tractor 'reads out' chemical conditions in a somewhat counterintuitive fashion.

The model as it stands requires that the tractor velocity gradient be maintained in order to sustain the deformation. However, when deformed, cells reorganize their cytoskeleton. Therefore, a deformation should 'fix', so that the the deformed state becomes the new 'rest state', or reference configuration. This effect is simply modeled by allowing the rest height, H_o to relax to the current deformed height, H:

$$\frac{dH_o}{dt} = \bar{k}(H_o - H) \tag{6}$$

Thus the equilibria on the (H, H_o) phase plane is the line $H = H_o$ (Fig. 4.) When held in the deformed configuration by the cortical tractor, the system slowly accomodates until the 'deformed' configuration is the new 'undeformed' configuration.

3 APPLICATIONS OF THE MODEL

By itself, the CT model will not produce interesting morphogenetic behavior, for it is simply a mechanism by which cells mechanically respond to local chemical conditions. However, by coupling the model to other processes, one can model a variety of morphogenetic events. Indeed, all of the epithelial foldings produced by Odell, et al. (1981) using an apical contraction model can be produced as well using the cortical tractor model (Jacobson, et al., 1986). However, we feel that the most important contribution of the CT model is to provide a mechanism by which epithelial cells may actively change neighbors, while maintaining the integrity of the cortical seal. This process of active interdigitation is coming to be recognized as one of the major cellular processes driving embryogenesis (Keller, 1985; Fristrom, 1976). A more complete discussion of this important process is given in Jacobson, et al. (1986) in their treatment of neural plate morphogenesis.

REFERENCES

1. Fristrom, D. (1976). The mechanism of evagination imaginal discs of *Drosophila melanogaster*: evidence of cell rearrangement. Devel. Biol. **54** :163-71.

2. Jacobson, A., G. Odell, G. Oster (1985). The cortical tractor model for epithelial folding: application to the neural plate. In: *Molecular Determinants of Animal Form.* G. Edelman (ed.). UCLA Sympos. Molec. Cell. Biol., New Series, Vol. **31**. New York: Alan R. Liss.

3. Jacobson, A., G. Odell, G. Oster, L. Cheng (1986). Neurulation and the cortical tractor model for epithelial folding. J. Embryol. exp. Morphol. (submitted).

4. Keller, R., M. Danilchik, R. Gimlich, J. Shih (1985). Convergent extension by cell intercalation during gastrulation of **Xenopus laevis**. In: *Molecular Determinants of Animal Form.* G. Edelman (ed.). UCLA Sympos. Molec. Cell. Biol., New Series, Vol. **31**. New York: Alan R. Liss.

5. Odell, G., J. Bonner (1986). How the *Dictyosteleum discoideum* grex crawls. Phil. Trans. Roy. Soc. (in press).

6. Odell, G., G. Oster, B. Burnside, P. Alberch (1981). The mechanical basis of morphogenesis I: Epithelial folding and invagination. Devel. Biol. **85**:446-62.

7. Oster, G. (1984). On the crawling of cells. J. Embryol. exp. Morphol. **83**, Suppl. 329-64.

8. Oster, G. (1984). Mechanics of Cytogels. In: *Modelling of Patterns in Space and Time.* W. Jager, J. Murray (eds.). Berlin: Springer-Verlag.

9. Oster, G., J. Murray, P. Maini (1985). A mechanochemical model for cartilage condensations in the vertebrate limb. J. Embryol. exp. Morphol. (in press).

10. Oster, G., A. Perelson (1984). A mathematical model for cell locomotion. J. Math. Biol. **21**:383-88.

11. Oster, G., G. Odell (1984). A mechanochemical model for plasmodial oscillations in Physarum. Cell Motility **4**:469-503.

An Equilibrium Theory of Cell Distribution in *Dictyostelium discoideum*

Ei Teramoto
Department of Biophysics, Kyoto University, Kyoto 606

INTRODUCTION

The phenomena observed at each stage of the life cycle of cellular slime molds have been widely investigated as a simple system relevant to the studies of chemotaxis of cell aggregation, regulatory cell differentiation and regenerative pattern formation. Here we shall discuss the cell distribution pattern in cellular slime molds *Dictyostelium discoideum* at the migrating slug stage, using a standard model of statistical physics.

Migrating slugs consist of two major presumptive types of cells. Under normal conditions, the cells in the anterior 25% of the slug are prestalk cells and cells in the posterior 75% are prespore cells, regardless of the size of slug. Even when this proportion is altered by cutting a slug into fragments, each fragment regenerates the normal pattern by cell type conversions (Bonner and Slifkin 1949, Bonner 1952). This pattern can be visualized by vital staining, since the prestalk cells are stained while prespore cells are unstained.

In addition, there is a smaller sub-population of cells (about 10%) in the posterior zone of slug which have the same staining property as anterior prestalk cells. Hence, they are called "anterior-like cells" (Durston and Volk 1979, Sternfeld and David 1981). Under normal conditions, the cell differentiation into two major types occurs at the stage just before the formation of a migrating slug, and it is observed that the proportion of anterior-like cells in the posterior zone remains constant during migration of the slug. These anterior-like cells are supposed to be of the third type of cell and to play some important role in the processes of cell sorting during the pattern formation of a slug. It is also assumed that in a contact slug they are prevented from entering the anterior zone by a certain diffusible substance secreted from the prestalk region (Sternfeld and David 1981).

Recently, however, Kakutani and Takeuchi(1986) have revealed the fact that, as far as the behavioral property of cell movement in the slug is concerned, the anterior-like cells cannot be distinguished from the prestalk cells. They observed minutely the movement of the anterior-like cells in a migrating slug and found the following:

(A) Anterior-like cells move randomly in the posterior zone of the slug. Observed data on their velocities show an approximately normal distribution, as do Brownian particles in a fluid.

(B) Anterior-like cells are sometimes substantially absorbed into the anterior prestalk zone, and, also, new prestalk cells leak out of the anterior zone. Namely, they can mutually exchange as water molecules at the surface of water.

On the basis of these findings, they performed a well designed experiment. The prestalk region of a slug stained with Nile blue sulfate was grafted onto the prespore region of another slug stained with neutral red. After several hours the prestalk/prespore boundary became purple due to mixing of the blue prestalk and red anterior-like cells. The purple region gradually expanded and finally resulted in the equilibrium purple pattern of composite slug after about a 25 hour migration. This kinetic process of cell mixing was quantitatively investigated by counting the number of red cells that permeated the prestalk region (Fig. 1).

These experimental results presumably suggest the applicability of thermodynamical considerations. Here we shall study the equilibrium composition of prestalk cells and anterior-like cells by using a two-phase model of statistical mechanics. The kinetic process of mutual exchange of the cells in a composite slug is also quantitatively analysed, using the parameter which characterizes the equilibrium cell composition of the migrating slug.

EQUILIBRIUM THEORY OF CELL DISTRIBUTION

We first consider the equilibrium spatial distribution of the cells in a slug at the normal condition. Here we shall assume the following:

1) Under normal conditions, the anterior zone is filled with prestalk cells and the posterior zone consists of prespore cells, including a sub-population of anterior-like cells. The total number of cells in a slug, $X + Y + x = N$, is fixed.

2) As far as the characteristic properties relevant to the present model are concerned, anterior-like cells do not constitute an independent population but are essentially the same as prestalk cells. Therefore, the prestalk cells dispersed in the posterior zone are regarded as anterior-like cells in the present model.

3) Exchange between anterior-like cells and prestalk cells freely occurs through their evaporation and absorption at the zone boundary, and anterior-like cells are randomly moving in the posterior zone like Brownian particles in a fluid.

Now, we introduce the cohesive energies of cell-cell interactions,

- ε_1 : PST (prestalk cell) - PST and PST-ALC (anterior-like cell)
- ε_2 : PSP (prespore cell) - PSP
- ε_{12} : PST - PSP and PSP - ALC

where we assumed that an ALC has same property as a PST with respect to the cellular interactions (Assumption 2). Then it is easily seen that the total energy of the slug's cellular interactions at the internal equilibrium condition can be written as

$$E = -\frac{a}{2}\{\varepsilon_1 X + (\frac{Y}{Y+x}\varepsilon_2 + \frac{x}{Y+x}\varepsilon_{12})Y + (\frac{Y}{Y+x}\varepsilon_{12} + \frac{x}{Y+x}\varepsilon_1)x\} \quad + \text{const.}$$

$$= -\frac{a}{2}\{\varepsilon_1(X + \frac{x^2}{Y+x}) + \varepsilon_2\frac{Y^2}{Y+x} + 2\varepsilon_{12}\frac{Yx}{Y+x}\} \tag{1}$$

where a is the average number of nearest neighbor cells. The additional constant term represents the correction for the boundary energy.

Here, in order to take into account the Brownian movement of ALC's in the posterior zone, we introduce the concept of configurational entropy, which is usually used in statistical thermodynamics. Considering all possible cellular configurations in the posterior zone, the entropy can be expressed, by using Stirling's formula, as

$$S = K \log\frac{(Y+x)!}{Y!x!} \simeq K\{(Y+x)\log(Y+x) - Y\log Y - x\log x\} \tag{2}$$

where K is considered to be a characteristic constant of biological cell aggregation systems and corresponds to Boltzmann's constant in the case of physico-chemical molecular systems.

In equations (1) and (2), we can presume that the proportion of prespore cells remains at the regulated value of cell differentiation, i.e. $Y/N = r$. According to Sternfeld and David (1982), this ratio r is 0.766. Thus the internal energy (1) and entropy (2) of a slug become functions of a single variable x. Using the context of statistical physics, we can find the equilibrium condition by minimizing the free energy $F(x) = E(x) - T^*S(x)$. Here a characteristic parameter T^*, which corresponds to the temperature in statistical physics, is introduced. From the equilibrium condition

$$\frac{dF(x)}{dx} = a\left(\frac{\varepsilon_1+\varepsilon_2}{2} - \varepsilon_{12}\right)\frac{Y^2}{(Y+x)^2} - KT^*\log\left(\frac{Y+x}{x}\right) = 0 \qquad (3)$$

we obtain the relation

$$\frac{\log z}{(1-z)^2} = -\frac{a\varepsilon}{\theta} \qquad (4)$$

where

$$z = \frac{x}{Y+x}, \quad \varepsilon = \frac{\varepsilon_1+\varepsilon_2}{2} - \varepsilon_{12} \quad \text{and} \quad \theta = KT^* .$$

Using the equilibrium condition (5) and the observed value of the ratio $z = 0.116$, we obtain $a\varepsilon/\theta = 2.757$.

In statistical physics, we can usually test the validity of the model with data from experiments run at different temperatures. However, in the present case, T^* is not a controllable variable and the parameter $\theta = KT^*$ can be assumed to be a measure of average energy related to the random movement of the cells and to be determined by the physiological state of the cells at the stage of the migrating slug.

EXCHANGE PROCESS OF PST'S AND ALC'S

The equilibrium relation (5) can also be derived from a simple kinetic model for the exchange process of PST's and ALC's at the zone boundary of slug. Suppose that, at the boundary layer, a prestalk cell leaks out of the anterior zone with probability k and an anterior-like cell is absorbed into the anterior zone with probability k' in the unit time interval. Then the number of cells which change from PST to ALC and from ALC to PST per unit time is given by $k\rho A$ and $k'\rho A\cdot x/(y+x)$, respectively, where $\rho = N/V$ (V: volume of slug) is the density of cells and A is the area of the zone boundary. It is also assumed that ALC's have a uniform distribution throughout the posterior zone.

The number of ALC's, x, can be described by a simple kinetic equation

$$\frac{dx}{dt} = k\rho A - k'\frac{x}{Y+x}\rho A . \qquad (5)$$

The stationary solution of this equation is obviously given by

$$\frac{x}{Y+x} = \frac{k}{k'} = \kappa . \qquad (6)$$

Here, if we adopt the proposition of statistical physics that the equilibrium constant κ is given by Boltzmann's factor, whose exponent is the energy difference of two transition states, we have

$$\kappa = \exp\left[-\frac{1}{\theta}\{E(x+1) - E(x)\}\right]$$

$$= \exp\left[-\frac{a\varepsilon}{\theta}\left(\frac{Y}{Y+x}\right)^2\right] . \quad (7)$$

It is seen that equations (6) and (7) give us again the equilibrium relation (4).

Now let us consider the exchange process of the cells in a composite slug which is obtained by grafting the prestalk region stained with Nile Blue sulfate onto the prespore region stained with neutral red. We assume that the composite slug maintains its equilibrium distribution while exchange between ALC's and PST's continuously occurs. Thus, the numbers of cells stained with Nile blue sulfate, X, neutral red, x, and prespore cells, Y, are assumed to be fixed at the equilibrium values. Let ξ be the number of red cells in the prestalk region at time t. Then, at time t, the prestalk region of composite slug consists of ξ red and $X-\xi$ blue cells and the prespore region contains $x-\xi$ red and ξ blue anterior-like cells.

Then, we can describe the exchange process of coloured cells by the equation

$$\frac{d\xi}{dt} = k'\frac{x-\xi}{Y+x}\rho A - k\frac{\xi}{X}\rho A$$

or

$$\frac{d\xi}{d\tau} = k\left[1 - \frac{X+x}{xX}\,\xi\right] \quad (8)$$

where $\tau = \rho A t$ and relation (6) is used. Under the initial condition $\xi(0) = 0$, the solution of equation (8) is given by

$$\frac{\xi}{X} = \frac{x}{X+x}\{1 - e^{-\beta t}\} \quad (9)$$

where $\beta = k\rho A(X+x)/xX$. ξ/X is the proportion of red cells in the anterior prestalk region, which finally approaches the stationary value $x/(X+x)$. Using the data of Sternfeld and David(1982), $X/N = 0.133$, $x/N = 0.101$ and $Y/N = 0.766$, we have $x/(X+x) = 0.432$.

In Fig. 1 the mean values of the proportions of observed cells stained with neutral red in the prestalk region of the composite slug are shown with the standard deviation of three experiments. The curve shows theoretical results obtained by using the parameter value $\beta = 0.192$ and assuming that the exchange starts after a time lag of 1.5hr.

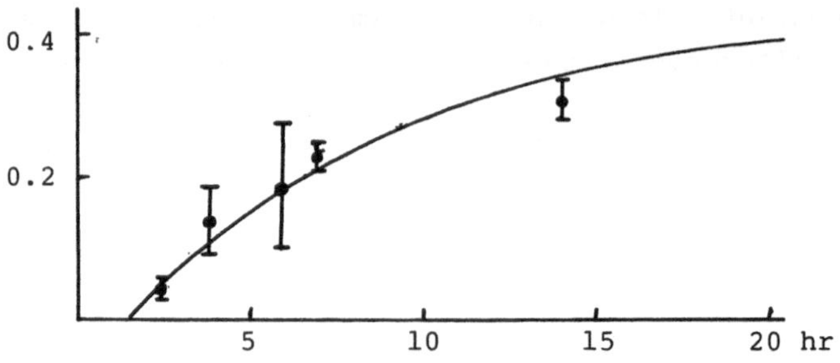

Fig. 1 Proportion of the cells stained with neutral red in the prestalk region of the composite slug. The mean values and the standard deviations of three experiments by Kakutani and Takeuchi(1986) are shown with our theoretical result.

DISCUSSION

We have considered a model for equilibrium distribution of the cells in the cellular slime mold *Dictyostelium discoideum*. Based on the experimental results of Kakutani and Takeuchi(1986), which disclosed that anterior-like cells randomly move in the posterior zone of the slug and that mutual exchange between prestalk cells and anterior-like cells occurs frequently, we have presented in this paper a statistical-physical model to discuss the equilibrium cell distribution pattern of the slug and to analyse quantitatively experimental data on cell exchange in a composite slug.

It has been shown that, as far as the equilibrium properties of cell distribution in the slug are concerned, the present model, based on principles of statistical physics, yields a plausible description of the experimental data. However, there seem to be doubts as to the applicability of the physical formula of molecular level. Needless to say, the dimensionality and the time scale of elementary processes are completely different from those of molecular systems; but if cellular system processes at equilibrium are statistically ergodic, it is not necessarily imprudent to apply the statistical-physical formula to the biological system by revising the meanings of characteristic quantities.

The most interesting theoretical problem is the dynamical description of regulated cell differentiation and pattern formation in cellular slime molds. We shall discuss this problem using a semiempirical model elsewhere in the near future.

ACKNOWLEDGMENT

The author wishes to thank Dr. Tetsuji Kakutani and Professor Ikuo Takeuchi for their kind offer of unpublished experimental results and useful suggestions.

REFERENCES

Bonner, J.T. and Slifkin, M.K. (1949). The proportion of stalk and spore cells. Am. J. Botan. **36**, 727-734.

Bonner, J.T. (1952). The pattern of differentiation in amoeboid slime mold. Am. J. Biol. **31**, 175-182.

Durston, A.J. and Volk, T. (1979). A cinematic study of the development of vitally stained *Dictyostelium discoideum*. J. Cell Sci. **36**, 261-279.

Kakutani, T. and Takeuchi, I. (1986). Characterization of anterior-like cells in *Dictyostelium* as analyzed by their movement. Dev. Biol. **115**, 439-445.

Sternfeld, J. and David, C.N. (1981). Cell sorting during pattern formation in *Dictyostelium discoideum*. Differentiation **20**, 10-21.

A MODEL FOR PATTERN FORMATION IN *DICTYOSTELIUM DISCOIDEUM*

Hans G. Othmer[1]
Department of Mathematics
University of Utah
Salt Lake City, Utah, 84112

and

E. F. Pate [2]
Department of Mathematics
Washington State University
Pullman, Washington 99163

1. Introduction

In many developing multicellular systems the location of a cell within the aggregate determines its developmental fate. In such systems the orderly specialization of cell structure and function and the arrangement of cells into tissues and organs require mechanisms for the spatial and temporal control of cellular activity. The concept of positional information [22], which is based on the supposition that a cell in a developing system must know where it is relative to other cells in order to follow the appropriate developmental pathway, has provided the framework for many analyses of developing systems. An alternative hypothesis to such direct linkage between the location of a cell and its subsequent fate is to suppose that cellular differentiation occurs independently of the spatial location of the cell, and that spatial patterns of cell type result from other processes such as cell sorting which occur after cell differentiation has occurred. That sorting of different cell types *in vitro* can produce nonuniform spatial distributions has long been known, but until recently there has been little evidence that cells *in vivo* can be determined at random locations in a developing tissue. However, recent experimental evidence [15,16,19] indicates that this can occur in the slug stage of the cellular slime mold *Dictyostelium discoideum*. A mathematical model based on these recent observations has been developed and analyzed elsewhere [14], and here we shall discuss some of the predictions of that model and compare them with recent experimental evidence.

An axial pattern of the two major presumptive cell types exists in the migrating slug, and under normal conditions the anterior 25% form the presumptive stalk zone while the posterior 75% form the presumptive spore zone. In addition, there is a smaller (approximately 10%) subpopulation of cells in the posterior portion of the slug that stains with vital dyes in a manner analogous to anterior, prestalk cells [15,16,19]. When the zone of anterior cells is removed from

[1]Supported in part by NIH Grant GM 29123.

[2]Supported in part by NIH Grant GM 29123.

a slug, these 'anterior-like' cells migrate to the anterior end of the posterior isolate, and the axial pattern (though not the proportions) of cell types is reestablished in approximately two hours. Experimental evidence strongly suggests that cell sorting is due to differential chemotaxis to cAMP [15,8,20,6], although differential adhesiveness of cell types may also be involved in cell sorting [7]. The anterior-like cell population in the transected slug only furnishes about 60% of the total number of anterior cells in a properly proportioned slug. The remainder apparently arise from the redetermination of cells previously determined to become prestalk cells by a regulation mechanism that reestablishes the 1/4 - 3/4 balance during the subsequent several hours. During this period the proper proportions of posterior and anterior-like cells are also restored in the posterior portion of the prespore isolate. In the proportioned slug, sorting of anterior-like cells to the anterior end of the slug is prevented by a diffusable inhibitor produced by the anterior cells [15].

In isolates comprised of anterior cells, cells with prespore characteristics appear at the posterior end of the regulating slug fragment within six hours following transection [17], but differentiated prespore cells do not appear within the prestalk zone [19]. This observation, coupled with the observation that posterior cells soon lose their prespore characteristics after crossing the prespore-prestalk boundary in migrating *D. mucoroides* slugs [4], suggests that the transition rates for conversion between the different cell types are not constant along the axis of the slug. Since the proper proportions of cell types are reestablished over a wide range of slug sizes, one can show that the mechanism by which these rates are controlled must exert its control in an essentially scale-invariant manner over a wide range of lengths [14].

Reaction-diffusion models of pattern formation originated with Turing [21], but have since been studied by many investigators [10]. These models involve two or more morphogens that react together and diffuse throughout the system. In a Turing model no cells are distinquished *a priori*; all can serve as sources or sinks of the morphogens. The most important fact about Turing systems in the context of pattern formation is that the stable morphogen distribution can be made spatially nonuniform rather than uniform by selecting the kinetic and diffusion parameters properly. However, the profiles in the standard Turing model are not strictly scale-invariant, nor even sufficiently close to it to reproduce the pattern invariance observed in the slug stage of *Dictyostelium discoideum* [13]. The regulation of morphogen profiles, and hence positional information, clearly requires a feedback mechanism by which information on the size of a system can affect its evolution in time. We have previously shown [11] that any desired degree of scale-invariance of the spatial profiles can be achieved simply by allowing the diffusion coefficients of the morphogens to depend on the concentration of a substance we call a control species. In [13] we showed that this model could account for many of the basic features of pattern regulation in the slug when only two types of cells are present and there is no cell sorting, and in [14] we showed that the model to be described later, which incorporates sorting and cell transitions, can account for the reported experimental observations. In the next section we will suggest a mechanism by which the postulated form of control can arise in the slug stage of *Dictyostelium discoideum*. As was pointed out in [11], scale-invariance can also be acheived by suitable modulation of the rate constants in the kinetic mechanism. However, this would lead to changes in the time scale of developmental processes when the spatial scale of the system is changed, and such changes in the time scale have not been observed. Thus schemes based on modulation of the reaction coefficients, as described in [5], seem less plausible at present.

2. The Mathematical Model

The major processes that must be incorporated in the model for the slug are as follows: (a) directed motion of the cells up the gradient of the chemotatic substance, (b) interconversion of the three cell types, (c) inhibition of chemotaxis of anterior-like and posterior cells by a substance released by the anterior cells, (d) regulation of the proportions of anterior and posterior cells over a wide range of slug sizes, and (e) morphallactic regeneration of a properly-proportioned slug following surgical modifications of the slug. Thus every model must at least contain equations that describe the time evolution of the three cell types, the chemotactic susbstance, and the inhibitor of chemotaxis. Since the pattern regulation mechanism is unknown there is more latitude in its choice. However, our previous model [13], in which scale-invariance is achieved via modulation of the diffusion coeficients of the morphogens, successfully reproduces many of the observations on regulation. For this reason we used the mechanism in that model for the expanded model developed in [14]. This mechanism employs two morphogen species, and thus the five equations mentioned previously must be augmented by equations for the morphogens and the control species. However, one of the morphogens can also serve as the chemotactic substance, and therefore the full model is described by a system of seven partial differential equations. Before displaying these equations we shall suggest a way in which the modulation of coefficients could arise in the slug.

Control of the permability or conductance of gap junctions via pH, calcium ions, and other chemical species is well-known [18,2,3], and in such cases it is straightforward to show that the diffusion coefficients in an averaged description depend on the concentration of the control species (cf. [12], wherein continuum descriptions of coupled cells are derived). Since there is no evidence as yet that cells in the slug form gap junctions, this derivation is not applicable here, and a more careful analysis of transport is required. The slug can be regarded as a two-phase mixture composed of a cell phase and the interstitial space between cells, the latter of which is filled with a viscous fluid. Although the fraction of interstitial space has not been measured, electron microscopy shows that the cells are densely packed in the slug [9], and therefore we expect that it is small. For simplicity we assume that it is constant throughout the slug. Material that is not localized in the cells can move through the slug by diffusion through the extracellular fluid, by cell-to-cell transport via small regions of fluid that are momentarily-trapped between cells, by a combination of these two processes, and by convection due to cell movement (Figure 1(a)). While convection might be important, there is no evidence for or against its inclusion, and we shall ignore it.

A simplified description of the major modes of transport is shown in Figure 1(b), wherein ϵ

is the fraction of interstitial space. Suppose that the diffusion from cell to fluid is modulated by another species whose concentration is W, and that the capacitance of the trapped fluid regions is negligible. If we assume that transport between the phases is a linear function of the

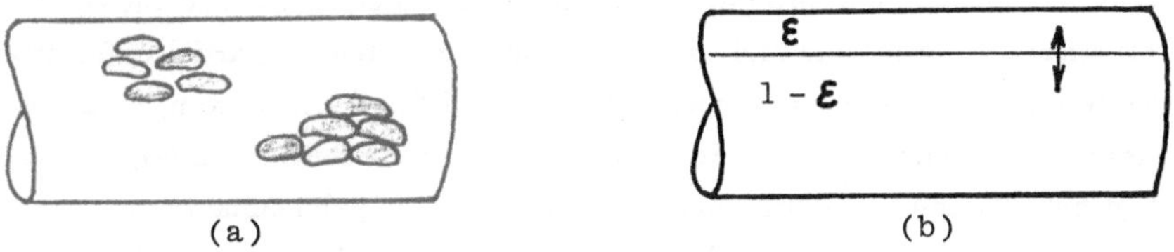

Figure 1. (a) A schematic of the slug, showing the various pathways for transport. (b) The simplified description used in deriving the equations for morphogen transport.

concentration difference between them, then the governing equations take the following form:

$$\epsilon\frac{\partial U}{\partial t} = \epsilon D_1 \nabla^2 U + H(W)(V-U)$$

$$(1-\epsilon)\frac{\partial V}{\partial t} = (1-\epsilon)\nabla\cdot(D_2(W)\nabla V) + H(W)(U-V)$$

Here U is the concentration in the interstitial fluid and V is the concentration in the cell phase. The sum of the equations yields

$$\frac{\partial}{\partial t}(\epsilon U + (1-\epsilon)V) = \nabla\cdot(\epsilon D_1 \nabla U + (1-\epsilon)D_2(W)\nabla V),$$

and if U $\equiv$ V one is led to the model of transport studied in [11] and applied to the slug in [13]. In general this will only be approximately true, but if ϵ is small and/or the diffusion coefficient in the interstitial fluid is small, transport via the connected pathway through the interstitial fluid will be negligible. In this case the single-phase description used in [11] will be a good approximation for the slug.

Since the length-to-diameter ratio of the slug is about five, we treat the slug as a one-dimensional system of length L and assume that the boundaries are impermeable to the two morphogens. The diffusion coefficients of the morphogens are assumed to depend upon the concentration of the control species [13]. We assume that the chemotactic flux of the ith species has the form

$$J_i^c = x_i \rho M_i(x_1, x_2, x_3) F_i$$

$$= x_i \rho \{\textstyle\sum_{k\neq i}(M_{ik} - M_{ki})x_k\}\nabla U$$

The x_i, i = 1,2,3, represent the fractions of anterior, posterior, and anterior-like cells, respectively, ρ is the total cell density, and U is the concentration of chemoattractant. M_{ik} is a measure of the ability of cells of type i to displace those of type k by virtue of their greater response to the gradient of the chemoattractant. The fluxes must satisfy the condition $\sum_{i=1}^{3} J_i^c = 0$. This condition stems from the assumption that the total cell density is constant in the slug. The more commonly-used equations for chemotaxis do not satisfy this condition and thus are not applicable to the slug.

When these facts are incorporated into the governing equations and the equations are cast into dimensionless form, one arrives at the following system of partial differential equations for the evolution of the chemical species and the cell fractions. Here u, v, and w denote the dimensionless concentration of the two morphogens and the control species, in that order, δ_1, δ_2 and δ_3 denote the respective diffusion coefficients, and the polynomial terms in the morphogen equations arise from a postive-feedback kinetic model [1]. The dimensionless concentration of the inhibitor of chemotaxis is denoted p, and s_{ik} represents the transition rate from cell type i to cell type k.

$$\begin{aligned}
\frac{\partial u}{\partial \tau} &= \delta_1 \frac{\partial}{\partial \varsigma}[w \frac{\partial u}{\partial \varsigma}] + k_0 - uv^2 - k_1 u \\
\frac{\partial v}{\partial \tau} &= \delta_2 \frac{\partial}{\partial \varsigma}[w \frac{\partial v}{\partial \varsigma}] + uv^2 + k_1 u - v \\
\frac{\partial w}{\partial \tau} &= \delta_3 \frac{\partial^2 w}{\partial \varsigma^2} + r_3 \\
\frac{\partial x_i}{\partial \tau} &= \frac{\partial^2 x_i}{\partial \varsigma^2} - \frac{\partial}{\partial \varsigma}[x_i \sum_{k \neq i}(m_{ik} - m_{ki}) x_k \frac{\partial u}{\partial \varsigma}] \\
&+ \sum_{k \neq i}(s_{ki} x_k - s_{ik} x_i) \qquad i = 1,2,3 \\
\frac{\partial p}{\partial \tau} &= \delta_p \frac{\partial}{\partial \varsigma}[w \frac{\partial p}{\partial \varsigma}] + e x_1 - f x_2
\end{aligned} \tag{1}$$

The boundary conditions used are as follows.

$$\begin{aligned}
\varsigma = 0: &\quad w\frac{\partial u}{\partial \varsigma} = w\frac{\partial v}{\partial \varsigma} = \frac{\partial p}{\partial \varsigma} = 0, \quad \delta_3 \frac{\partial w}{\partial \varsigma} = \theta w \\
\varsigma = 1: &\quad w\frac{\partial u}{\partial \varsigma} = w\frac{\partial v}{\partial \varsigma} = 0, \quad \frac{\partial w}{\partial \varsigma} = \frac{\partial p}{\partial \varsigma} = 0
\end{aligned} \tag{2}$$

The reasons for these choices are discussed in [13].

3. Numerical Results

Some numerical solutions for this model are summarized in Figure 2, which shows how the pattern of cell types and their proportions are reestablished in the posterior half of a surgically-transected slug as a function of time. Parameter values and additional details concerning the numerical procedure are given in [14]. Figure 2(A) shows the steady-state distributions of anterior, posterior, and anterior-like cells, and Figure 2(a) displays the steady-state distributions

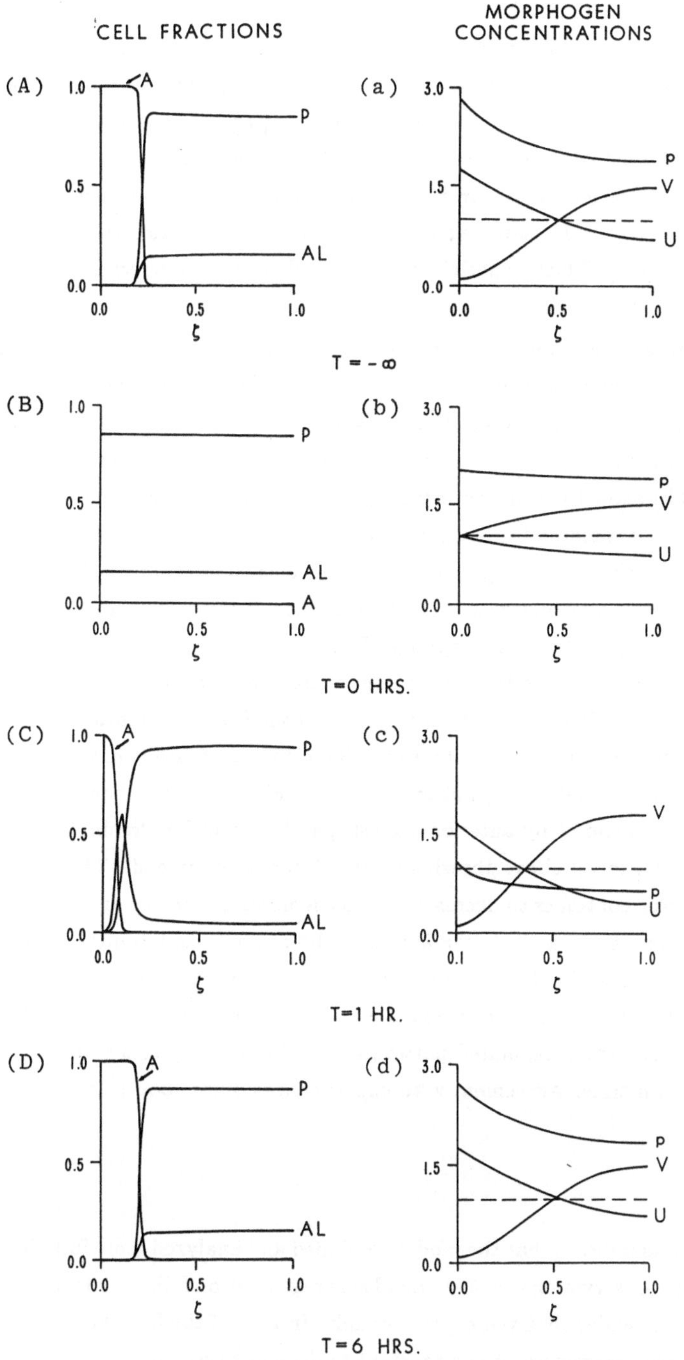

Figure 2. Left column: Fractions of anterior (A), posterior (P), and anterior-like (AL) cells. Right column: The concentration of U, V and P. $\varsigma = 0$ is the anterior end, $\varsigma = 1$ the posterior end. (A) and (a): Steady-state distributions; (B) and (b): Distributions immediately following cut; (C) and (c): One hour after the cut; (D) and (d): Six hours after the cut.

of the morphogens and the inhibitor of chemotaxis. The proportions and spatial distribution of the three cell types are maintained by several processes. The morphogen U serves as the chemoattractant species. All cells are assumed to be chemotactic in the absence of the diffusable inhibitor of chemotaxis produced by the anterior cells [15] and degraded by the posterior cells, but anterior cells can displace posterior and anterior-like cells. The chemotactic properties of anterior cells are unchanged in the presence of the inhibitor, but we assume that there is a threshold value of P ($p_T = 1$) above which anterior-like and posterior cells do not respond to the chemoattractant [15]. Thus a spatial pattern of cell types will emerge, even in the absence on interconversion between cells, as a result of cell sorting.

Cell proportions are maintained by two factors. It is asssumed that when the concentration of the morphogen is above a fixed threshold ($v_T = 0.35$), anterior cells differentiate into prespore cells, and that when $v < v_T$ anterior-like and posterior cells differentiate into prestalk cells. In addition, there is a random, position-independent interconversion of anterior-like and posterior cells. Figures 2(B,b) show the cell and morphogen distributions in the rear portion of the slug (rescaled) following a simulated transection experiment. In the absence of anterior cells, posterior cells degrade the chemotactic species and its concentration rapidly falls to a subthreshold value. Anterior-like cells now migrgate to the anterior end, preferentially displace posterior cells, in the reestablishing gradient of the morphogen U. Although the morphogen V is initially above the threshold v_T along the entire transected half, a subthreshold value soon appears at the anterior end, and anterior-like cells which have moved to the front differentiate into anterior cells and begin to secrete the chemotaxis inhibitor. These processes are summarized in in Figures 2(C,c), which shows the distributions after 1 hour. In Figure 2(C), a band of anterior cells is visible, followed by a band of by anterior-like cells, and then by predominately posterior cells. The chemotaxis inhibitor is above threshold only at the anterior end. Thus the anterior-like population in the rear continues to decrease due to chemotaxis, and the anterior zone grows due to cellular conversion. Eventually (after 2 hours), the chemotaxis inhibitor is above threshold everywhere. The increase in the anterior zone due to chemotaxis of anterior-like stops, and proportions consistent with the distribution of V are reestablished. Figures 2(D,d) demonstrate the reestablished, proportion-regulated pattern after 6 hours. Figure 3 shows the evolution of the cell fractions with time. Agreement with experiment is quite good (see [15], Fig. 2).

4. Discussion

The preceding results indicate that the model developed and analyzed more fully in [14] is able to reproduce the major observations on slime mold slug regeneration following transection, and that it does so on the time scales observed experimentally. In view of the fact that the model contains numerous adjustable parameters, it is important to determine to what extent the experimental observations place constraints on these parameters. The major observable processes are cell-cell interconversion and chemotaxis, and the time scales on which these occur are known. This sets some fairly tight constraints on the parameters involved in these processes. For instance, we are

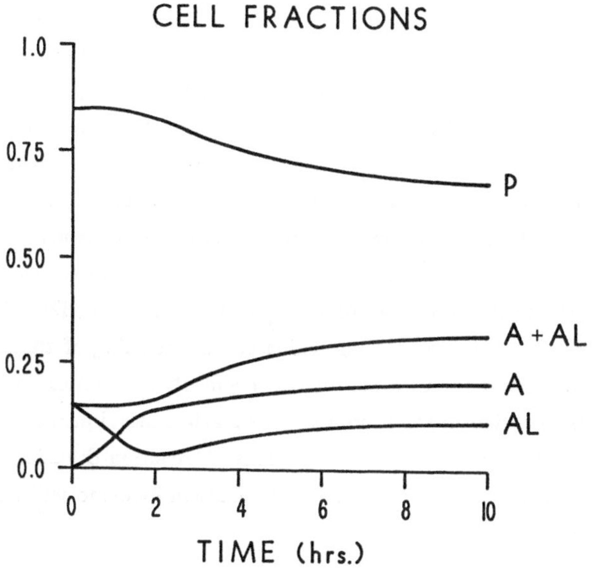

Figure 3. Numerically-computed fractions of anterior (A), posterior (P), anterior-like (AL) cells, and the sum of anterior and anterior-like (A + AL) cells as a function of the time elapsed since the cut.

able to show analytically that the transition rates between cell types cannot be constant along the slug, simply by arguments based on the observed time scales [14]. Moreover, this conclusion depends only on the assumption that the transitions are first order, and is independent of the details of the other processes involved. Similarly, the observed times for cell sorting provide the orders of magnitude of the differences of the chemotaxis coefficients. Parameters such as the diffusion coefficients can be more variable, but the values we use here and in [14] are within the observed range of diffusion coefficents of small molecules in solution. In particular, we are careful to ensure that the apparent diffusion coefficients of the morphogens do not exceed such bounds, even at the highest levels of the control species.

We have suggested a plausible mechanism by which the modulation of the diffusion coefficients can arise, but as we indicated earlier, the pattern regulation mechanism is unknown at present. Therefore our results concerning this aspect of the problem only demonstrate that our proposed mechanism is feasible. However, one can say that an alternate mechanism for producing scale-invariance, one in which the coefficients in the reaction mechanism for the morphogens scale inversely with L^2, is less plausible. This mechanism would require that whenever the characteristic length of the slug increases ten-fold the characteristic times in development increase one hundred-fold. To our knowledge, this has not been observed for the slug.

There are numerous ways in which the basic model could be altered without altering the predictions in an essential way. Firstly, we used the morphogen U as the chemotactic species, but this is purely for reasons of simplicity. It is easy to devise alternate schemes for creating a cAMP gradient along the length of the slug. For instance, the tip could function as a source of cAMP, therby serving as a classical organizer. This would require inclusion of another partial differential equation in the model, and this is not warranted at present. Secondly, we have assumed, following [15], that anterior-like cells constitute a distinct class of cells that arise from posterior cells. Instead, one could assume that anterior-like cells are simply anterior cells that have somehow been displaced from the anterior zone and have subsequently lost their ability to chemotact in the presence of the inhibitor[6]. Were anterior-like cells to arise in this way, it would not affect our conclusions concerning reproportioning of the slug. Finally, the basic conclusions do not depend on our choice of positive feedback kinetics for the morphogens. These kinetics are a simple caricuture that embody the mathematical properties necessary to produce the spatially-nonuniform morphogen distributions. If reaction-diffusion mechanisms are in fact the origin of the spatial pattern, the kinetic mechanism involved will undoubtedly prove to be much more complicated.

5. Bibliography

[1] M. Ashkenazi and H. G. Othmer, J. Math. Biol. **5** (1978), 305.

[2] M. V. L. Bennett, D. C. Spray, and A. L. Harris, in Perspectives in Differentiation and Hypertrophy, (W. Anderson and W. Sadler, eds.), Elsevier, New York (1982).

[3] S. Caveny, Ann. Rev. Physiol. **47** (1985), 319.

[4] J. Gregg and R. Davis, Differentiation **21** (1982), 200.

[5] L. G. Harrison, in Developmental Order: Its Origin and Regulation, (S. Subtelny and P. B. Green, eds.), New York: A. R. Liss (1982), 3.

[6] W. F. Loomis, S. A. Wheeler, W. R. Springer, and S. H. Barondes, Dev. Biol., **109** (1985), 111.

[7] H. MacWilliams and J. Bonner, Differentiation, **14** (1979), 1.

[8] S. Matsukuma and A. J. Durston, J. Embry. Exp. Morph., **50** (1979), 243.

[9] E. H. Mercer and B. M. Shaffer, J. Biophys. Biochem. Cytol., **7** (1960), 353.

[10] H. G. Othmer, in Lectures on Mathematics in the Life Sciences, **9** (1977), 57.

[11] H. G. Othmer and E. Pate, Proc. Nat. Acad. Sci., **77** (1980), 4180.

[12] H. G. Othmer, J. Math. Biol., **17** (1984), 351.

[13] E. Pate and H. G. Othmer, Differentiation **28** (1984), 1.

[14] E. Pate and H. G. Othmer, J. Theor. Biol. (to appear) (1986).

[15] J. Sternfeld and C. N. David, Differentiation **20** (1981), 10.

[16] J. Sternfeld and C. N. David, Devel. Biol. **93** (1982), 111.

[17] Y. Sakai and I. Takeuchi, Develop. Growth Differ. **13** (1971), 231.

[18] D. C. Spray, A. L. Harris and M. V. L. Bennett, in Cellular Communication During Ocular Development, (J. B. Sheffield and S. Hilfer, eds.), Springer-Verlag, New York (1982).

[19] I. Takeuchi, M. Tasaka, M. Oyama, A. Yamamoto, and A. Amagai, in Embryonic Development, Part B, Cellular Aspects, (M. M. Burger and R. Weber, eds.), New York: A. R. Liss (1982), 283.

[20] M. Tasaka and I.Takeuchi, J. Embry. Exp. Morph. **49** (1979), 89.

[21] A. M. Turing, Phil. Trans. Roy. Soc. Lond. **B237** (1952), 37.

[22] L. Wolpert, J. Theor. Biol. **25** (1969), 1.

A Density Dependent Model for Prestalk/Prespore Pattern Formation in Dictyostelium discoideum
I. Basic Mathematical Framework

Youichi Kobuchi

Department of Biophysics, Faculty of Science,
Kyoto University, Kyoto, 606 JAPAN

1. Introduction

Pattern formation is one of the central issues in the study of developmental biology. Although its implications are diverse, it is related mainly to cell differentiation and morphogenesis. In this respect, cellular slime molds seem to be quite handy as a model for the aggregation process, various typical morphogenetic changes, and the formation and regulation of slug pattern, to name a few (Bonner, 1967; Loomis, 1982). This note treats the slug stage of the cellular slime mold Dictyostelium discoideum and proposes a simple mathematical model to explain the prestalk/prespore pattern formation through density dependent cell movement and differentiation. The model will have good regulation capability and generate size independent typical slug patterns.

A D. discoideum slug has a patterned configuration : about 20-25% prestalk cells (T cells, for short) are in the anterior part, and in the remaining posterior prespore cell (P cell) region, there are some 10 % randomly-distributed anterior-like cells (A cells) (Loomis, 1982; Sternfeld and David, 1981). T and A cells have virtually identical staining and microscopic characteristics. The above mentioned slug pattern is well regulated and is restored without cell proliferation if part of the slug is excised. Both the proportions of the three cell types and their pattern are thus stable in that they are largely independent of slug size (i.e., the total number of cells) (MacWilliams and Bonner, 1979).

Regulation of proportions, however, also seems to occur within clumps formed in a shaken liquid culture of cells from a dissociated slug or an earlier aggregation stage. (Oyama, Okamoto, and Takeuchi,

1983) In this case, none of the patterns typically seen in the slugs are observed, which suggests that the above-mentioned T/P(A) pattern formation is not necessarily a prerequisite for the regulation of proportion. If this is the case, the basic proportion regulation mechanism should also work when T, A, and P cells are randomly distributed in an aggregate.

On the other hand, increasing attention has been paid to the roles of A cells in pattern formation and regulation processes. It is known that A cells migrate to the anterior end to form a new prestalk region when the anterior part of an intact slug is cut off (Sternfeld and David, 1982). Kakutani and Takeuchi(1986) analysed the dynamical behavior of each A cell in a whole slug and found some striking facts: contrary to customary understanding, an A cell under observation seems to move randomly relative to slug migration, and it sometimes enters the prestalk region, which suggests that T and A cells are in constant exchange.

The model presented in this note thus hypothesizes the following to explain the slime mold pattern formation and regulation:

1) A and T cells are interchangeable depending on their location. That is, we assume that the A cell is but another name for the T cell when its density is low.

2) T and A cells move around randomly (relative to slug migration) in the population of P cells with density dependent deviation.

3) Any changes of cell types are density dependent.

2. Density dependent cell movement

Consider a one-dimensional compartmental system with n compartments, where n is an arbitrary integer. The compartments, numbered 1 through n from left to right, represent the corresponding portions of a slug. The i-th compartment is denoted as c_i and we assume (without loss of generality) that c_1 is the anterior end portion of the slug and c_n the posterior end portion.

Since we have assumed that T and A cells differ only in their modes of existence, we hereafter refer to them as "T" cells and consider the density of "T" cells in each compartment. We assume that all compartments have the same fixed cell capacity and that each compartment is filled with "T" and P cells. We therefore treat a slug as a long cylinder divided into n equal parts, as is shown in Fig.1.

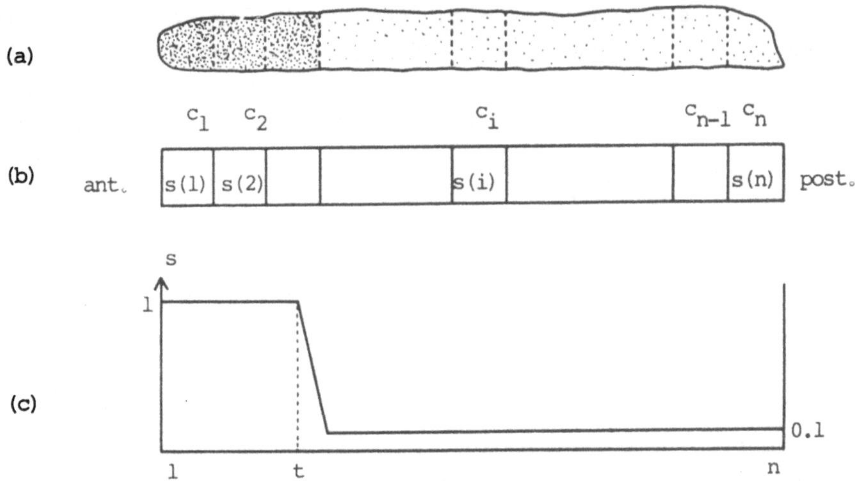

Fig. 1. Slug as a compartmental system

(a) A slug (b) Corresponding compartmental system (c) "T" cell density pattern for a typical slug For the symbols, see the text.

Let s(i) denote the density of "T" cells in c_i over the cell capacity. By definition, the density of P cells equals 1 - s(i). Let t be an integer close to 0.2n. Then for a typical D. discoideum slug, s(i)=1 for $1 \leq i \leq t$ (i.e., when c_i is in the prestalk region) and s(i) = 0.1 for $t < i \leq n$, which corresponds to the existence of A cells in the prespore region. Thus we can specify any T(A), P slug pattern (to a certain degree, since we assume random distribution of the cells within each compartment) by giving the value s(i) for every i (i = 1,2,...,n) in our compartmental system framework. Because we have assumed a fixed cell capacity for every compartment, larger n means a longer slug and vice versa.

Now we give a simple cell movement rule to explain the formation and maintenance of a typical slug pattern.

Cell movement rule

Let α, δ, and γ be the parameters such that $0 < \alpha < 1 - \gamma/\delta$ and $0 < \gamma < \delta < 1/2$.

i) If $s(i) \leq \alpha$, then "T" cells in c_i go to the neighboring

c_{i-1} and c_{i+1} with equal probability δ.

ii) If $s(i) > \alpha$, then "T" cells in c_i go to the anterior c_{i-1} with probability $\delta + \gamma$ and to the posterior c_{i+1} with probability $\delta - \gamma$.

We note that, because of the limited capacity of the compartments, a "T" cell can move into a new compartment if and only if a P cell moves out of the compartment, and thus it cannot move into a compartment c_i where $s(i)=1$. If we write the average value of $s(i)$ at the next time step as $s'(i)$, then we have the following relation:

$$s'(i)=(\delta-\Gamma(i-1))s(i-1) + (1-2\delta)s(i) + (\delta+\Gamma(i+1))s(i+1)$$

where

$$\Gamma(j) = \begin{cases} \gamma & \text{if } s(j) > \alpha \\ 0 & \text{otherwise} \end{cases}$$

with Neumann boundary condition $s(0) = s(1)$, $s(n+1) = s(n)$, $\Gamma(0) = -\Gamma(1)$, and $\Gamma(n+1) = -\Gamma(n)$.

The boundary condition simply states that there are no cells going outside of the system. Since we are not considering cell type conversion at this stage, this means that $\sum_{i=1}^{n} s(i)$ is kept constant all the time. If we are given the set of values $s(i)$ as "T" cell density in c_i $(i=1,2,...,n)$, then we can calculate the values $s'(i)$ $(i=1,2,...,n)$ by the above formula. This means that a slug pattern is changed to another after a unit time lapse by the proposed cell movement. Note, in passing, that if $s(i-1)$, $s(i)$, and $s(i+1)$ are all less than or equal to 1, then so is $s'(i)$ for $i=2,3,...,n$ because of the relationship assumed to hold among the parameter values. By the same formula, however, it is possible that $s'(1) > 1$, in which case $s'(1)$ should be 1 because of the capacity restriction, and the rest (i.e., $s'(i) - 1$) should be allocated to c_2 and to still more distant c_i's if necessary.

Let $s^*(i)$ $(i=1,2,...,n)$ denote the steady values of the system governed by the above mentioned transition rule.

First, assume that $s^*(1)=s^*(2)=...=s^*(t)=1$ and $\alpha < s^*(t+1) < 1$ for some nonnegative integer t. Let $M = s^*(t+1)$ and $R = (\delta-\gamma)/(\delta+\gamma)$. Then we have the following relationship, assuming that $R^{p-1}M > \alpha$ where p is the largest such integer:

$$s^*(t+1) = M > R$$

$$s^*(t+2) = RM$$

$$\vdots$$

$$s^*(t+p)=R^{p-1}M$$

Let $L = s^*(t+p+1) < \alpha$. Then we have

$$s^*(t+p+1) = L = R^{p-1}(1-\gamma/\delta)M,$$
$$s^*(t+p+2) = L,$$
$$\vdots$$
$$s^*(n) = L.$$

If $s^*(1) \leq \alpha$, then $s^*(i) = s^*(1)$ for $i=1,2,\ldots,n$.

Thus, if there are enough "T" cells such that the average "T" cell ratio $a = (1/n)\sum_{i=1}^{n} s(i)$ is greater than α, then there exist t and p such that:

$$t + \sum_{i=1}^{p} R^{i-1}M + (n-t-p)L = an$$

where

$$1 > M \geq R = (\delta-\gamma)/(\delta+\gamma)$$
and
$$R^{p-1}M > \alpha > L = R^{p-1}(1-\gamma/\delta)M.$$

Thus

$$\frac{t}{n} = \frac{a-L}{1-L} - \frac{\frac{\delta+\gamma}{2\gamma}M - (p+\frac{\delta}{2\gamma})L}{(1-L)n},$$

and the proportion t/n tends to (a-L)/(1-L) as n becomes large.

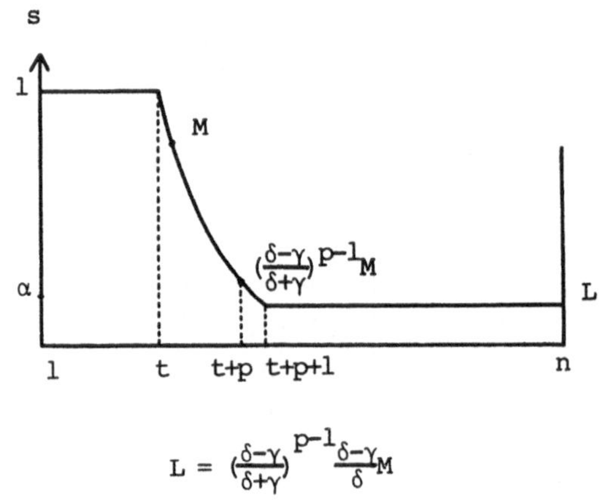

Fig. 2. An equilibrium pattern defined by the cell movement rule

We have shown what amounts to the following: a simple density dependent cell movement rule generates a stable slug-like T/P(A)

pattern whose proportion is almost independent of slug size.

Our premise so far has been that no change in cell type occurs during the pattern formation process, which might be unrealistic for some cases. We turn to this cell type conversion problem in the following section.

3. Density dependent cell type change

How a cell determines or changes its type and what information it uses in doing so constitute one of the central incompletely resolved questions in biology. In the case of cellular slime molds there has been, and still is, a controversy as to which of two mechanisms is operative: position-dependent differentiation as opposed to sorting-out after position-independent differentiation. (See, for example, Morrissey, 1982 for review.) Many recent experiments, however, seem to favor the latter. That is, cells differentiate randomly in the aggregate before forming a typical slug pattern (Tasaka and Takeuchi, 1981). Supporting evidence is provided by the observation that, as has been noted in section 1, the proportion regulation mechanism is also working when "T" and P cells are randomly distributed in an aggregate (Oyama et al., 1983). If we assume that the basic principle determining cell type is the same for these processes, it would be natural to first analyse the simplest ideal case as follows.

Let there be an aggregate composed of T and P cells and let s denote the proportion of T cells. We assume that T and P cells are so randomly distributed that each cell is able to "guess" the value of s by "looking around" its neighborhood. If θ is the target value of the variable s, then the simplest mechanism for each cell to attain or keep s value at θ would be as follows.

<u>Simple cell type change rule</u>

i) If $s > \theta$, then T cells change to P cells with rate c_{TP}.

ii) If $s < \theta$, then P cells change to T cells with rate c_{PT}.

By this rule, the T cell proportion s changes to s' at the next time step where

$$s' = \begin{cases} (1-c_{TP})s & \text{if } s > \theta, \\ s & \text{if } s = \theta, \\ s+c_{PT}(1-s) & \text{if } s < \theta. \end{cases}$$

In general, the rates of cell type change c_{TP} and c_{PT} might be functions of, say, $(s-\theta)$. We assume that they are constants and are reasonably small for the sake of simplicity. Then, if $s > \theta$, the proportion decreases exponentially until it reaches θ and, conversely, if $s < \theta$, the proportion increases exponentially to θ. The s value does not change when $s = \theta$. These patterns of change do not seem to be very unrealistic in light of experimental observations when there are no T/P patterns in an aggregate (Oyama et al., 1983).

What if we apply this simple rule for change in cell type to the case in which the typical T/P(A) slug pattern can arise ? That is, we combine the cell movement rule described in section 2 with this rule for change in cell type to see what will happen. In so doing, we assume that any changes in the state of a cell depend on the s value (i.e., the proportion of "T" cells) of the compartment to which the cell belongs. Thus each cell in a model slug changes its location and/or its state according to the local "T" cell density. It is important to note the relative rate differences of the events which these two independent rules represent. A biologically plausible assumption is that any changes in cell type occur far more slowly than cell movement, which may well be caused by random cell motion, differential chemotaxis, differential cell adhesion, etc. (cf. Sekimura and Kobuchi, 1986) In terms of the rule parameters, this amounts to saying that c_{TP} and c_{PT} are much smaller than δ and $\delta \pm \gamma$. Then, roughly speaking, a typical stable T/P(A) pattern calculated in the previous section can be maintained if the following relation holds:

$$c_{TP}(t/n + p/2n) = c_{PT}(1-L)(1-t/n-p/2n).$$

This is obtained by approximating the stable pattern with a step function which takes the value 1 for $1 \leq i \leq t + p/2$ and L otherwise. The approximation in turn can be condoned by supposing that the stable pattern takes a value close to θ at location i where i is the closest integer to t+p/2.

We can interpret, conversely, the above relation as it specifies the ratio of c_{TP} and c_{PT} in order to maintain a typical slug pattern characterized by the numbers t/n, p/n, and L.

Thus, for a suitable parameter set, we can regulate the proportion of T(A) and P cells both for patterned and non-patterned cell aggregates. In particular, the typical slug pattern can be restored if we start simulation from an anterior T cell segment, a posterior P(A) cell segment, or any combination of portions, provided the anterior-posterior axis is maintained. (Some of these simulations are shown in Fig. 3.)

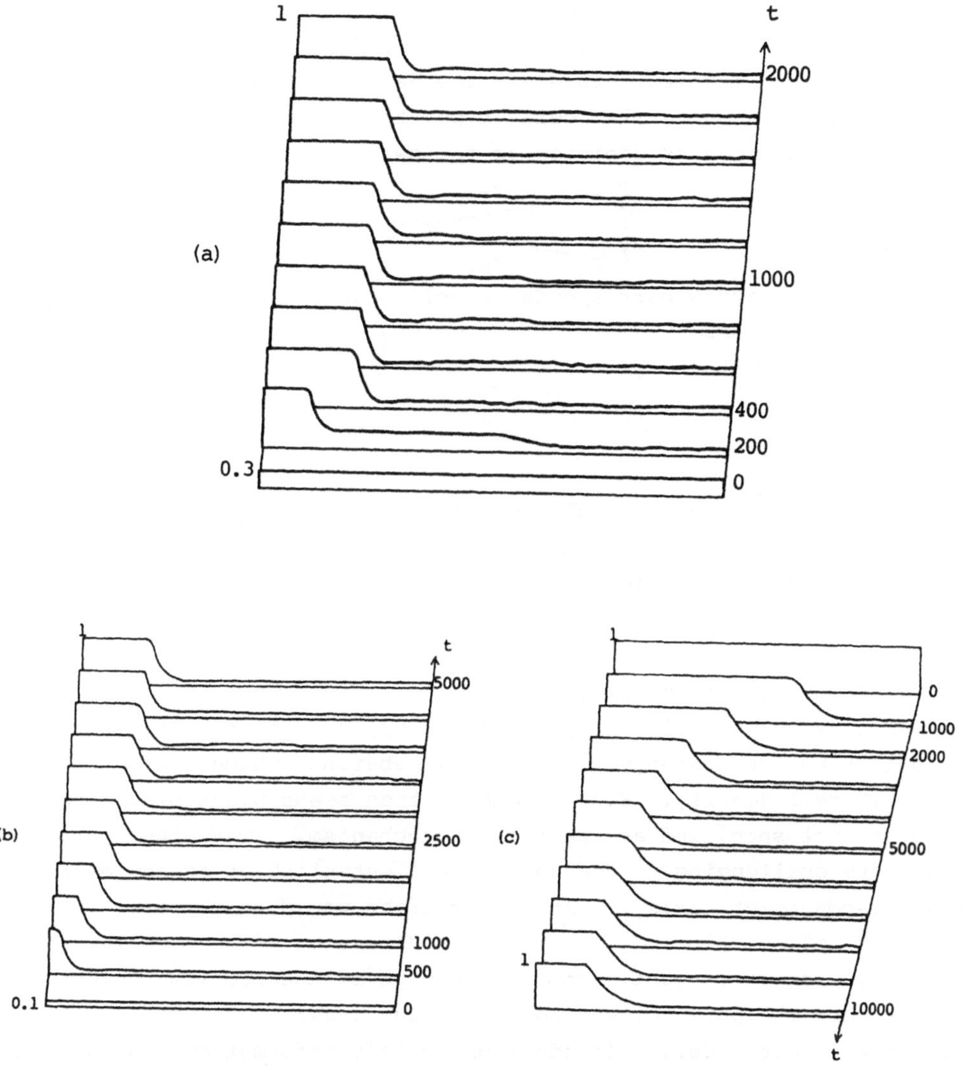

Fig. 3. Simulation of a basic density dependent model
Cell movement parameters: $\alpha = 0.12$, $\delta = 0.3$, $\gamma = 0.075$.
Cell type change parameters: $\theta = 0.3$, $c_{TP} = 0.0004$, $c_{PT} = 0.00014$.

(a) Pattern formation from 30% randomly distributed "T" cells (n = 100)

(b) Pattern regulation from prespore isolates (with 10% anterior like cells) (n = 70)

(c) Pattern regulation from prestalk isolates (n = 30)

It seems that basic pattern formation and regulation in a D. discoideum slug are thus accomplished by simply combining the density dependent cell movement rule and the density dependent cell type change rule. An overall cell type transformation of the kind described so far may be depicted as follows:

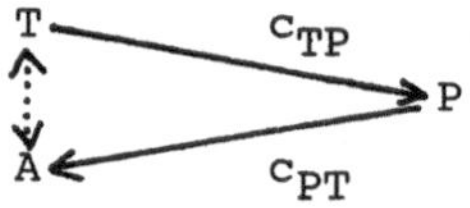

where a dotted arrow corresponds to cell movement and a solid one to change in cell type.

4. Discussion

We first assumed a simple cell movement rule by which "T" cells basically move randomly (relative to slug migration) subject to the condition that more "T" cells move to an anterior compartment than to a posterior one when the "T" cell density is high. Two points are crucial here : how to know the "T" cell density and how to establish the anterior-posterior axis. A "T" cell could sense the local "T" cell density by directly recognizing the types of surrounding cells or by detecting the concentration of some substance which reflects the local "T" cell density. Such an explanation seems plausible, although we have not specified any particular mechanism. The other problem looks more challenging. How does a cell establish or acknowledge its anterior-posterior axis ? An easy retreat would be to rely on the existence of a tip and the migration direction of the whole slug. This is, however, just a cosmetic change cf the problem unless we explain how a tip is formed and how a slug crawls -- problems we did not treat in our model. In addition to this information, the driving force of deviated random T or A cell movement can be accounted for by differential chemotaxis and cell adhesion, as has been done elsewhere (Sekimura and Kobuchi, 1986).

In our model, cell movement (random or deviated) plays an essential part in both pattern formation and proportion regulation. According to our density-dependent differential cell movement and cell type conversion hypotheses, each cell's behavior is determined solely by the local density of other cells of the same type. The almost stable slug pattern we obtain in the end is in dynamic equilibrium. Thus we have presented some experimentally testable hypotheses which, as a first approximation, successfully explain the formation and regulation of slime mold slug pattern. Few of the characteristic features of our model are shared with hitherto proposed models. (See MacWilliams and Bonner, 1979 and Morrissey, 1982 for review.)

There are, however, some problems in this seemingly plausible scenario. For example, cell type conversions between "T" and P are incessantly occurring everywhere even after the slug pattern is established. Although the actual cell type conversion processes in this case are not known well enough as to settle the point, it is possible that our rule is too simplistic. We can elaborate the cell type conversion scheme a little to accord with more subtle observation, and this will be reported elsewhere.

Acknowledgement

The author would like to thank Profs. Teramoto, Takeuchi, and members of their laboratories for their interest and encouragement.

References

Bonner, J.T. (1967) The Cellular Slime Molds. (2nd edition) Princeton University Press.

Kakutani, T. and Takeuchi, I. (1986) Dev. Biol. 115, 439-445.

Loomis, W.F. (1982) (ed.) The Development of Dictyostelium discoideum. Academic Press.

MacWilliams, H.K. and Bonner, J.T. (1979) Differentiation 14, 1-22.

Morrissey, J.H. (1982) in The Development of Dictyostelium discoideum. (ed. Loomis, W.F.) Academic Press. 411-449.

Oyama, M., Okamoto, K., and Takeuchi, I. (1983) J. Embryol. exp. Morph. 75, 293-301.

Sekimura, T. and Kobuchi, Y. (1986) J. theor. Biol. 122, 325-338.

Sternfeld, J. and David, C.N. (1981) Differentiation 20, 10-21.

Sternfeld, J. and David, C.N. (1982) Dev. Biol. 93, 111-118.

Tasaka, M. and Takeuchi, I. (1981) Differentiation 18, 191-196.

ORIGIN OF BURSTING AND BIRHYTHMICITY IN A MODEL FOR CYCLIC AMP OSCILLATIONS IN *DICTYOSTELIUM* CELLS

J.L. MARTIEL and A. GOLDBETER

Faculté des Sciences, Université Libre de Bruxelles
Campus Plaine, C.P. 231, B-1050 Brussels, Belgium

1. Introduction.

The periodic generation of cyclic AMP (cAMP) pulses during aggregation of the slime mold *Dictyostelium discoideum* is one of the best-known examples of periodic behavior at the cellular level (Gerisch and Wick, 1975; Goldbeter and Caplan, 1976). We have recently analyzed a model for this phenomenon based on receptor modification (Goldbeter, Martiel and Decroly, 1984; Martiel and Goldbeter, 1984). In the course of this analysis, we found that in addition to simple periodic oscillations, the model is capable of more complex modes of oscillatory behavior, namely, bursting and chaos (Martiel and Goldbeter, 1985) as well as birhythmicity (Goldbeter and Martiel, 1985). The latter phenomenon refers to the coexistence between two simultaneously stable periodic regimes (Decroly and Goldbeter, 1982). It is the rhythmic counterpart of the more common mode of bistability in which two stable steady states coexist in the same conditions.

Although birhythmicity has been observed in some oscillatory chemical reactions (Alamgir and Epstein, 1983), it has not yet been found experimentally in biological systems. Bursting, on the other hand, has not been observed in *D. discoideum* but is a common phenomenon in neurobiology (Johnston and Brown, 1984), as exemplified by the R15 neurone in *Aplysia* (Adams and Benson, 1985).

We analyze here the origin of bursting and birhythmicity in the model for cAMP oscillations in *Dictyostelium* cells. Bursting phenomena have been analyzed theoretically in a number of other biological systems (Chay and Keizer, 1983; Kopell and Ermentrout, 1986; Plant, 1981). Our qualitative analysis is closely related to that proposed by Rinzel et al. for the mechanism of bursting in the Belousov-Zhabotinsky reaction (Rinzel and Troy, 1981) and in models of excitable membranes (Rinzel, 1985; Rinzel and Lee, 1985).

2. Bursting and birhythmicity.

The model for cAMP oscillations is governed by three nonlinear ordinary differential equations (Goldbeter and Martiel, 1985):

$$\frac{d\rho_t}{dt} = -\rho_t\mu(\gamma) + (1-\rho_t)\eta(\gamma)$$

$$\frac{d\gamma}{dt} = \sigma^*\Phi(\rho_t,\gamma,\alpha) - k_e\gamma$$

$$\frac{d\alpha}{dt} = \varepsilon^*\{v^* - \sigma^*\Phi(\rho_t,\gamma,\alpha)\} \qquad (1)$$

with $\mu(\gamma) = \dfrac{k_1 + k_2\gamma^2}{1 + \gamma^2}$; $\eta(\gamma) = \dfrac{k_1L_1 + k_2L_2c^2\gamma^2}{1 + c^2\gamma^2}$

$$\Phi(\rho_t,\gamma,\alpha) = \frac{\alpha\,(\lambda\theta + \varepsilon\rho_t\gamma^2/(1+\gamma^2))}{1 + \alpha\theta + (1+\alpha)\varepsilon\rho_t\gamma^2/(1+\gamma^2)}$$

$$\varepsilon^* = \frac{h(k_i+k_t)}{k_tq} ; \quad \sigma^* = \frac{\sigma k_tq}{h(k_i+k_t)} ; \quad v^* = \frac{vk_tq}{h(k_i+k_t)}$$

The various parameters are defined in previous publications (Goldbeter et al., 1984; Martiel and Goldbeter, 1984) where parameter c should read $c=(K_R/K_D)^{1/2}$.

In the above equations, ρ_t is the fraction of cAMP receptor in the active state (binding of cAMP to this state of the receptor elicits cAMP synthesis through coupling with adenylate cyclase); γ is the normalized concentration of extracellular cAMP, and α is the normalized concentration of ATP, the substrate of adenylate cyclase.

Whereas eqs.(1) generally admit a single, stable periodic solution whenever the unique steady state is unstable, numerical integration of these equations shows that for appropriate parameter values the system may evolve to either one of two stable limit cycles, depending on initial conditions (Fig.1). As shown by the phase space representation (Fig.2), the larger limit cycle is of the busting type. The second cycle is characterized by smaller variations in α and γ.

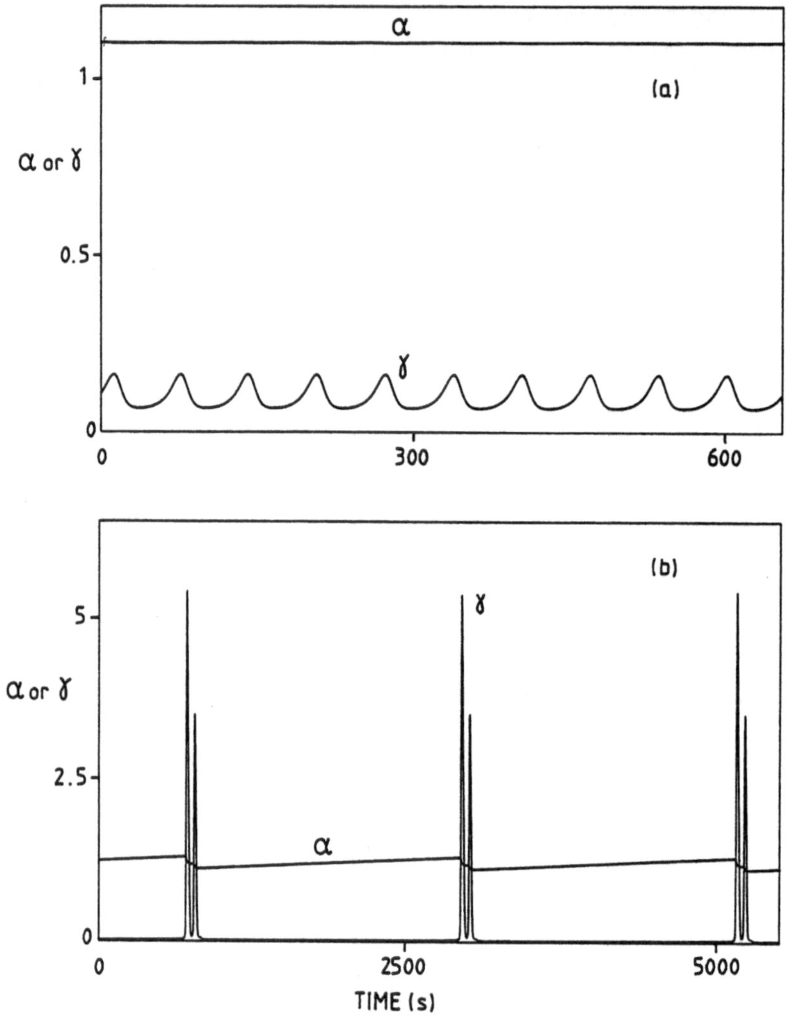

Fig.1: Birhythmicity. Two different oscillatory regimes are obtained by numerical integration of eqs.(1) for the same set of parameter values: $\lambda=\theta=0.01$, $\sigma=0.1\ s^{-1}$, $c=100$, $L_1=85.507$, $L_2=8.5507\times10^{-3}$, $k_1=0.141\ s^{-1}$, $k_2=0.0564\ s^{-1}$, $\varepsilon=0.2$, $h=5$, $q=4000$, $k_e=0.35\ s^{-1}$, $k_i=0.6\ s^{-1}$, $k_t=0.4\ s^{-1}$, $v^*=3.15\times10^{-2}s^{-1}$. Appropriate initial conditions for the two modes of oscillations are $\alpha=1.24$, $\rho_t=0.951$, $\gamma=1.84\times10^{-2}$ for (b) and $\alpha=1.095$, $\rho_t=0.308$, $\gamma=0.118$ for (a).

Bursting may also occur in the absence of birhythmicity (Martiel and Goldbeter, 1985).

The passage from the small to the large cycle can be achieved upon addition of a pulse of γ with the right magnitude at the appropriate phase; the reverse transition from the bursting regime to the small-amplitude cycle can sometimes be obtained by means of similar perturbations (Goldbeter and Martiel, 1985) but, as will be explained below, this transition cannot occur for the parameter values considered in Fig.1.

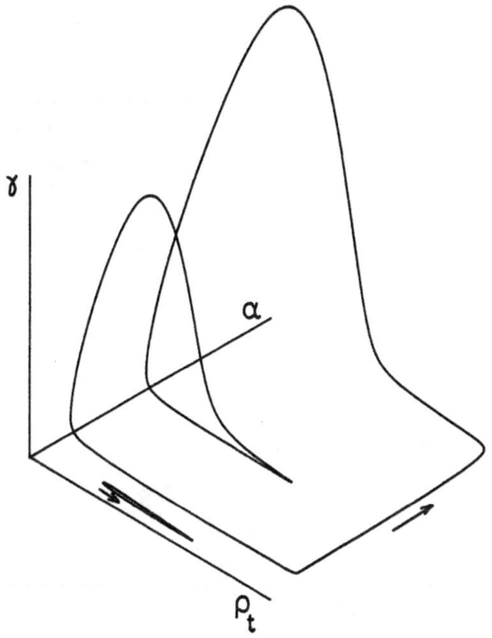

Fig.2: Phase space representation of birhythmicity. The two coexisting limit cycles correspond to the two stable modes of oscillations shown in Fig.1. The range of variation is 1-1.3 for α, 0-1 for ρ_t, and 0-4 for γ.

3. Origin of bursting and birhythmicity.

The behavior shown in Figs. 1 and 2 can be comprehended qualitatively by analyzing first a two-variable reduction of system (1) in which α is treated as a parameter. Letting α vary allows, in a second stage, to account for the behavior of the three-variable system.

3.1. Two-variable analysis.

Parameter ε^* in eqs.(1) is much smaller than unity, given that the experimental value of q is of the order of 10^3. In the limit $\varepsilon^* \to 0$, eqs.(1) reduce to the two-dimensional system:

$$\frac{d\rho_t}{dt} = -\rho_t\mu(\gamma) + (1-\rho_t)\eta(\gamma)$$

$$\frac{d\gamma}{dt} = \sigma^*\Phi(\rho_t,\gamma,\alpha) - k_e\gamma \qquad (2)$$

where α is now a parameter.

A schematic bifurcation diagram for this reduced system is shown in Fig.3. Here, the steady state value of γ in system (2), together with the mean value $\langle\gamma\rangle$ over a period of oscillation, are plotted as a function of α. This diagram shows the possibility of multiple steady states produced by saddle node bifurcations in S_1 and S_2.

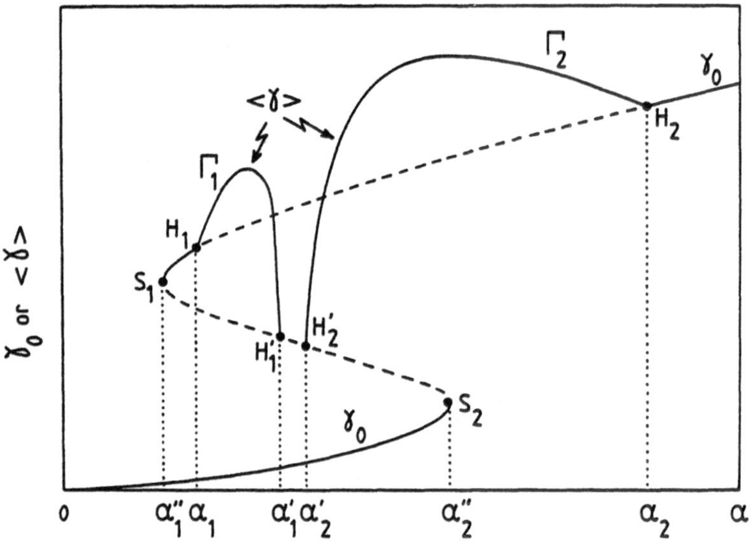

Fig.3: Bifurcation diagram showing the origin of bursting and birhythmicity. The diagram is established for the reduced two-variable system governed by eqs.(2). For the sake of clarity, data are represented in a schematic manner. Shown are the steady state concentration of extracellular cAMP, γ_0, and the mean value $\langle\gamma\rangle$ over a period in the domain of existence of the two branches of periodic solutions Γ_1 and Γ_2 (see text). For the parameter values of Fig.1, the abscissas of the critical points are $\alpha_1''=1.074$, $\alpha_1=1.084$, $\alpha_1'=1.101$, $\alpha_2'=1.126$, $\alpha_2''=1.287$, $\alpha_2=3.934$.

The branch of steady state solutions characterized by higher values of γ is unstable for $\alpha_1<\alpha<\alpha_2$. Two families of periodic solutions, denoted Γ_1 and Γ_2, emanate through Hopf bifurcations in H_1 and H_2. The two families of periodic solutions disappear, respectively, in H_1' and H_2' where eqs.(2) admit a homoclinic solution (Guckenheimer and

Holmes, 1983). The above situation occurs in the case considered in Figs. 1 and 2 but is by no means the rule as for other parameter values a continuous family of periodic solutions extends from H_1 to H_2.

As the parameter α tends to the values α_1' and α_2' the mean value $\langle\gamma\rangle$ approaches the saddle steady state in H_1' and H_2' since the system spends an infinite amount of time near this state. For $\alpha_1' < \alpha < \alpha_2'$ there is no periodic solution and the unique attractor of eqs.(2) is the stable steady state in which γ is low.

3.2. Dynamics of the complete system.

We can use the bifurcation diagram of Fig.3 to describe the behavior of the three-variable system. An important property of eqs.(1) is that they admit a unique steady-state solution which for γ takes the simple form $\gamma_0 = v^*/k_e$. Thus the steady-state values of α and γ in Fig.3 lie at the intersection of the S-shaped curve with a line parallel to the α axis; the ordinate of this line, v^*/k_e, increases with the substrate input v^*.

For small values of ε^*, the three-variable system continues to admit Hopf bifurcation points near H_1 and H_2, whereas two new Hopf bifurcations will appear in the vicinity of S_1 and S_2. The phenomenon of bistability indeed transforms into a phenomenon of oscillations once α becomes a variable (For simplicity, we shall consider in the following discussion that the four Hopf bifurcation points in the three-variable system correspond to the points H_1, H_2, S_1, S_2 in Fig.3).

As long as the steady state is located between the origin and S_2, it remains stable. Between S_2 and S_1 it is unstable. The dynamics of the system can then be understood in a manner analogous to that proposed by Rinzel (1985). Starting at the origin, the system moves slowly toward S_2 as the substrate α accumulates: the constant input v^* indeed exceeds substrate consumption through the enzyme reaction since adenylate cyclase functions at a basal rate (only at higher values of γ is the latter enzyme activated). In S_2, the reduced system (ρ_t,γ) no longer admits a stable steady state and is attracted by the stable solution that belongs to the branch of periodic solutions Γ_2: oscillations begin on this branch.

Given that the level of γ is now higher, adenylate cyclase is fully activated by extracellular cAMP so that the substrate consumption by the enzyme reaction exceeds the constant supply through v^*. The variable α therefore decreases progressively in the course of oscillations. Each peak in γ removes a roughly constant amount $\Delta\alpha$ from the substrate pool. As soon as α drops below α_2', the reduced system no longer admits any stable periodic solution and is attracted by the steady state corresponding to a lower value of γ. In this point, however, v^* once again exceeds substrate consumption so that α rises towards S_2. The three-variable system thus undergoes bursting oscillations, roughly in the range $\alpha_2' < \alpha < \alpha_2''$, as shown by the large-amplitude periodic solution in Figs. 1 and 2.

When v^* is such that the steady state of the complete system lies between S_1 and H_1, this state is stable. For appropriate initial conditions, the bursting regime nevertheless persists as the system, starting on the lower branch, cannot reach this stable steady state owing to the existence of the homoclinic orbit in H_2'. Hard excitation obtains in these conditions.

When v^* is slightly larger so that the steady state lies just to the right of H_1, a small-amplitude periodic solution belonging to the branch Γ_1 appears in the reduced system. This periodic solution, when it exists in the complete system, coexists with the bursting solution described above. Birhythmicity develops in these conditions, as illustrated in Figs. 1 and 2.

A stable periodic solution of small amplitude occurs in the complete system only if the substrate input v^* counterbalances exactly the substrate consumption over a period. As shown below, the condition of existence of such periodic solution can be obtained by perturbation analysis. Our analysis is based on the assumption that the time variation of α is much slower than that of ρ_t and γ; hence it is not valid in the vicinity of homoclinic orbits.

We expand small-amplitude periodic solutions of eqs.(1) as a function of the small parameter ε^*:

$$\begin{aligned} \rho_t &= \bar{\rho}(t,\bar{\alpha}) + \varepsilon^* \rho_1(t,\tau) + \dots \\ \gamma &= \bar{\gamma}(t,\bar{\alpha}) + \varepsilon^* \gamma_1(t,\tau) + \dots \\ \alpha &= \bar{\alpha} + \varepsilon^* \alpha_1(t,\tau) + \dots \end{aligned} \tag{3}$$

where $\tau=\varepsilon^* t$ is the slow time scale; $(\bar{\rho}(t,\bar{\alpha}), \bar{\gamma}(t,\bar{\alpha}))$ stands for a solution of eqs.(2) where $\bar{\alpha}$ is the value of the bifurcation parameter α in the reduced system (at this stage, $\bar{\alpha}$ is an unknown constant). When the expansion (3) is inserted into eqs.(1), the equations for $d\rho_t/dt$ and $d\gamma/dt$ are trivially satisfied at order zero. A necessary condition for periodicity of $\alpha(t)$ is given by (Nayfeh, 1981):

$$\int_0^{T(\bar{\alpha})} \{v^* - \sigma^* \Phi(\bar{\rho}(t,\bar{\alpha}), \bar{\gamma}(t,\bar{\alpha}), \bar{\alpha})\} \, dt = 0 \tag{4}$$

This solvability condition expresses that the mean value for $\alpha(t)$ over the period $T(\bar{\alpha})$ is $\bar{\alpha}$. Given that $\bar{\rho}_t$ and $\bar{\gamma}$ is a periodic solution of the reduced system, the integration of $d\gamma/dt$ over a period yields:

$$\sigma^* \int_0^{T(\bar{\alpha})} \Phi(\bar{\rho}(t,\bar{\alpha}),\bar{\gamma}(t,\bar{\alpha}),\bar{\alpha})\, dt \; - \; k_e \int_0^{T(\bar{\alpha})} \bar{\gamma}(t,\bar{\alpha})\, dt = 0 \qquad (5)$$

Hence, combining eqs.(4) and (5) we obtain

$$\frac{1}{T(\bar{\alpha})} \int_0^{T(\bar{\alpha})} \bar{\gamma}(t,\bar{\alpha})\, dt = \langle\gamma\rangle = \frac{v^*}{k_e} = \gamma_0 \qquad (6)$$

We see that condition (6) selects among the possible periodic solutions of eqs.(2) those corresponding to a value $\bar{\alpha}$ for which the mean value of γ over a period is just equal to the steady-state value of γ in the complete system. Such periodic solutions correspond in the three-variable system to oscillations in ρ_t and γ accompanied by a small variation in α around $\bar{\alpha}$, in the range $\alpha_1 < \alpha < \alpha_1'$ (see the small-amplitude periodic solution in Figs. 1 and 2).

Upon further increasing v^*, the amplitude of the limit cycle on the branch Γ_1 increases until it reaches a maximum. Beyond this point, the periodic solution on Γ_1 becomes unstable as it separates a stable solution of Γ_1 from the stable bursting behavior.

3.3. <u>Switching between the two oscillatory regimes</u>.

In a previous study (Goldbeter and Martiel, 1985), we showed the possibility of switching reversibly between the two simultaneously stable periodic regimes. Here, numerical simulations indicate so far that the transition occurs only in a single direction. In the conditions of Figs. 1 and 2, a pulse in γ at the appropriate phase of the small-amplitude oscillations brings the system into the bursting mode once the perturbation exceeds a threshold. This behavior can be comprehended from Fig.3: a large pulse in γ activates the cyclase to such extent that α drops below α_1'' and bursting develops. If the pulse is too small, the subsequent consumption of α fails to bring the substrate level below α_1''. Then the system remains on the upper branch, to the right of S_1; α rises up to the value $\bar{\alpha}$ satisfying eq.(6), and the small-amplitude oscillation resumes.

It is more difficult – and sometimes impossible – to achieve the passage from bursting to the small-amplitude limit cycle. Such transitions may occur in some cases just after the last burst in γ, when the magnitude of the perturbation is comprised between two critical values (see Fig.4 in Goldbeter and Martiel, 1985). The reason why this transition appears to be ruled out in the conditions of Fig.1 is made clear by the diagram of Fig.3. When the system undergoes the last burst, α is close to α_2'. An addition of γ will produce a decrease in α. There exists two values of the perturbation between which α is brought down to a level comprised between α_1 and α_1'. Then the system is trapped into the small-amplitude oscillatory regime. If the perturbation is too small or too large, α will decrease to a value above α_1' or below α_1, respectively, and bursting will resume.

Parameter values in Fig.1 are such that the net substrate consumption $\Delta\alpha$ due to the perturbation is smaller than $(\alpha_2' - \alpha_1')$: the system remains trapped in the basin of attraction of the bursting regime.

4. Discussion.

The present analysis of bursting and birhythmicity leads to a number of predictions. First, birhythmicity should disappear altogether upon disappearance of the two homoclinic orbits of the reduced system, just beyond the coalescence of H_1' and H_2' after Γ_1 and Γ_2 have merged. The disappearance of the homoclinic orbits in the three-variable system closely follows their disappearance in the reduced system. Birhythmicity may in fact occur for parameter values in the intermediary range, when the homoclinic orbits have vanished in the two-variable reduction

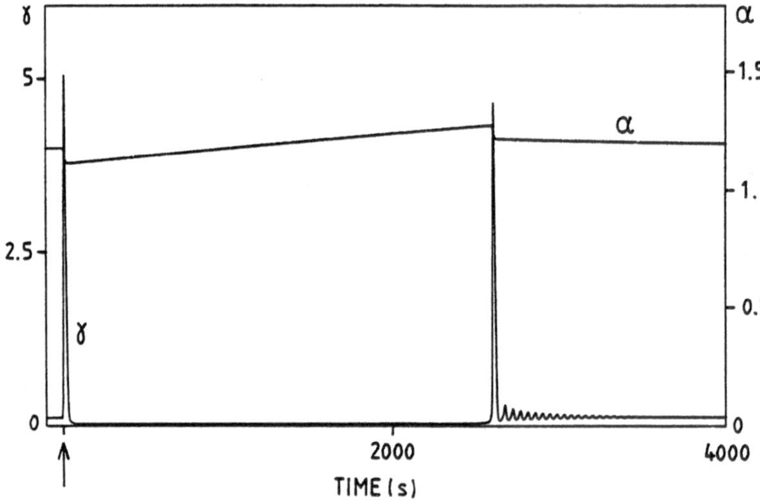

Fig.4: Complex excitability. Upon stimulation by a pulse of extracellular cAMP at time zero (arrow) the system amplifies the perturbation before returning to the stable steady state in a complex manner. Initial conditions are α=1.217, ρ_t =0.498 (steady-state values) and γ=0.15. Here L_1=63, L_2=0.0063, v^* = 0.02775 s^{-1}, k_1=0.15/s, k_2=0.06/s.

but not in the three-variable system, as exemplified in the case considered in a previous publication (Goldbeter and Martiel, 1985).

The disappearance of the homoclinic orbits also bears on the characteristics of bursting. When v^* is such that the steady state lies between S_1 and S_2, the number of peaks in the bursting phase should increase abruptly upon disappearance of the homoclinic orbits. The bursting mode in which α varies between α_2' and α_2'' would indeed give rise to a mode in which α varies in a larger range between α_1'' and α_2''.

Another prediction is that of a coexistence between two non-bursting oscillatory regimes, one belonging to Γ_1 and the other to Γ_2. So far we have not observed this situation in our numerical simulations.

That birhythmicity originates here from the breaking of an oscillatory domain into two parts by homoclinic orbits markedly bears on the possibility of switching from one periodic regime to the other upon chemical perturbation. In the present model, such transition may sometimes occur only in a single direction. This situation has to be contrasted with that encountered in a simpler two-variable model in which it is always possible to switch back and forth, at the appropriate phase, between a small-amplitude limit cycle and a second, enclosing cycle (Moran and Goldbeter, 1985).

The present analysis based on the bifurcation diagram of Fig.3 is supported by the observation of a phenomenon of complex excitability in the dynamics of the three-variable system (Fig.4). Such phenomenon occurs when the steady state lies between S_1 and H_1, in the absence of homoclinic orbits. Then the addition of a pulse of γ is followed, after the initial peak, by a phase of quiescence (corresponding to the increase in α from α_1'' to α_2'' on the lower branch of the hysteresis loop in Fig.3), a second peak in γ (due to the jump when α exceeds α_2'') and small-amplitude oscillations as the system returns to the stable steady state. The latter oscillations are damped as the value of v^* is too low to stabilize any periodic regime on the upper branch.

During the process of modeling the dynamics of cAMP signaling in Dictyostelium amoebae, we have uncovered closely intertwined scenarios for the onset of bursting and birhythmicity. Of novelty is the observation that the suppression of the homoclinic trajectories may affect both phenomena in an abrupt manner. Thus, when v^* is such that the steady state lies to the right of H_1, the disappearance of the homoclinic orbits results in the evolution to a unique non-bursting periodic regime characterized by small-amplitude variations in α. At the same time as birhythmicity vanishes, bursting disappears through a global bifurcation.

REFERENCES

Adams, W.A. and Benson, J.A. (1985) The generation and modulation of endogenous rhythmicity in the Aplysia bursting pacemaker neurone R15. Progr. Biophys. Mol. Biol. 46: 1-49.

Alamgir, M. and Epstein, I.R. (1983) Birhythmicity and compound oscillations in coupled chemical oscillators: chlorite-bromate-iodide system. J. Am. Chem. Soc. 105: 2500-2501.

Chay, T.R. and Keizer, J. (1983) Minimal model for membrane oscillations in the pancreatic β-cell. Biophys. J. 42: 181-190.

Decroly, O. and Goldbeter, A. (1982) Birhythmicity, chaos, and other patterns of temporal self-organization in a multiply regulated biochemical system. Proc. Nat. Acad. Sci. USA 79: 6917-6921.

Gerisch, G. and Wick, U. (1975) Intracellular oscillations and release of cyclic AMP from Dictyostelium cells. Biochem. Biophys. Res. Commun. 65: 364-370.

Goldbeter, A. and Caplan, S.R. (1976) Oscillatory enzymes. Ann. Rev. Biophys. Bioeng. 5: 449-476.

Goldbeter, A., Martiel, J.L. and Decroly, O. (1984) From excitability and oscillations to birhythmicity and chaos in biochemical systems. In: Dynamics of Biochemical Systems (eds. Ricard, J. and Cornish-Bowden, A.), pp 173-212. Plenum, New York.

Goldbeter, A. and Martiel, J.L. (1985) Birhythmicity in a model for the cyclic AMP signalling system of the slime mold Dictyostelium discoideum. FEBS Lett. 191: 149-153.

Guckenheimer, J. and Holmes, P. (1983) Nonlinear Oscillations, Dynamical Systems, and Bifurcations of Vector Fields. Springer, New York.

Johnston, D. and Brown, T.H. (1984) Mechanisms of neuronal burst generation. In: Electrophysiology of Epilepsy, pp 277-301. Academic Press, London.

Kopell, N. and Ermentrout, G.B. (1986) Subcellular oscillations and bursting. Math. Biosci., in press.

Martiel, J.L. and Goldbeter, A. (1984) Oscillations et relais des signaux d'AMP cyclique chez Dictyostelium discoideum: Analyse d'un modèle fondé sur la modification du récepteur pour l'AMP cyclique. C.R. Acad. Sci. (Paris) Sér. III 298: 549-552.

Martiel, J.L. and Goldbeter, A. (1985) Autonomous chaotic behaviour of the slime mould Dictyostelium discoideum predicted by a model for cyclic AMP signalling. Nature 313: 590-592.

Moran, F. and Goldbeter, A. (1984) Onset of birhythmicity in a regulated biochemical system. Biophys. Chem. 20: 149-156.

Nayfeh, A.H. (1981) Introduction to Perturbation Techniques. Wiley, New York.

Plant, R.E. (1981) Bifurcation and resonance in a model for bursting nerve cells. J. Math. Biol. 11: 15-32.

Rinzel, J. (1985). Bursting oscillations in an excitable membrane model. In: Ordinary and Partial Differential Equations (eds. Sleemann, B.D., Jarvis, R.J. and Jones, D.S.). Springer.

Rinzel, J. and Lee, Y.S. (1986) On different mechanisms for membrane potential bursting. In: Nonlinear Oscillations in Biology and Chemistry (ed. Othmer, H.G.). Springer.

Rinzel, J. and Troy, W.C. (1982) Bursting phenomena in a simplified Oregonator flow system model. J. Chem. Phys. 76: 1775-1789.

III. THEORETICAL NEUROSCIENCES AND RELATED PROBLEMS IN PHYSIOLOGY

Neurosciences

Physiology and Related Problems

MATHEMATICAL MODELLING OF MACROSCOPIC BRAIN PHENOMENA

Robert M. Miura

Departments of Mathematics and Pharmacology & Therapeutics
Institute of Applied Mathematics
University of British Columbia
Vancouver, B.C. Canada V6T 1Y4

1. INTRODUCTION

Theoretical studies of the brain have concentrated mainly on microscopic phenomena at the membrane and cellular levels. At the macroscopic level, there are "gross" brain phenomena, e.g., some kinds of epilepsy, that occur on space scales which are large compared to cell size and on time scales which are long compared to time constants associated with molecular and cellular events. Because of the large number of nerve and glial cells involved (on the order of 10^{11} (Hubel, 1979)), it is difficult to construct a mathematical model of such phenomena by starting with individual cells, connecting them into small circuits, and then combining these circuits into a coherent model for a particular brain structure.

In this paper, we review an alternative approach to modelling macroscopic brain phenomena. This approach treats the brain as a continuum and describes macroscopic events while accounting for physiological mechanisms that occur at the cellular and membrane levels. To illustrate the continuum approach, we indicate how it has been applied to a steady progressing wave phenomenon called spreading cortical depression (SD for short) (Tuckwell & Miura, 1978) which has received considerable attention experimentally (Bures et al., 1974). Also, we consider a two-dimensional rotating wave phenomenon associated with SD waves (Shibata & Bures, 1972, 1974) and present some new results on an analog model for an excitable medium.

2. MODELLING THE BRAIN AS A CONTINUUM

In studying phenomena such as SD, we confine the modelling to the cortex of various brain structures. The cortex, or grey matter, is a thin layer of tissue at the surface (analogous to the rind of an orange but not as distinct) wherein the neuronal cell bodies and dendritic branches lie. Below this cortex is the white matter consisting of axons covered with myelin sheaths made up of Schwann cells. In viewing the brain as a continuum, the underlying assumption is that the length scale of the phenomenon of interest (for SD, on the order of millimeters) is long compared to a typical length scale for neurons (on the order of 50 - 100 microns for cell bodies). This continuum approach sets up two or more overlapping spaces, corresponding to extracellular space and intracellular spaces of various cells (neurons and glial cells) or subunits of cells (cell bodies, dendrites, and presynaptic terminals), within and between which ions can move. Ionic concentrations are the basic dependent variables and nonlinear diffusion equations are derived for their variations in time and space. The values of the extracellular and intracellular ionic concentrations at a point are to correspond with the extracellular and intracellular concentrations nearest that point.

For simplicity, we consider only one space dimension, namely the direction parallel to the surface of the cortex and only two compartments for ions, namely, extracellular and intracellular spaces (for SD this intracellular space consists of the insides of the presynaptic terminals located at the ends of axons; note that between 1000 and 10,000 synapses impinge on the dendritic tree of a single neuron). The extracellular and intracellular ionic concentrations of the jth ion are denoted by C_j^o and C_j^i, respectively, and we determine their time evolution. The extracellular space is completely connected and ions diffuse according to Fick's law. Also, there is exchange of ions through the cellular membrane between the extracellular and intracellular spaces via the mechanisms of electrodiffusion (channels) and energy consumption (pumps). On the other hand, the intracellular space is not connected and in order for an ion to move to a distant intracellular point, it must first become an extracellular ion, undergo a random walk, and then go back into the intracellular space, and possibly repeat this process to reach the remote point. Thus we ignore intracellular diffusion. The macroscopic model equations are then given by

$$\frac{\partial C_j^o}{\partial t} = D_j \frac{\partial^2 C_j^o}{\partial x^2} + I_j + P_j \quad ,$$

$$\frac{\partial C_j^i}{\partial t} = -\alpha(I_j + P_j) \quad .$$

where D_j is the diffusion coefficient for the jth ion in aqueous solution (Nicholson & Phillips, 1981) and I_j and P_j correspond to ionic movements via electrodiffusion and energy consumption, respectively, and alpha accounts for the difference between the extracellular and intracellular volumes (the extracellular space is approximately 20% of the total volume (Nicholson & Phillips, 1981)). Thus for each ionic species, the concentrations are governed by a nonlinear diffusion equation and an ordinary differential equation. Complete specifications of these equations require detailed descriptions of the terms I_j and P_j.

3. WAVES OF SPREADING CORTICAL DEPRESSION

Spreading cortical depression is a pathological brain phenomena which was discovered in 1944 by the Brazilian physiologist, A.A.P. Leao (1944). (The name, spreading cortical depression, comes from the depression of the electroencephalogram (EEG) signal and not from psychological depression.) These slowly propagating waves can be instigated by chemical, mechanical, and electrical stimuli in the cortex of various brain structures in a variety of animals (Bures et al., 1974) and can propagate at speeds ranging from 1 up to 11 mm per minute (Tobiasz & Nicholson, 1982). One notable feature of SD is that during its propagation, ionic concentrations in the extracellular space reach pathological levels and thus it provides a controlled paradigm for detailed studies of some brain mechanisms, notably those controlling (large) movements of ions across cell membranes (Kraig & Nicholson, 1978; Nicholson et al., 1978). The importance of ionic concentrations as basic dependent variables is thus clear. Other manifestations of SD are depolarization of nerve cell membranes and small negative DC surface cortical potentials. Although primarily a phenomenon observed experimentally, recently it has been conjectured that an SD wave in the visual cortex produces the visual aura reported by patients suffering from classical migraine (Gardner-Medwin, 1981; Olesen, 1985). This aura takes about a half hour to spread and goes away prior to the onset of a painful headache.

A simplified model for SD waves in which only the two major ions, potassium and calcium, were assumed to undergo changes was derived by Tuckwell & Miura (1978; see also Miura, 1981). The other two major ions, sodium and chloride, were assumed to have fixed concentrations. This model contains a conjectured mechanism for SD propagation. In particular, an increase in the extracellular potassium ion concentration depolarizes presynaptic terminal membranes in the apical dendritic region and permits the influx of calcium ions which in turn release neurotransmitter substance. This neurotransmitter produces conductance changes on postsynaptic membrane and thereby more potassium ions are released into the extracellular space which then diffuse into regions of lower potassium ion concentration. This process repeats itself continuously as the SD wave propagates. These model equations, consisting of two nonlinear diffusion equations coupled to two ordinary differential equations, were solved numerically and it was shown that a solitary SD wave could be generated and that a collision of two solitary SD waves annihilated each other. These solutions exhibit features qualitatively similar to the experimental data. (See Tuckwell & Miura (1978) for details and equations.) Although the speed of the solitary SD wave was about 1 mm/min, which is in the physiological range, such a comparison is premature at this time since it is necessary to account for other mechanisms.

In addition to accounting for those ions which have been assumed not to undergo concentration changes in the simplified model, there are other mechanisms which affect the speed and shape of an SD wave. Although one might expect that action potentials are essential for SD waves to exist and that they have a significant effect on ionic concentrations, recent experiments show that they are not essential for SD propagation (Sugaya et al., 1975) and have little effect on the speeds of SD waves (Tobiasz & Nicholson, 1982). On the other hand, there are other mechanisms which can have effects on the instigation and propagation of SD waves. The problem of instigation may be distinct from the problem of propagation. The fact that multiple SD waves which can be generated experimentally from a single stimulus could not be reproduced using the simplified SD model indicates that other mechanisms such as volume changes and water movement may be important. For propagation, one such mechanism called the "spatial buffer mechanism" (Orkand et al., 1966; Gardner-Medwin, 1983) is conjectured to rapidly transport potassium ions from one point in the extracellular space to another point. This mechanism utilizes the fact that glial cells, which do not fire actions potentials, are close to being potassium electrodes, i.e., locally their membrane potentials are determined by the potassium ionic concentrations on the two sides of the membrane. Incorporating this mechanism into the simplified model may lead to a significant change in the speeds of propagation.

4. TWO-DIMENSIONAL ROTATING WAVES

One of the most interesting phenomena associated with SD waves is the occurrence of two-dimensional rotating waves (Shibata & Bures, 1972). Experimentally, these rotating waves have been shown to persist for as many as 50 cycles around an artificially created lesion in the cortex (Shibata & Bures, 1975). It was not determined whether the natural termination of these rotating waves was due to metabolic exhaustion or due to some other inhibiting process. The method used for generating these rotating waves was originally proposed by Wiener & Rosenblueth (1946) in their studies of a network of connected excitable elements, in particular, with application to cardiac muscle. A point stimulus in a homogeneous excitable medium generates an outgoing circular wave and a delayed application of a second point stimulus at an appropriate location just behind this wave (accounting for refractoriness) will generate a single front semi-circular wave propagating in the opposite direction. By inserting a localized obstacle in the medium, Wiener & Rosenblueth (1946) were able to show that a rotating wave could be established. Shibata & Bures (1972) showed that rotating SD waves around a lesion in the cortex could be generated experimentally.

Our interest in this phenomenon was to determine if such rotating waves can be generated in a region without a lesion, i.e., is a lesion required for a rotating SD wave to persist in the cortex. A study of rotating waves in an excitable medium without internal obstacles was carried out by Winfree (1974) using a two-dimensional piecewise linear version of the FitzHugh-Nagumo equations consisting of a nonlinear diffusion equation coupled to an ordinary differential equation in time. (These equations can be viewed as an analog model of SD.) He showed that a persistent rotating wave in an excitable medium could be established by choosing appropriate initial distributions for the two variables. To mimic the experiments of Shibata & Bures (1972) and study rotating waves in lesion-free cortex, Miura & Plant (1981) performed various numerical experiments on the system studied by Winfree. Our goal was to show that persistent rotating waves could be generated using physiological type stimuli, i.e., not by setting up special initial conditions. The computations were performed in a circle to avoid the effects of corners and we were able to show that a rotating wave can be generated (see Miura & Plant, 1981). Computations were carried out for hundreds of time steps but the rotating wave did not stop and disappear nor did it appear to be tending strongly towards a rigid body rotation as might be expected for steady state motion in a circle. The lack of a strong tendency towards rigid body rotation can be accounted for partly by the use of a homogeneous Neumann boundary condition around the edge of the circle.

Shibata & Bures (1975) studied various ways of terminating a rotating SD wave around a lesion. To terminate a rotating wave in a region without an obstacle, Plant & Miura (unpublished) generated a counter-rotating wave which collided with the already circulating wave. This annihilation of the rotating wave is shown in the sequence of pictures in Figure 1. However, if the second applied stimulus is not placed properly, then the resulting pattern can become more complicated than the simple single rotating wave. Indeed, as shown in Figure 2, one might get two adjacent counter-rotating waves. If the rotating wave occurred in some clinical application, one would have to take great care in placement of the terminating stimulus to avoid catastrophic results.

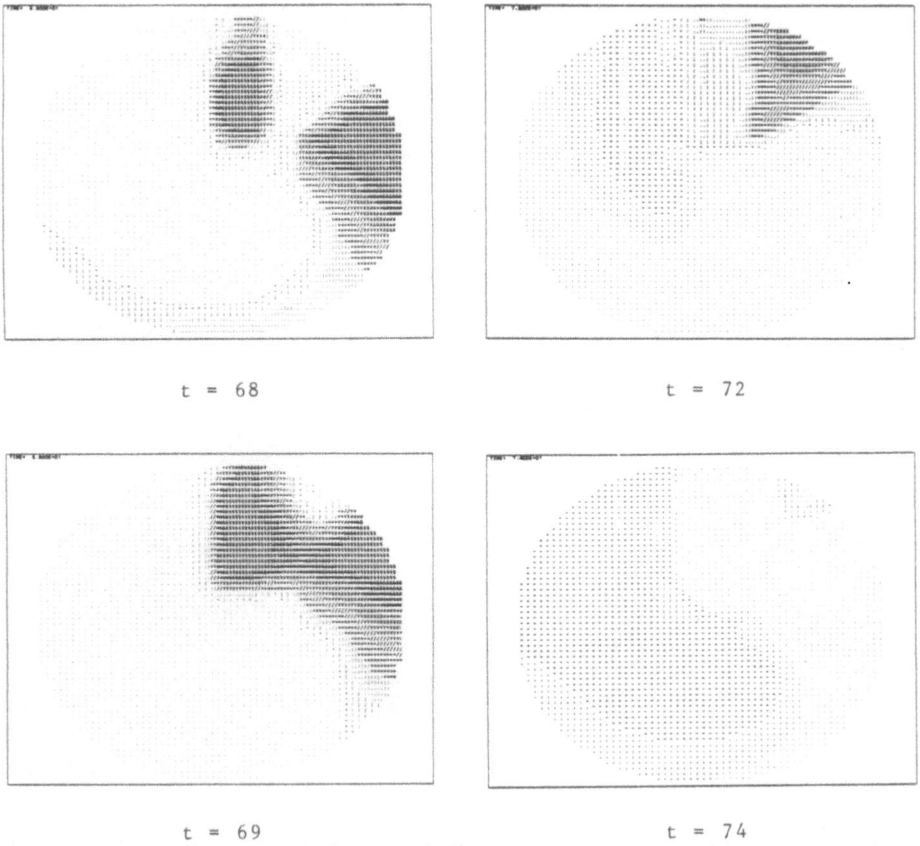

Figure 1. Termination of a rotating wave using the piecewise linear FitzHugh-Nagumo model. A counter-clockwise rotating wave on the right at t = 68 is terminated by a wave stimulated at the top right. Note that the region to the left of the stimulus is in a refractory state.

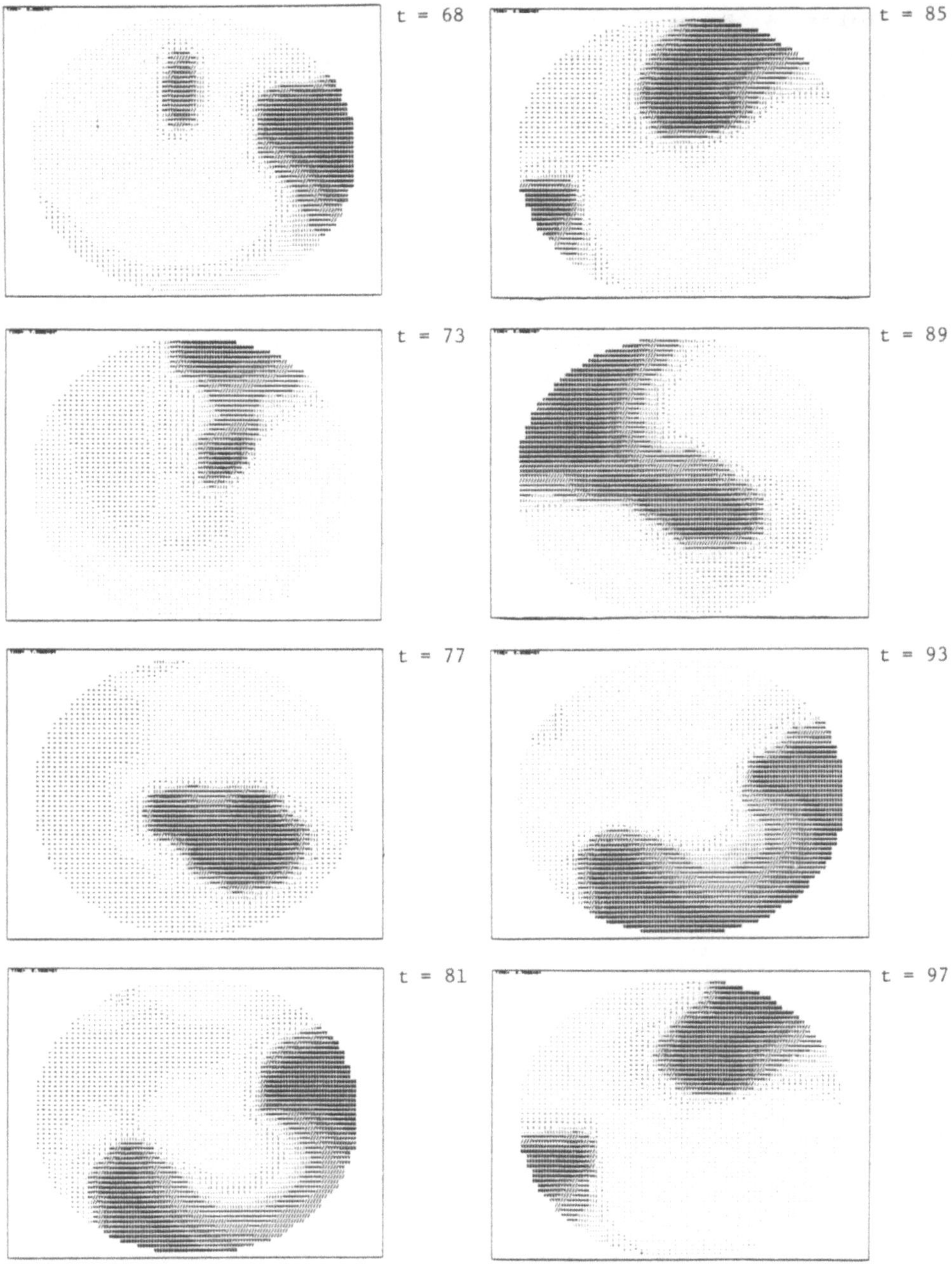

Figure 2. Unsuccessful termination of a rotating wave medium using piecewise linear FitzHugh-Nagumo model. Generation of two counter-rotating waves. A slight change in position of the stimulus at t = 68 results in the escape of a bolus of excitation at t = 73. At t = 81 this excitation splits into two counter-rotating waves. The rotation cycle starting at t = 81 is completed at t = 93.

5. CONCLUSION

Studies of pathological brain phenomena can lead to better understanding of normal brain function by providing a wider range of data from which important parameters can be evaluated and mechanisms can be conjectured. Our goals in this paper have been to review the continuum approach to modelling macroscopic phenomena in the brain and indicate how it has been applied to SD waves. The simplified SD model needs further study and improvements. This approach may be applied to other macroscopic brain phenomena such as epilepsy. Also, we presented some new numerical experiments on successful and unsuccessful termination attempts of two-dimensional rotating waves.

ACKNOWLEDGMENT. We thank Mr. John Ko who skillfully carried out the numerical calculations shown in Figures 1 and 2. This work was supported in part by the Natural Sciences and Engineering Research Council Canada under Grant No. A4559. Additional support was provided by the Department of Applied Mathematics at the University of Washington and by the Department of Mathematics at the University of California at Los Angeles.

REFERENCES

BURES, J., BURESOVA, O., & KRIVANEK, J. (1974). The Mechanism and Applications of Leao's Spreading Depression of Electroencephalographic Activity. Prague: Academia.

GARDNER-MEDWIN, A.R. (1981). Possible roles of vertebrate neuroglia in potassium dynamics, spreading depression and migraine. J. Exp. Biol. 95, 111-127.

GARDNER-MEDWIN, A.R. (1983). A study of the mechanisms by which potassium moves through brain tissue in the rat. J. Physiol. 335, 353-374.

HUBEL, D.H. (1979). The brain. Scientific American 241, 45-53.

KRAIG, R.P. & NICHOLSON, C. (1978). Extracellular ionic variations during spreading depression. Neurosci. 3, 1045-1059.

LEAO, A.A.P. (1944). Spreading depression of activity in the cerebral cortex. J. Neurophysiol. 7, 359-390.

MIURA, R.M. (1981). Nonlinear waves in neuronal cortical structures. In Nonlinear Phenomena in Physics and Biology, eds. ENNS, R.H., JONES, B.L., MIURA, R.M., & RANGNEKAR, S.S., pp. 369-400. New York: Plenum.

MIURA, R.M. AND PLANT, R.E. (1981). Rotating waves in models of excitable media. In Differential Equations and Applications in Ecology, Epidemics, and Population Problems, eds. BUSENBERG, S.N. & COOKE, K.L., pp. 247-257. New York: Academic Press.

NICHOLSON, C. & PHILLIPS, J.M. (1981). Ion diffusion modified by tortuosity and volume fraction in the extracellular microenvironment of the rat cerebellum. J. Physiol. 321, 225-257.

NICHOLSON, C., TEN BRUGGENCATE, G., STOCKLE, H., & STEINBERG, R. (1978). Calcium and potassium changes in extracellular microenvironment of cat cerebellar cortex. J. Neurophysiol. 41, 1026-1039.

OLESEN, J. (1985). Migraine and regional cerebral blood flow. Trends in NeuroScience 8, 318-321.

ORKAND, R.K., NICHOLLS, J.G. & KUFFLER, S.W. (1966). Effect of nerve impulses on the membrane potential of glial cells in the central nervous system of amphibia. J. Neurophysiol 29, 788-806.

SHIBATA, M. & BURES, J. (1972). Reverberation of cortical spreading depression along closed-loop pathways in rat cerebral cortex. J. Neurophysiol. 35, 381-388.

SHIBATA, M. & BURES, J. (1974). Optimal topographic conditions for reverberating cortical spreading depression in cats. J. Neurobiol. 5, 107-118.

SHIBATA, M. & BURES, J. (1975). Techniques for termination of reverberating spreading depression in rats. J. Neurophysiol. 38, 158-166.

SUGAYA, E., TAKATO, M., & NODA, Y. (1975). Neuronal and glial activity during spreading depression in cerebral cortex of cat. J. Neurophysiol. 38, 822-841.

TOBIASZ, C. & NICHOLSON, C. (1982). Tetrodotoxin resistant propagation and extracellular sodium changes during spreading depression in rat cerebellum. Brain Research 241, 329-333.

TUCKWELL, H.C. & MIURA, R.M. (1978). A mathematical model for spreading cortical depression. Biophys. J. 23, 257-276.

WEINER, N. & ROSENBLUETH, A. (1946). The mathematical formulation of the problem of conduction of impulses in a network of connected excitable elements, specifically in cardiac muscle. Arch. Inst. Cardiol. Mex. 16, 205-265.

WINFREE, A.T. (1974). Rotating solutions to reaction-diffusion equations in simply-connected media. SIAM-AMS Proceedings 8, 13-31.

A Formal Classification of Bursting Mechanisms in Excitable Systems

J. Rinzel
Mathematical Research Branch, NIDDK
National Institutes of Health
Bethesda, MD 20892, U.S.A.

1. Introduction.

Burst activity is characterized by slowly alternating phases of near steady state behavior and trains of rapid spike-like oscillations; examples of bursting patterns are shown in Fig. 2. These two phases have been called the silent and active phases respectively [2]. In the case of electrical activity of biological membrane systems the slow time scale of bursting is on the order of tens of seconds while the spikes have millisecond time scales. In our study of several specific models for burst activity we have identified a number of different mechanisms for burst generation (which are characteristic of classes of models). We will describe qualitatively some of these mechanisms by way of the schematic diagrams in Fig. 1.

The basic idea is that there are slow processes which modulate the faster spike generating dynamics. For the models we describe here, the slow subsystem, however, does not act independently as a forcing function to the fast subsystem. There are two way interactions between the fast and slow subsystems; the fast variables play a crucial role in the slow dynamics. For membrane electrical activity, the membrane potential is an important shared variable between the fast and slow processes.

Our understanding of these systems has come from extensive numerical calculations and investigation of broad parameter ranges in a number of explicit models [22-27]. For the most part, these models were formulated to mimic experimental data in a semi-quantitative way. For example, rates of slow processes were chosen to match the time scales of burst patterns. The fast variables account for macroscopic features of spike shape and frequency while some aspects of spike shapes were disregarded to prevent introducing nonessential details which frequently do not alter qualitative properties. Also, each such detail places a further burden on numerical calculations.

In Section 2 we will present schematically our qualitative view of a number of different mechanisms for bursting. To understand these mechanisms we exploit the different time scales. We first identify the fast and slow subsystems. Then the fast dynamics are considered with the slow variables treated as parameters. A full description of the steady state and periodic solution sets to the fast subsystem yields the slow manifold; i.e., this step is essentially a global bifurcation analysis of the fast subsystem with the slow variables as parameters. We find several different bifurcation structures with which we identify and correlate features of different observed burst patterns. For example, the various types of transition behavior between steady state and oscillation branches: subcritical Hopf bifurcation, large amplitude homoclinic orbits, and degenerate homoclinic orbits (which contact saddle-node singularities) lead to different spike frequency characteristics at the beginning or end of the active phase. To complete this lowest order approximation we then consider the flow of the slow dynamics on the branches of the slow manifold. By varying parameters of the slow dynamics one obtains a variety of burst patterns and other activity which correspond to various experimental findings. Our formal analysis is essentially the first step in a systematic singular perturbation treatment of these complex oscillators.

In Section 3, we will discuss some of the explicit models for excitable membrane behavior and offer biophysical interpretations of the theoretical results.

2. A Qualitative Catalog of Burst Generating Mechanisms.

For this discussion we will consider a model of the form

$$\dot{x} = F(x,y) \quad , \quad x \in R^n \qquad \text{(FAST)}$$

$$\dot{y} = \varepsilon\, G(x,y) \quad , \quad y \in R^m \,. \qquad \text{(SLOW)}$$

in which there are n fast variables x and m slow variables y; here, $0 < \varepsilon << 1$. We suppose that the components of G and F are O(1) except in small neighborhoods around zeroes of these vector functions. Our qualitative catalog with graphical representations is based on the limit of ε near zero.

In this limit, we focus on the steady state, $x_{ss}(y)$, and periodic solutions, $x_{osc}(t;y)$, to FAST (with y as a vector of parameters) which satisfy:

$$0 = F(x_{ss},y) \,,$$

and

$$\dot{x}_{osc} = F(x_{osc},y) \,, \qquad x_{osc}(t+T;y) = x_{osc}(t;y) \,,$$

respectively, where $T = T(y)$ is the period of x_{osc}.

Specific model equations, with identification of physiological slow and fast variables, are given in Section 3. For example, the Chay-Keizer equations (which correspond to the schematic of Fig. 1B) are a model for the electrical activity of insulin-secreting β-cells of the pancreas [6,7].

In some types of bursters the fast dynamics exhibit bistability in which, for certain ranges of the slow variables, there are two different attracting branches of the slow manifold. For example, a (pseudo) steady state and an oscillation which would correspond to the silent and active phases respectively. In such models, the slow dynamics cause y to sweep back and forth through this regime and the burst trajectory essentially traces a hysteresis loop.

The simplest example of such hysteresis behavior is illustrated in Figs. 1A and 2A. In this case there is one slow variable, $y \in R^1$, and the fast subsystem exhibits three steady states over a certain range of y: two, upper and lower, which are stable, and one unstable saddle. Without being explicit, let us suppose that the slow dynamics are such that when x is in the upper state then $\dot{y} > 0$ (y is produced slowly) and when x is in the lower state then $\dot{y} < 0$ (y is depleted slowly). The full system then generates a relaxation oscillation (Fig. 1A (right) and Fig. 2A). We might identify this waveform as a burst without spikes or slow wave. It is important to realize that we have not hypothesized a mechanism which requires the slow system to oscillate independently through the hysteresis zone. The sign of $\dot{y}$ depends crucially on the values of x, and the dependence is autonomous. Without the hysteresis or bistability of FAST this slow wave would not exist; if instead of the Z-curve relation, x_{ss} vs y were monotonically decreasing, then we would not have bursting. In this sense the fast subsystem, and its hysteresis, drives the slow variable y so as to produce the slow oscillation.

A minimal model for the slow wave of Fig. 1A, might involve only a single fast variable, $x \in R^1$. In this case the full system would be a classical second-order relaxation oscillator. Such a model could not account for an active phase with spiking. For this, the fast subsystem must have at least two variables. For example, suppose again that $y \in R^1$ and that FAST still had a steady state Z-shaped bifurcation diagram; for comparison purposes, let it be identical to that in Fig. 1A. But, suppose the upper state, instead of being stable, is unstable and is "surrounded" by an oscillation over some y-range (Fig. 1B). An illustrative example of a phase portrait (in the case $x \in R^2$) is shown in Fig. 3. In this model, the branch of oscillations emerges via Hopf bifurcation at HB and terminates at HC. At termination, the periodic orbit makes contact with the saddle point of the middle branch of the Z-curve; the period becomes infinite at HC and the orbit is called homoclinic. Beyond HC, the lower steady state is globally attracting. Let us again hypothesize slow dynamics. Suppose, as above, that y is decreasing when x is in the lower steady state, and, analogous to the above, when x is in the upper attractor, i.e., the oscillatory mode, we suppose that y experiences a net increase for each cycle of the fast x-oscillation x_{osc}. (This means that $\bar{G}(x_{osc},y)$ is O(1) and positive, where $\bar{G}(x_{osc}, y) = T^{-1} \int_0^T G(x_{osc}(t;y),y)dt$, $T = T(y)$). One can intuitively predict that under these hypotheses the full system generates a burst pattern (schematized in the right panel of Fig. 1B) with the time course of Fig. 2B. The upper oscillation branch of FAST generates the repetitive spike pattern of the burst's active phase. The fork (maximum and minimum of x_{osc} over a period) of the bifurcation diagram becomes the envelope for this train of spikes. In comparing Figs. 2A and 2B (and recalling that the Z-curves are identical), we see that the spikes of the active phase (left panels) in Fig. 2B oscillate around the plateau level in Fig. 2A; also, the y excursion is larger in Fig. 2A than in Fig. 2B since the slow wave trajectory progresses to the right knee of the Z-curve (cf., Fig. 1A) but the burst trajectory (cf., Fig. 1B) does not. We may view the burst of Figs. 1B, 2B as a generalized relaxation oscillation, i.e. repetitive visitation to overlapping coexistent branches of the slow manifold. But, in contrast to the classical relaxation oscillator in which both branches are steady state branches, here one branch is oscillatory. From Fig. 1B, we reach a qualitative conclusion about the spike pattern during the burst: the instantaneous spike frequency drops dramatically at the end of the active phase (see Fig. 2B) as the trajectory passes near the homoclinic orbit of FAST.

Variations on this general hysteresis-based mechanism will lead to different burst patterns. For example, if the supercritical Hopf bifurcation point were rightward of the Z-curve's left knee then the active phase would exhibit an initial portion of near (upper) steady state behavior followed by growing oscillations (around the upper steady state) whose frequency drops suddenly as the active phase ends. One could also predict qualitatively the burst pattern in case the Hopf bifurcation is subcritical.

We mentioned for the case of Fig. 1A, that if hysteresis in FAST is destroyed, say, by stretching the "Z" out of the slow manifold, then the slow (relaxation) wave is lost. Similarly here, if hysteresis is precluded between the upper oscillatory and low steady state attractors of FAST then bursting (in a robust way) is lost. For example, by adjusting parameters of FAST the homoclinic termination of the oscillation branch can be moved leftward until it meets the knee of the Z-curve (Fig. 1C); in Section 3, we describe how this can be accomplished in the Chay-Keizer model. In this situation, FAST has a unique global attractor for each value of y; hysteresis is lost. The response will be either time independent, if $G(x_{ss},y) = 0$, and $\partial G/\partial y < 0$, for some y rightward of HC with x_{ss} on the lower steady state branch, or continuous periodic spiking, if $\bar{G}(x_{osc},y) = 0$, and $\partial\bar{G}/\partial y < 0$, for some y leftward of HC (with x_{osc} on the oscillatory branch). The time courses of two such different response patterns are shown in Fig. 2C, dashed and solid, respectively.

For the fast dynamics represented by Fig. 2C, the full model would exhibit no robust parameter range for bursting as parameters of SLOW are tuned through a range which includes steady state behavior at one extreme to continuous spiking at the other. On the other hand, if such tuning were done (dynamically) in a smooth repetitive manner then a burst pattern would be generated nonautonomously. This burst trajectory would cross the HC boundary both at the beginning and at the end of the active phase and, consequently, a plot of instantaneous spike period versus spike number would appear as a concave-upward parabola. In the neurobiology literature such a waveform has been called a parabolic burst pattern [1]. To generate this behavior in an autonomous manner requires at least two slow variables. In the case $y \in R^2$, the slow manifold may be represented schematically as the surface in Fig. 1D; it includes both steady state and oscillation branches. With the additional degree of freedom in y, one can formulate slow dynamics which lead to autonomous oscillatory

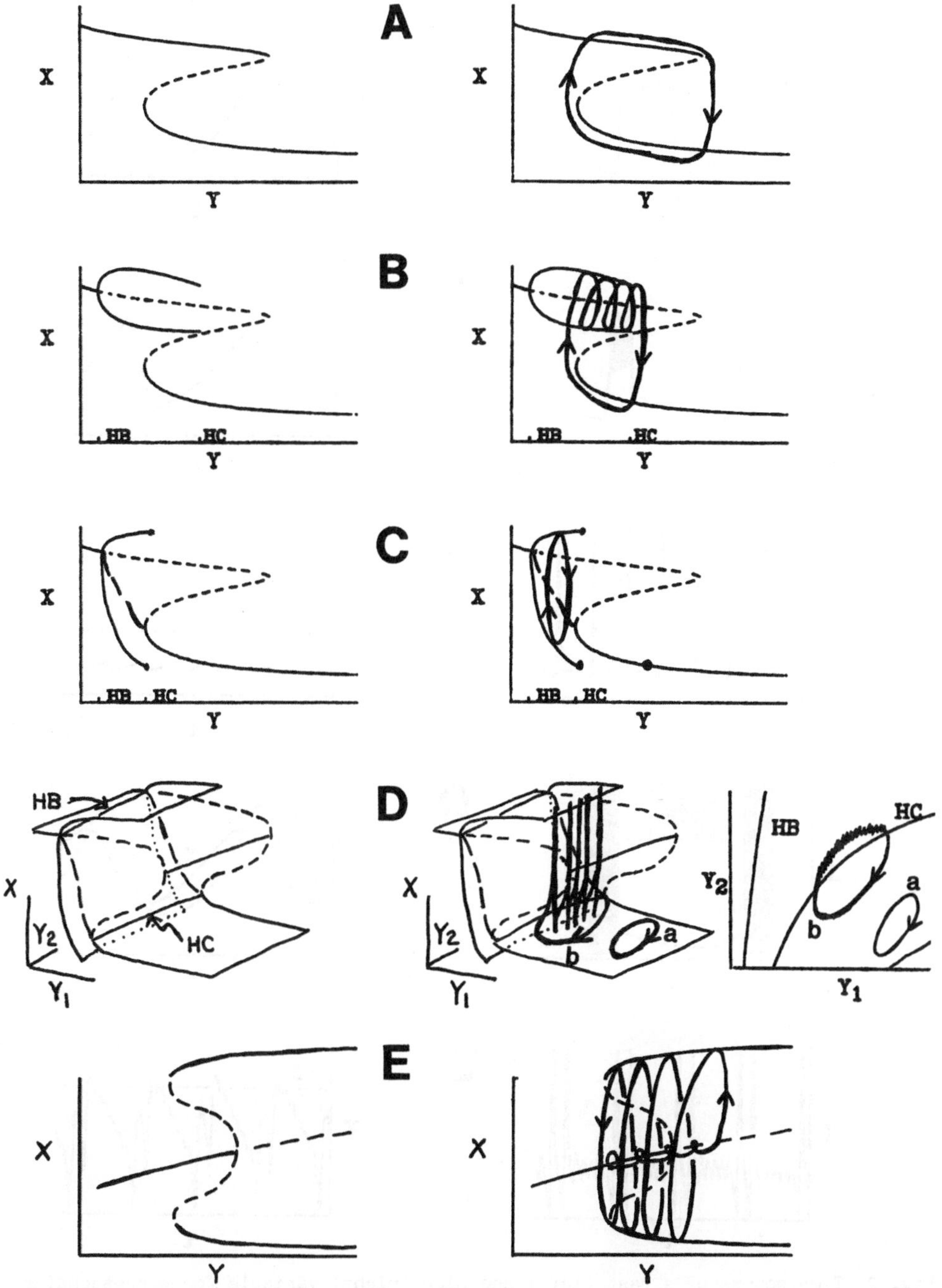

Fig. 1 Left: Bifurcation diagram (compact description) of periodic and steady state solutions to fast subsystem (FAST) with slow variables as parameters. Maximum and minimum values of some solution component, or its time average over a period (long dashes), is plotted. Unstable solutions indicated by short dashes. Right: Schematic representation (heavy curves) of slow wave, burst, or continuous spiking trajectory as projected on bifurcation diagram (and its projection, case D, far right) of corresponding left panel.

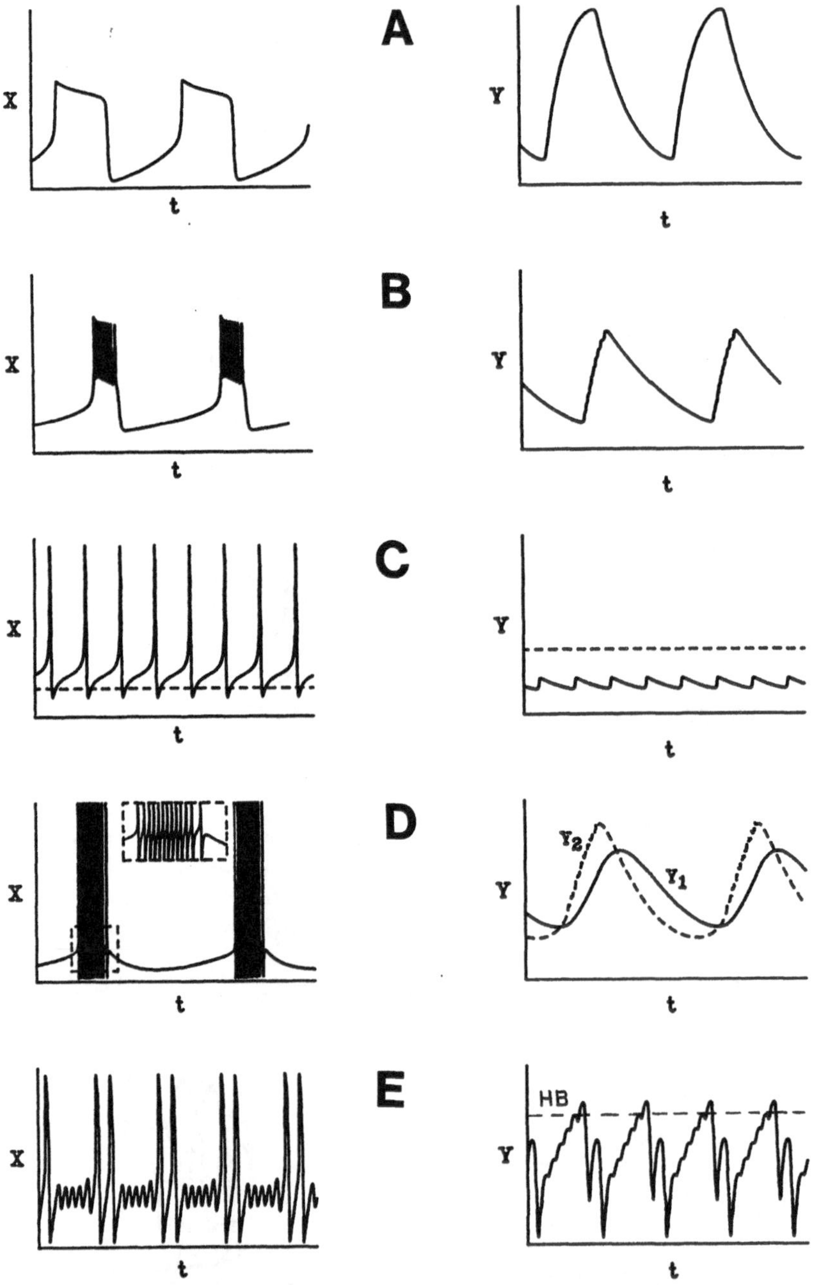

<u>Fig. 2</u> Time course of a fast (left) and slow (right) variable for corresponding schematics of Figure 1. These are for computed solutions of specific excitable membrane models. Cases A, B, C: V and Ca from eqns. (3.1)-(3.3) with parameter changes: $\lambda_n = 2.0$, $f = 0.08$ (A); $\lambda_n = 0.1$, and $k_{Ca} = 0.006$ (solid) or $k_{Ca} = 0.0005$ (dashed) (C). Case D: V, Ca (right, solid), and slow calcium conductance, x (right, dashed) for Plant's model (as given in [25] but with $\tau_x = 1.88 \times 10^4$, $K_C = 5.525 \times 10^{-3}$, $\rho = 0.9 \times 10^{-4}$). Case E: v and y for eqns. (3.4)-(3.6) with $\varepsilon = 0.0008$ (note, ε not too small here, so only 2 pulses per burst and y-increments not small).

behavior of y even when x is restricted to the lower steady state branch of the slow manifold. In this case, we would find a slow wave (Fig. 1D, right panels, case (a)) but it would not be of relaxation type as in Fig. 1A; both x and y would vary smoothly on the slow time scale.

To understand how a burst pattern may be generated, we note first that variations in the parameters of SLOW do not alter the slow manifold (the surfaces in Fig. 1D) in any way. Thus, the parameters of G could be varied appropriately to move the slow wave toward the HC boundary. With sufficient parameter variation, a portion of the slow wave would just cross into the oscillatory regime of FAST and a parabolic burst trajectory would be obtained (Fig. 1D, right, case (b) and Fig. 2D). This intuitive description supposes that the trajectory will return to the steady state branch and then recycle through the silent phase. This is reasonable to expect. Since, for y near the HC boundary, periodic solutions to fast spend most of their time near the HC steady state, the (fast time) "averaged" flow of $\bar{G}(x_{osc},y)$ will be continuously extended off of the steady state branch and therefore will inherit the (slow) oscillatory properties which G exhibits on the lower steady state branch. In this type of burst mechanism one can identify the underlying slow wave; it appears to drive the spikes. We note, again, that this mechanism for bursting does not require the fast subsystem to exhibit multiple <u>stable</u> states for any y in the bursting region.

A common feature of the above examples is that, over (at least part of) the y-range for bursting, the fast subsystem exhibits multiple steady states. In the preceding example of Fig. 1D, at most one steady state is stable. A different burst generating mechanism, illustrated in Fig. 1E, does not require multiple steady states and still can be realized with only a single slow variable. It is based on subcritical Hopf bifurcation in FAST, so it relies on hysteresis. To obtain bursting, we hypothesize for SLOW that y exhibits a net decrease over a cycle of the large amplitude fast oscillation, i.e., $\bar{G}(x_{osc},y) < 0$ for $x_{osc}(t;y)$ on the "outer" branch. Thus during the active phase y decreases toward the turning point or knee of the fast oscillation branch where the stable and unstable fast periodic solutions coalesce and disappear. When y reaches the knee, x falls into the domain of attraction of the stable steady state.

In a neighborhood of this steady state we suppose that y, on the average, increases. Thus we have slow rightward movement along the steady state branch during the silent phase. If y passes inside HB, this pseudo steady state becomes unstable and x cannot continue to track it. Therefore x returns to the oscillatory mode of FAST to initiate the active phase. We have found that the silent phase does not necessarily end immediately when y passes inside HB but, that there may be some delay as small oscillations in x grow slowly; we have estimated analytically this escape time [27]. This behavior is seen in the schematic phase space projection of Fig. 1E (right) and in the time courses in Fig. 2E. In this type of burst pattern one does not see the underlying smooth slow wave as in Figs. 1D, 2D nor the relaxation pattern of Figs. 1B, 2B. Here, the fast oscillations of the active phase surround the steady state of the silent phase. Also, since the pseudo steady state may behave as a damped oscillation in the entire overlap range of the subcritical Hopf bifurcation, one may expect to find small decaying and then growing oscillations in the silent phase.

3. Bursting in Excitable Membrane Systems.

Models for excitable membrane electrical activity usually involve modifications to the classical Hodgkin-Huxley description [15] of action potential generation in squid axon membrane. In models of bursting, the fast subsystem is based on such modified Hodgkin-Huxley spike dynamics. The associated fast variables include the membrane potential V, and V- and t-dependent activation/inactivation variables for ionic channel currents. These variables describe membrane properties. The fast subsystem may involve two or more variables. The rate-limiting dynamics of the slow processes typically involve at least one non-membrane quantity, e.g., intracellular free calcium concentration Ca.

A biophysical description, e.g. in the case of insulin-secreting β-cells [2], would be the following. The membrane has ion-selective channels which activate, and possibly inactivate, to V-dependent levels and with V-dependent rates. For example, calcium channels allow inward current flow which increases $\dot{V}$ while potassium currents are outward and their activation decreases $\dot{V}$. When inward current kinetics are fast and outward currents are slower, then these two opposing currents can lead to oscillatory spike activity. In the β-cell system, these dynamics are such that, depending on Ca, the membrane can remain with V at the lower silent phase potential ($\doteq$ -55 mV) or oscillate with V in a range (-40 to -25 mV) well above the silent phase potential. During the active phase of a burst, each spike causes a small net increase in Ca; it is small because only a fraction f of the entering calcium is free - most of it is bound rapidly to high affinity binding sites inside the cell. Here, f determines the slow time scale of the burst pattern. To modulate the fast membrane variables, there must be a feedback site for calcium at the membrane. Such feedback to the fast dynamics gives the bifurcation structure of Fig. 1B (with V on the ordinate and Ca along the abscissa). In a number of models, this site is hypothesized to be a calcium-activated potassium channel and it is usually considered distinct from the more classical Hodgkin-Huxley K^+-channel. Thus, as Ca slowly accumulates, it activates the conductance g_{K-Ca} (which is often treated as V-independent). As g_{K-Ca} rises during the active phase, so does the threshold for spike activity. This threshold corresponds to the saddle point on the middle branch of the Z-curve in Fig. 1B. This rising threshold meets the trajectory for repetitive spike activity, and the active phase terminates at this point; the trajectory falls below threshold and V drops to the lower (pseudo steady state) silent phase potential. At this low V, the calcium channels are not active, and so there is no influx of calcium. The removal of calcium dominates during the silent phase; g_{K-Ca} slowly decreases, while the pseudo steady state V increases and the threshold falls. When V meets the threshold then the active phase is reentered and the cycle repeats.

Explicit equations for the β-cell system were formulated originally by Chay and Keizer [6] (based upon the biophysical model of [2]) and subsequently modified [7]. A FAST/SLOW analysis of the original model was presented in [23] and a simplified model was treated in [24]. The equations of the simplified model are given by:

$$C_m\dot{V} = -\bar{g}_{Ca}m_\infty^3(V)h_\infty(V)(V-V_{Ca}) - [\bar{g}_K n^4+\bar{g}_{K-Ca}\frac{Ca}{1+Ca}](V-V_K) - \bar{g}_L(V-V_L) + I_{app} \quad , \tag{3.1}$$

$$\dot{n} = \lambda_n\,[n_\infty(V)-n]/\tau_n(V) \quad , \tag{3.2}$$

$$\dot{Ca} = f[\alpha\bar{g}_{Ca}m_\infty^3(V)h_\infty(V)(V_{Ca}-V) - k_{Ca}Ca] \quad , \tag{3.3}$$

where,

$$j_\infty(V) = \alpha_j(V)/[\alpha_j(V)+\beta_j(V)] \quad , \qquad j = m,\ h,\ \text{or}\ n$$

$$\tau_n(V) = 1/[\alpha_n(V)+\beta_n(V)] \quad ,$$

$$\alpha_m(V) = 0.1(-V-25)/[\exp\{0.1(-V-25)\}-1] \quad , \qquad \beta_m(V) = 4\exp\{(-V-50)/18\} \quad ,$$

$$\alpha_h(V) = 0.07\exp\{(-V-50)/20\} \quad , \qquad \beta_h(V) = 1/[\exp\{0.1(-V-20)+1] \quad ,$$

$$\alpha_n(V) = 0.01(-V-20)/[\exp\{0.1(-V-20)\}-1] \quad , \qquad \beta_n(V) = 0.125\exp\{(-V-30)/80\} \quad .$$

The variables n and Ca are dimensionless; t, V's, and $\bar{g}$'s have units of msec, mV, mmhos/cm^2, respectively. Parameter values used here are: $C_m = 1[\mu F/cm^2]$, $V_{Ca} = 100$, $V_K = -75$, $V_L = -40$, $\bar{g}_{Ca} = 1.79934$, $\bar{g}_K = 1.69765$, $\bar{g}_{K-Ca} = 0.0104998$, $\bar{g}_L = 0.00698514$, $\lambda_n = 0.3$, $k_{Ca} = 0.00513$, $f = 0.0058$, and $\alpha = 0.0259102$ (which involves Faraday's constant and the cell radius).

In this simplification, the calcium current depends instantaneously upon V while the potassium current activation, n, has the time constant $\tau_n(v)$. Here, the V-n equations form the fast subsystem in which the slow variable Ca appears. The phase portrait of FAST with Ca fixed at a value in the bursting range is shown in Fig. 3. With the parameters given above, this model has FAST/SLOW dynamics and burst behavior as described by Figs. 1B, 2B. Several features of experimentally observed burst patterns are consistent with the model. The spike frequency is seen to drop sharply near the end of the active phase. Changes in glucose do not alter spike envelopes or the silent phase potentials [5]. This is found theoretically when the effect of increasing glucose is modeled by increasing k_{Ca}; such changes affect only SLOW but not FAST. On the other hand, glucose alters the relative duration of the active and silent phases [5]; very low glucose leads to a stable rest state at low V and very high glucose yields continuous spiking. This behavior can be understood theoretically by superposing the Ca-nullcline onto Fig. 1B (see [23,24]) and noting that changing k_{Ca} repositions the nullcline relative to the oscillatory and steady state branches. Other biophysical insights from the model's burst behavior are described in [4,6,7,23]. An idealized model (without identifiable biophysical variables) for neuronal burst activity of this general class has also been formulated and studied by Hindmarsh and Rose [14]. Also, Martiel and Goldbeter [19] have applied our formalism to understand complex oscillations in a model for cyclic AMP signaling in Dictyostelium amoebae.

If, in (3.1)-(3.3), n acts too rapidly (i.e., λ_n large enough) then the upper steady state, for fixed Ca, is stable and the model behaves as in Figs. 1A, 2A. This parameter range has not been found experimentally; relaxation slow waves have not been observed. Finally, if n acts too slowly, then the FAST dynamics of the Chay-Keizer model correspond to Figs. 1C, 2C. This would predict that, as glucose increases, the response would evolve from steady state (Ca to right of HC) to continuous spiking (average Ca to left of HC) without passing through a bursting regime. Such conditions have been induced experimentally with TEA (see Fig 4 of [3]).

A well studied neuronal pacemaker exhibits parabolic bursting [1]. Plant [21] formulated a model for such a system; our FAST/SLOW analysis [24,25] reveals that its structure is that of Figs. 1D, 2D. In this model, the inward current for fast spike generation is carried by sodium ions rather than calcium ions. There is also experimental evidence for an additional, and very slow, inward current with a substantial calcium component. In the model, this means there are two slow variables: the above mentioned slow membrane conductance for calcium (an equation like $\dot{n}$ but with a much slower rate) and the intracellular free calcium concentration (with dynamics as in (3.3)). Parameter adjustments in the model (in particular, merely in the slow subsystem, as described qualitatively in Section 2) can lead to bursting, to smooth slow wave activity without spiking, or to continuous spiking. In Plant's formulation, as well as some others (see [1]), the biophysical mechanism for calcium feedback was modeled by a calcium-activated potassium (outward) channel. This view has been reconsidered and an alternative hypothesis (see [1], and its references) involves calcium inactivation of the slow inward current. We have also shown how this second hypothesis can be formulated, and explored, without altering the fast subsystem [25].

Plant's original model was also studied by Honerkamp et al. [16]. Kopell and Ermentrout [18] also consider a mechanism for parabolic bursting in which a degenerate homoclinic connection plays a key role. Their hypotheses however are not satisfied by Plant's model and they did not require feedback from FAST to produce an oscillation in SLOW (in this sense, the slow subsystem would generate an oscillation independently and thus it acts more like a driving force for FAST).

Models of bursting of the type represented by Figs. 1E, 2E have been studied in the context of the Belousov-Zhabotinskii oscillating chemical reaction (see [24,25] and references therein). Such bursting has not been exposed for a quantitative biophysical model or widely studied experimental preparation for excitable membrane. On

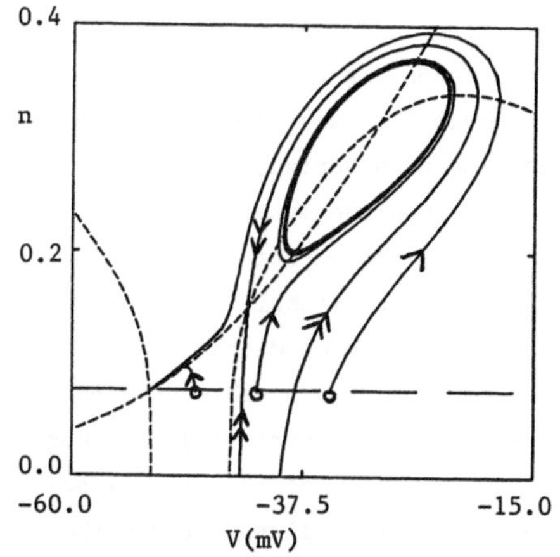

Fig. 3 Phase plane portrait for fast subsystem, eqns. (3.1)-(3.2), of Chay-Keizer model for a fixed value of Ca, 0.7 μM. V and n-nullclines are shown dashed. Three singular points. Double arrowheads denote stable manifolds of saddle point. Responses for three different initial conditions (open circles) show bistability with stable lower steady state and stable oscillation surrounding unstable upper steady state.

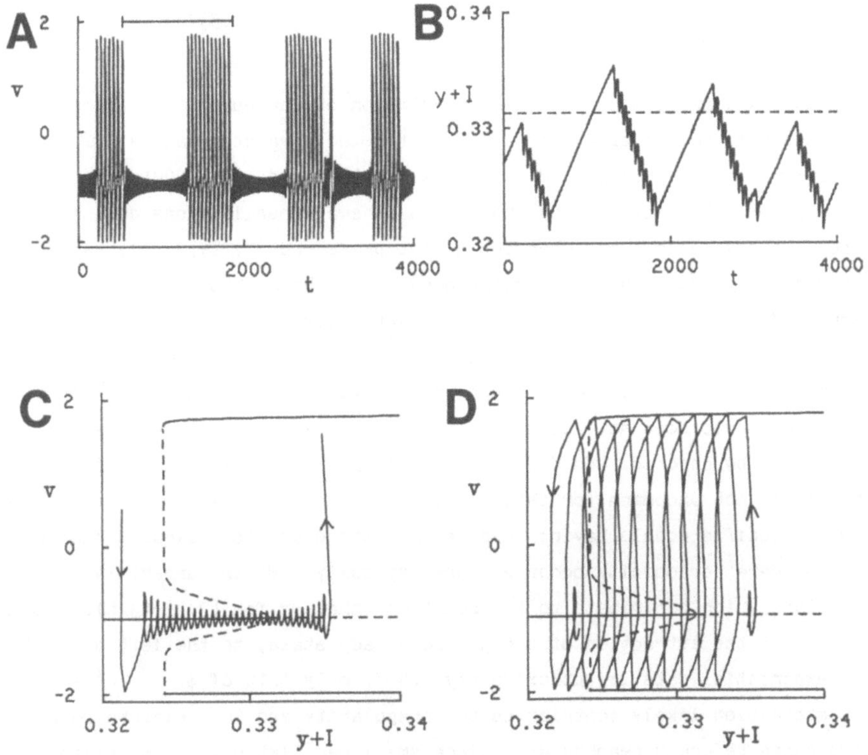

Fig. 4 Solution to the bursting model, eqns. (3.4)-(3.6); parameters given in text. (A), (B): time course of v (fast variable) and y (slow variable), respectively. Projection of silent phase (C) and active phase (D) of a burst (time interval indicated by bar in (A)) onto bifurcation diagram of fast subsystem (cf. Fig. 1E).

the other hand, the classical Hodgkin-Huxley model exhibits a bifurcation structure like Fig. 1E (left), with y representing the applied external current [22]. Experiments show that squid axon behaves this way in appropriate regimes [11]. This bifurcation structure can be seen also with variations in other parameters of the Hodgkin-Huxley model, and presummably if such parameters can be appropriately treated as slow autonomous, dynamic variables then such bursting could be generated. The "skip runs" observed experimentally by Guttman and Barnhill [12] are similar in appearance to Fig. 2E, and are suggestive of this type of mechanism.

An idealized nerve membrane model which exhibits bursting as represented by Figs. 1E, 2E was formulated and studied numerically (FitzHugh and Rinzel, 1976, unpublished). The fast subsystem for this model is the classical FitzHugh-Nagumo equation [10,20]. For appropriate parameter values, it has the subcritical Hopf bifurcation structure of Fig. 1E (left) with applied current I as the parameter [17,30]. By formulating a suitable slow dynamics for I, one generates the desired burst behavior. The model takes the following form:

$$\dot{v} = v - v^3/3 - w + y + I, \qquad (3.4)$$

$$\dot{w} = \phi(v + a - b\, w), \qquad (3.5)$$

$$\dot{y} = \varepsilon(-v + c - d\, y), \qquad (3.6)$$

in which I is fixed and y describes slow modulation of the current. Figures 4A, B illustrate the time course of a burst pattern for the parameter values: $I = 0.3125$, $a = 0.7$, $b = 0.8$, $c = -0.775$, $d = 1.0$, $\phi = 0.08$, and $\varepsilon = 0.0001$. These plots illustrate the characteristic decay and growth of small oscillations during the silent phase. Observe also that, when the silent phase ends, y typically has progressed considerably above the value corresponding to the Hopf bifurcation (represented by horizontal dashed line in Fig. 4B) in the fast subsystem. Projection of an active and silent phase of this solution onto the y-v plane (Figs. 4C, 4D) yields a comparison with the FAST/SLOW prediction. Notice also, over the time interval for Figs. 4A, 4B, that the response does not appear to be periodic; its period could be long or the solution may be chaotic. We did not explore this in detail, but we found that such behavior is not uncommon for this model. This contrasts with our experience [26,27] with specific models having this general structure for Belousov-Zhabotinskii oscillations where periodicity occurred more typically. We are uncertain about the primary factors which contribute to the solution behavior for (3.4)-(3.6). One possibility is that the attraction of the pseudo steady state, to the left of HB, is weak; its exponential rate is approximately 0.008, only 1/10 of ϕ, the rate of w. This weak attraction likely contributes to irregularity and to premature reentry of the active phase before y reaches HB. Since small oscillations of the silent phase are not damped rapidly, the v-w trajectory (as y moves rightward toward HB) may cross the unstable periodic solution (dashed in Figs. 4C, D) of FAST and then proceed to the "outer" stable oscillatory attractor before y reaches HB; this is premature reentry. (Note, corresponding phenomena were observed in the Chay-Keizer model, an example of the class represented by Figs. 1B, 2B.) The model, (3.4)-(3.6), can also exhibit bistability in the burst response, i.e. two different stable burst patterns for the same parameter values. One can hope that this model is simple enough so that further insight might be obtained analytically. Honerkamp et al. [16] have also studied a model of this sort.

4. Discussion.

We have outlined our formal approach to understand qualitatively the different mechanisms for bursting in a number of models. Our FAST/SLOW dissection is essentially the first step in a systematic singular perturbation treatment of these complex oscillations. We have computed numerically, for some explicit models, the bifurcation structures (slow manifolds) corresponding to each of the cases schematized in Fig. 1. In such efforts the automatic capabilities of AUTO [9] for branch-tracking, bifurcation and stability analysis of stable and unstable, steady state and/or periodic, solutions have been extremely useful. Solutions to the full models were usually obtained by Gear's method of numerical integration for stiff equations [13]. We hope that these explicit examples will motivate analysts to provide a more rigorous basis and insight for our formal approach. The early results of Tikonov [29] and others (see [31] for an introduction and references), provide some groundwork for relaxation oscillators like that of Fig. 1A. Such work must be extended to cases in which the slow manifold has more complex solution sets. (For recent analytic work in this direction see [28]). In Figs. 1B, C, D, the branch of oscillatory solutions terminates in a homoclinic orbit. One could imagine even more complex attractors for FAST. The examples we have discussed are minimal in this regard. Also, we have not addressed the details of solution structure to the full model as parameters are varied, for example, of how the transition is made of a burst pattern of n-spikes to one of (n+1)-spikes, of how bursting goes into continuous spiking, or, of how chaos arises (see, for example, [8]).

We do not view our catalog of bursting mechanisms as complete. Also, we appreciate that the classifications are not rigorous but that they convey the qualitative essence. Furthermore, our experience shows, for a model with given nonlinearities, by varying the time scale of merely one of the fast variables that different response classes can be obtained. For example, different values of λ_n in (3.1)-(3.3) can yield the behavior of Figs. 1A, B, or C.

The qualitative categorizations of Figs. 1 and 2 allow us to realize better the limitations and richness of specific models. By identifying the fast and slow time scales, and the underlying structure, we can identify parameters which affect certain aspects, but not others, of solution behavior. The work does not stop here, however. For explicit models one can ask about absolute quantities and values. How small must ε be to match the time scale of experimentally observed patterns and to guarantee that our FAST/SLOW analysis is valid? How robust is the bursting mechanisms to parameter variations in the physiologic range?

Acknowledgement
I thank Steven M. Baer for computational help with Fig. 4 and for reading the manuscript carefully.

1. Adams, W. B., and J. A. Benson. 1985. The generation and modulation of endogenous rhythmicity in the Aplysia bursting pacemaker neurone R15. Prog. Biophys. Molec. Biol. 46:1-49.
2. Atwater, I., C. M. Dawson, A. Scott, G. Eddlestone, E. Rojas. 1980. The nature of the oscillatory behavior in electrical activity for pancreatic β-cell. J. of Hormone and Metabolic Res., Suppl. 10:100-107.
3. Atwater, I., B. Ribalet, and E. Rojas. 1979. Mouse pancreatic β-cells: tetraethylammonium blockage of the potassium permeability increase induced by depolarization. J. Physiol. 288:561-574.
4. Atwater, I., and J. Rinzel. The β-cell bursting pattern and intracellular calcium. In Ionic Channels in Cells and Model Systems (ed., R. Latorre). Plenum, New York, 1986.
5. Beigelman, P. M., B. Ribalet, and I. Atwater. 1977. Electrical activity of mouse pancreatic beta-cells II. Effects of glucose and arginine. J. Physiol., Paris 73:201-217.
6. Chay, T. R., and J. Keizer. 1983. Minimal model for membrane oscillations in the pancreatic β-cell. Biophys. J. 42:181-190.
7. Chay, T. R., and J. Keizer. 1985. Theory of the effect of extracellular potassium on oscillations in the pancreatic β-cell. Biophys. J. 48:815-827.
8. Chay, T. R., and J. Rinzel. 1985. Bursting, beating, and chaos in an excitable membrane model. Biophys. J. 47:357-366.
9. Doedel, E. J. Software for Continuation Problems in Ordinary Differential Equations. Tech. Report, Applied Math. Dept., Cal Tech., 1986.
10. FitzHugh, R. 1961. Impulses and physiological states in models of nerve membrane. Biophys. J. 1:445-466.
11. Guttman, R., S. Lewis, and J. Rinzel. 1980. Control of repetitive firing in squid axon membrane as a model for a neuroneoscillator. J. Physiol. 305:377-395.
12. Guttman, R., and R. Barnhill. 1970. Oscillation and repetitive firing in squid axons: Comparison of experiments with computations. J. Gen. Physiol. 55:104-118.
13. Hindmarsh, A. C. Ordinary differential equations systems solver. Report UCID-30001. Lawrence Livermore Lab. Livermore, CA, 1974.
14. Hindmarsh, J. L., and R. M. Rose. 1984. A model of neuronal bursting using three coupled first order differential equations. Proc. R. Soc. Lond. B 221:87-102.
15. Hodgkin, A. L., and A. F. Huxley. 1952. A quantitative description of membrane current and its applicaton to conduction and excitation in nerve. J. Physiol. (Lond) 117:500-544.
16. Honorkamp, J., G. Mutschler, and R. Seitz. 1985. Coupling of a slow and a fast oscillator can generate bursting. Bull. Math. Biol. 47:1-21.
17. Hsu, I., and N. K. Kazarinoff. 1976. An applicable Hopf bifurcation formula and instability of small periodic solutions of the Field-Noyes model. J. Math. Anal. App. 55:61-89.
18. Kopell, N., and G. B. Ermentrout. 1986. Subcellular oscillations and bursting. Math. Biosci. 78:265-291.
19. Martiel, J. L., and A. Goldbeter. Origin of bursting and biorhythmicity in a model for cyclic AMP oscillations in Dictyostelium cells. In Proceedings of Kyoto International Symposium on Mathematical Biology (ed., E. Teramoto). Springer, to appear.

20. Nagumo, J. S., S. Arimoto, and S. Yoshizawa. 1962. An active pulse transmission line simulating nerve axon. Proc. IRE. 50:2061-2070.
21. Plant, R. E. 1981. Bifurcation and resonance in a model for bursting nerve cells. J. Math. Biol. 11:15-32.
22. Rinzel, J. 1978. On repetitive activity in nerve. Federation Proc. 37:2793-2802.
23. Rinzel, J. Bursting oscillations in an excitable membrane model. In Ordinary and Partial Differential Equations (eds., B. D. Sleeman and R. J. Jarvis). Lecture Notes in Mathematics 1151, Springer-Verlag, New York, 1985.
24. Rinzel, J., and Y. S. Lee. On different mechanisms for membrane potential bursting. In Nonlinear Oscillations in Biology and Chemistry (ed., H. G. Othmer). Springer, New York, to appear.
25. Rinzel, J., and Y. S. Lee. Dissection of a model for neuronal parabolic bursting, preprint.
26. Rinzel, J., and I. B. Schwartz. 1984. One variable map prediction of Belousov-Zhabotinskii mixed mode oscillations. J. Chem. Phys. 80:5610-5615.
27. Rinzel, J., and W. C. Troy. 1982. Bursting phenomena in a simplified Oregonator flow system model. J. Chem. Phys. 76:1775-1789.
28. Siska, J., L. Kubinova, and I. Dvorak. Time hierarchy in systems with general attractors. In Proceedings of the Fourth International Conference on Mathematical Modeling (Zurich, 1983). Pergamon Press, to appear.
29. Tikhonov, A. N. 1952. Systems of differential equations containing a small parameter multiplying the highest derivatives. Mat. Sb. NS (31) 73:575-585. (in Russian)
30. Troy, W. C. 1974. Oscillation phenomena in nerve conduction equations. Doctoral thesis, State Univ. of New York at Buffalo.
31. Wasow, W. Asymptotic Expansions for Ordinary Differential Equations. Wiley-Interscience, New York, 1965, pps. 297-303.

ON THE TOPOLOGICAL REPRESENTATION OF SIGNALS IN SELF-ORGANIZING NERVE FIELDS

Shun-ichi Amari
Faculty of Engineering, University of Tokyo
Minoru Maruyama
Central Research Laboratory, Mitsubishi Electric Corp.

1. Introduction

The brain can self-organize its structure based on environmental information. More specifically, when a set of signals is applied repeatedly to a nerve system, for each signal in the set, the system forms by self-organization sets of representative cells that are excited in response to particular signals but are not excited by any other signals. Such a representation may be regarded as a model of the outer world formed in the nervous system. This is one simple aspect of self-organization taking place in the brain. Physiologists have so far found hypercolumnar and microcolumnar structures in the primary visual cortex in which orientation-detecting cells, i.e., cells representative of lines of various orientations, are formed and fixed by self-organization. Moreover, hypercolumns are arranged retinotopically, and orientation-detecting cells are arranged in each hypercolumn in the order of preferable orientations. Physiologists have also found in various parts of the cerebrum cells which are responsive to specific shapes of objects, specific types of motions, and, in particular, faces of men or monkeys.

The present paper analyzes a simple mathematical model of a nerve field for the purpose of studying a plausible neural mechanism of self-organization which accounts for formation in the brain of a model of the outer world. When a set X of signals is received from the outer world, an inner model is generated by self-organization in a neural field F in the form of representative cells. This is a model of the outer world. When X has a topological or metrical structure, it is of interest to know how the topology of the signal space X is preserved in the model formed in the nerve field F. In other words, it is of interest to know how and where the representative cells are formed in the nerve field F. This problem has been studied by Willshaw and Malsburg [1974], Amari [1980,1983], Takeuchi and Amari [1979], Kohonen [1984], etc. The problem is closely related to retinotopical map and columnar microstructure formation in developmental biology. We give a mathematical formulation and a detailed analysis of this problem by using a simple model in continuation of the work of Amari [1980, 1983] and Takeuchi and Amari [1979].

2. Fundamental Equations

Let F be a model of a nerve field in which neurons are arranged to form a one-dimensional continuum. Let $u(\xi, t)$ be the average membrane potential of the neurons at spatial position ξ at time t, where ξ is a coordinate system specifying the position in the neural continuum (Fig. 1). The firing rate of the neurons at ξ is given in terms of their average potential $u(\xi, t)$ as

$$z(\xi, t) = f[u(\xi, t)],$$

where f is a monotone increasing function satisfying $f(u) = 0$ when $u < 0$. The neurons are recurrently interconnected within the field. Let $w(\xi, \xi')$ be synaptic efficacy of the connections from neurons at ξ' to those at ξ. Since the field is assumed to be homogeneous and isotropic, we can put $w(\xi, \xi') = w(|\xi - \xi'|)$. It is customary to assume that $w(\xi)$ is positive when $|\xi|$ is small and is otherwise negative.

The dynamical equation for neural excitation in the field is written as

$$\varepsilon \partial u(\xi, t)/\partial t = - u(\xi, t) + w * f(u) + S(\xi), \qquad (1)$$

where $S(\xi)$ is the total sum of input stimuli applied from the outside to the neurons at ξ and

$$w * f(u) = \int w(\xi - \xi') f[u(\xi', t)]d\xi'$$

Fig. 1

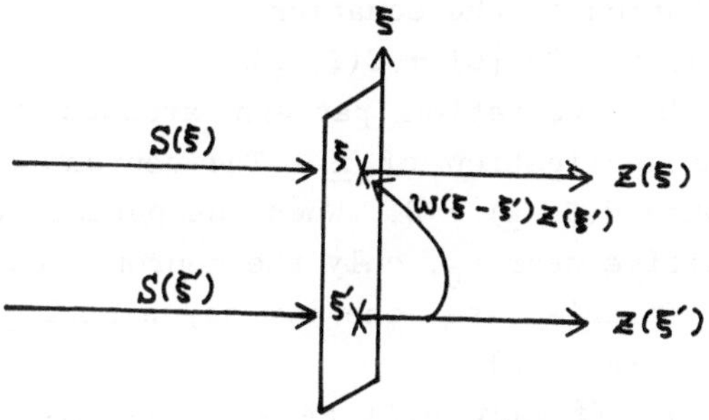

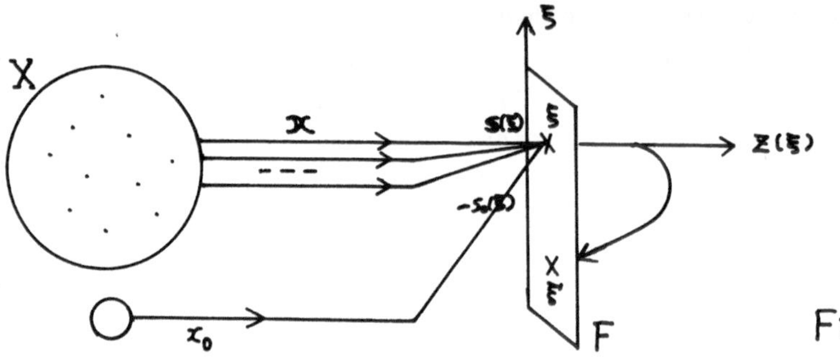

Fig. 2

is the total sum of recurrent stimuli from the other positions in the field.

Let X be a signal space consisting of vector signals $\underline{x} = (x_1, \ldots, x_n)$. A vector signal $\underline{x}$ is applied commonly to every part of the field at a time. Since the synaptic efficacy of input signal $\underline{x}$ depends on the position ξ, different positions receive different total sums of stimuli when a common $\underline{x}$ is applied to F. Let

$$\underline{s}(\xi) = [s_1(\xi), \ldots, s_n(\xi)],$$

be the vector denoting the synaptic efficacy at position ξ. Then the total sum of stimuli for $\underline{x}$ is represented by the inner product $\underline{s}(\xi)\cdot\underline{x}$. In addition to the excitatory signal $\underline{x}$, the field receives inhibitory inputs. We summarize all the inhibitory inputs to a single component x_0 and denote its synaptic efficacy by $s_0(\xi)$ (Fig. 2). For the sake of simplicity, x_0 is assumed to be constant. Therefore, when an input $\underline{x}$ is applied, the total sum of stimuli at position ξ is written as

$$S(\xi; \underline{x}) = \underline{s}(\xi)\cdot\underline{x} - s_0 x_0. \qquad (2)$$

Let $u(\xi, t; \underline{x})$ be the potential distribution in the field when $\underline{x}$ is applied from X. Let $U(\xi; \underline{x})$ be the equilibrium solution of (1) when $\underline{x}$ is applied from X, i.e., when $S(\xi) = S(\xi; \underline{x})$. Then, it is given by the solution of the equation

$$U(\xi; \underline{x}) = w * f[U] + S(\xi; \underline{x}). \qquad (3)$$

It represents the excitation pattern aroused in the field F in response to the application of $\underline{x}$. The neurons at ξ are excited in response to $\underline{x}$ when $U(\xi; \underline{x}) > 0$. When the pattern $U(\xi; \underline{x})$ is unimodal on F and is positive near ξ_0, only the neurons near ξ_0 are excited in response to $\underline{x}$. In this case, one may say signal $\underline{x}$ is represented by the neurons at ξ_0 (Fig. 3).

The synaptic efficacy $\underline{s}(\xi)$ as well as $s_0(\xi)$ is assumed to be modifiable, so that the response pattern $U(\xi; \underline{x})$ to $\underline{x}$ is changed as

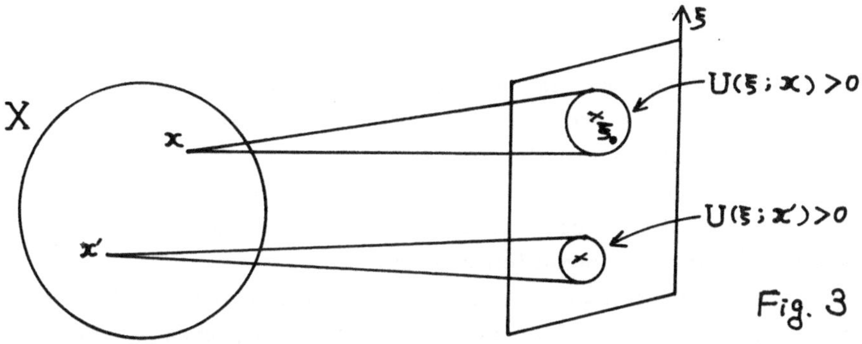

Fig. 3

synaptic efficacy changes. The Hebbian rule is assumed so that the dynamical equations for the synaptic modification are

$$\tau \partial \underline{s}(\xi, t)/\partial t = - \underline{s}(\xi, t) + c\ f[u(\xi, t; \underline{x})]\underline{x}, \qquad (4)$$

$$\tau \partial s_0(\xi, t)/\partial t = - s_0(\xi, t) + c'\ f[u(\xi, t; \underline{x})]x_0, \qquad (5)$$

where τ is a time constant much larger than ε and c, c' are also constants. The equations show that the synaptic efficacy $\underline{s}(\xi)$, $s_0(\xi)$ increases in proportion to the intensity of the input $\underline{x}$, x_0 when, and only when, the neurons at ξ are excited (i.e., $z(\xi, t) > 0$).

The environment signal space X would have a certain structure to which the nerve field adapts by self-organization. For example, some signals appear very frequently and hence are important to us, while others seldom appear and are not important. Such a structure is represented by a probability distribution $p(\underline{x})$ over X. In this simple case, we may take the ensemble average of (4) and (5), and replace $\underline{s}(\xi)$ and $s_0(\xi)$ in (1) or (2) by their ensemble averages. We then have the averaged equation

$$\tau \partial \underline{s}/\partial t = - \underline{s} + c\langle f[u(\xi, t; \underline{x})]\underline{x}\rangle, \qquad (6)$$

$$\tau \partial s_0/\partial t = - s_0 + c'\langle f[u(\xi, t; \underline{x})]x_0\rangle, \qquad (7)$$

in which $\underline{s}$ and s_0 are replaced by the averaged ones, and $\langle\ \rangle$ denotes the ensemble average, e.g.,

$$\langle a(\underline{x})\rangle = \int a(\underline{x})p(\underline{x})\, d\underline{x}.$$

The total stimulus $S(\xi; \underline{x})$ given by (2) changes as the synaptic efficacy changes according to (6), (7). By differentiating (2) with respect to t and substituting (6) and (7) in it, we have

$$\tau \partial S(\xi, t; \underline{x})/\partial t = - S + k(\underline{x}, \underline{x}') \circ f[u(\xi, t; \underline{x})],$$

where

$$k(\underline{x}, \underline{x}') = c\underline{x} \cdot \underline{x}' - c_0 x_0^2$$

denotes the similarity of two signals $\underline{x}$ and $\underline{x}'$ and

$$k \circ f[u] = \int k(\underline{x}, \underline{x}')f[u(\xi, t; \underline{x}')]p(\underline{x}')\, d\underline{x}'.$$

We thus have the following two equations

$$\varepsilon \partial u/\partial t = - u + S + w * f(u) \qquad (8)$$

$$\tau \partial S/\partial t = - S + k \circ f(u) \qquad (9)$$

which together describe the neural excitation and self-organizing behavior of the nerve field F under the application of signals in X. This may be regarded as an activator-inhibitor system, where the inhibitory effect is brought in through w * and k o. If the convolution operators w * and k o are replaced by diffusion terms, it reduces to a non-linear reaction-diffusion system, which has various pattern formation abilities. This suggests that our system also has an interesting pattern formation ability.

Since the time constant ε of neural excitation is far smaller than the time constant τ of synaptic modification, when we solve (9), we may replace $u(\xi, t; \underline{x})$ by its equilibrium solution $U(\xi; \underline{x})$. We then have

$$U(\xi; \underline{x}) = S(\xi; \underline{x}) + w * f(U), \tag{10}$$

$$\tau \partial S/\partial t = - S + k \circ f(U). \tag{11}$$

The final equilibrium $\bar{U}(\xi; \underline{x})$ is given by the equation

$$\bar{U} = k \circ f(\bar{U}) + w * f(\bar{U}). \tag{12}$$

3. The Equilibrium Solution and Topographic Organization

The primary visual cortex and the somato-sensory cortex have a topographic organization in the sense that the representative neurons of the positions of the retina or body surface are arranged in the fields continuously with respect to the retina or the body surface. This suggests that the cells representing signals in X are formed and arranged in F in accordance with the topology of X. In some cases, the arrangements in F have microstructures of the kind seen in the visual cortex. When X has a higher-dimensional topology, it is impossible to map X topologically to F. Hence, such a microstructure will help the representation of X in F to be as faithful as possible. In this case, the neurons in F are divided into a mosaic of topologically arranged groups, and a certain topological microstructure also appears within each group.

We now study a plausible mechanism accounting for the formation of such a topographic organization in a nerve field. For the sake of simplicity, we consider a simple case where F is one-dimensional, and the frequently presented signals also have a one-dimensional structure in X. More specifically, let $\underline{x}(\eta)$ be a curve in S parametrized by a scalar η, and let $p(\eta)$ be the probability density of signal $\underline{x}(\eta)$ (Fig. 4). It is assumed that the probability of a signal not in the curve is negligibly small.

We put

$$k(\eta, \eta') = \underline{x}(\eta) \cdot \underline{x}(\eta') - c_0 x_0^2,$$

$$U[\xi; \underline{x}(\eta)] = U(\xi, \eta), \quad S[\xi; \underline{x}(\eta)] = S(\xi, \eta).$$

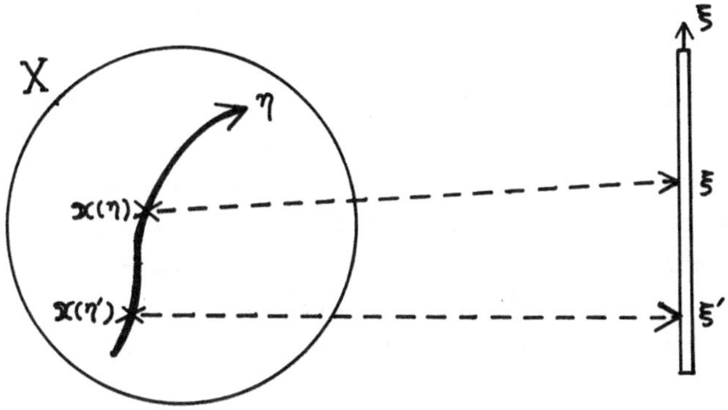

Fig. 4

Then, $S(\xi, \eta)$ denotes the total sum of stimuli which the neurons at ξ receive when $\underline{x}(\eta)$ is input, and $U(\xi, \eta)$ denotes the average final potential of the neurons at ξ when $\underline{x}(\eta)$ is input. These quantities change by self-organization, and their equations are the same as (10) and (11), where the operators $*$ and k o should be interpreted as

$$w * f(U) = \int w(\xi - \xi')f[U(\xi', \eta)]d\xi'$$

$$k \circ f(U) = \int k(\eta, \eta')f[U(\xi, \eta')]p(\eta')d\eta'.$$

When the one-dimensional structure in X is homogeneous in the sense that $p(\eta') = \text{const.}$ and

$$k(\eta, \eta') = k(\eta - \eta'),$$

it is expected that the final equilibrium solution is also homogeneous, satisfying

$$\bar{U}(\xi, \eta) = \bar{U}(\xi - \eta) = a(\xi - \eta),$$

by making an adequate scaling of ξ and η. If the function $\bar{U}$ or a is close to the delta function, the signal $\underline{x}(\eta)$ is represented by an excitation of the neurons near $\xi = \eta$, thus forming a natural correspondence $\eta \mapsto \xi = \eta$ (Fig. 5a). However, this natural homogeneous solution becomes unstable in some cases, and a non-homogeneous pattern will emerge in the nerve field (Fig. 5b).

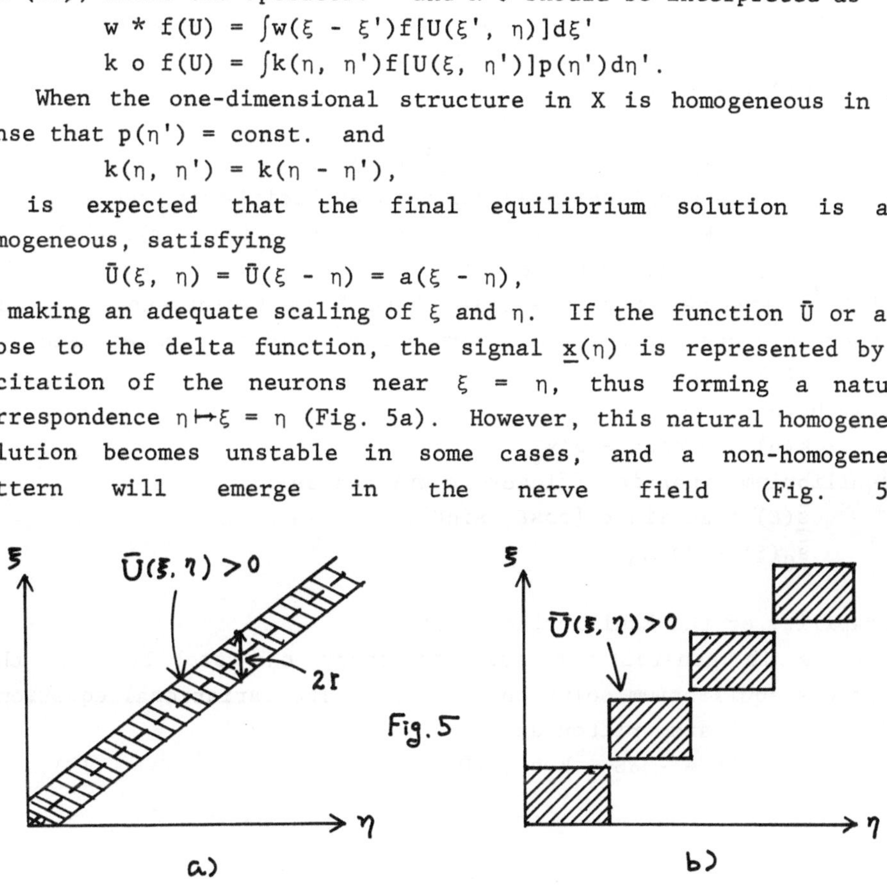

Fig. 5

This is considered the mechanism of the formation of a microstructure in the nerve field.

When the function $f(u)$ is approximated by the step function $1(u)$ ($1(u) = 1$ when $u > 0$ and otherwise $1(u) = 0$, i.e., neurons fire at the maximum rate when $u > 0$ and never fire when $u < 0$), we have the explicit form of the equilibrium solution. We can assume that the equilibrium excitation pattern with $\bar{U}(\xi - \eta) = a(\xi - \eta)$ is unimodal if the recurrent connection function $w(\xi)$ is of a strong lateral inhibitory type (Amari [1977], Kishimoto and Amari [1979]). Then for some constant r,

$$\bar{U}(\xi - \eta) > 0$$

hold for $|\xi - \eta| < r$ and $\bar{U}(\xi - \eta) < 0$ for $|\xi - \eta| > r$. (Obviously, $\tilde{U}(\xi, \eta) = \bar{U}(\xi - \eta - c)$ is also an equilibrium solution for any c. We put here $c = 0$). The explicit form of $\bar{U}$ is then written from (10) as

$$\bar{U}(\xi - \eta) = \int_{|\xi-\eta'|<r} k(\eta - \eta')d\eta' + \int_{|\xi'-\eta|<r} w(\xi - \xi')d\xi'.$$

In this representation, a signal $\underline{x}(\eta)$ is represented by the neurons in the interval $\xi = \eta - r \sim \eta + r$. The constant r is obtained from $\bar{U}(\pm r) = 0$. We thus have the equation to determine r,

$$W(2r) + K(2r) = 0, \qquad (13)$$

where

$$W(r) = \int_0^r w(\xi)d\xi, \quad K(r) = \int_0^r k(\eta)d\eta.$$

Using this r, the equilibrium solution is explicitly written as

$$a(\xi) = K(\xi + r) - K(\xi - r) + W(\xi + r) - W(\xi - r). \qquad (14)$$

$$\bar{U}(\xi, \eta) = a(\xi - \eta + c).$$

Let us consider a simple case where X is two-dimensional, and the frequently presented signals form a circle in it parametrized by

$$\underline{x}(\eta) = [\cos\eta, \sin\eta].$$

Then, we have

$$k(\eta) = c\cos\eta - c'x_0^2.$$

The equilibrium synaptic efficacy is written as

$$\bar{\underline{s}}(\xi) = 2c\sin r\,[\cos\xi, \sin\xi],$$

$$\bar{s}_0(\xi) = 2c'x_0.$$

4. Stability of the Equilibrium Solution

It is in general not easy to study the stability of the homogeneous equilibrium solution $\bar{U}(\xi - \eta)$. The variational equations of (10) and (11) are written as

$$\tau\partial\delta S/\partial t = -\delta S + k \circ \delta f(\bar{U}), \qquad (15)$$

$$\delta\bar{U} = \delta S + w*\delta f(\bar{U}). \tag{16}$$

In the present case, when f(u) is replaced by the step function 1(u), we have a tricky method of analyzing the variational equation (cf. Amari [1977], Takeuchi and Amari [1979]). Since the equilibrium solution $\bar{U}(\xi, \eta)$ satisfies

$$\bar{U}(\xi, \eta) > 0$$

in the region

$$R = \{(\xi, \eta) \mid |\xi - \eta| < r\},$$

$f[\bar{U}(\xi, \eta)] = 1$ in R and is otherwise equal to 0. Since k o and w * are integral operators, the effect of the variation $\delta f[\bar{U}]$ is expressed by integration over the region δR, where $R + \delta R$ is the region on which $\bar{U} + \delta\bar{U}$ is positive.

Let us represent the region in the form

$$R + \delta R = \{(\xi, \eta) \mid -r + \delta_1 r(\xi) < \xi - \eta < r + \delta_2 r(\xi)\},$$

where $\delta_1 r(\xi)$ and $\delta_2 r(\xi)$ denote the variations in the excited region. From the implicit function theorem, we have the following relation

$$\begin{bmatrix} \alpha - w(0) & w(r) \\ w(r) & \alpha - w(0) \end{bmatrix} \begin{bmatrix} \delta_1 r(\xi) \\ \delta_2 r(\xi) \end{bmatrix} = \begin{bmatrix} -\delta S(\xi, & \xi - r) \\ \delta S(\xi, & \xi + r) \end{bmatrix},$$

where

$$\alpha = -\partial\bar{U}(\eta + r, \eta)/\partial\xi = \partial\bar{U}(\eta - r, \eta)/\partial\xi,$$

and α is the slope of the waveform $a(\xi)$ at $\xi = \pm r$. Let us put

$$X_1(\xi) = \delta S(\xi, \xi - r),$$
$$X_2(\xi) = \delta S(\xi, \xi + r).$$

Then, we have the variational equations

$$\tau\dot{X}_1(\xi) = w(2r)\delta_1\ell(\xi) + k(2r)[\delta_2 r(\xi) - \delta_1 r(\xi)],$$
$$\tau\dot{X}_2(\xi) = w(2r)\delta_2\ell(\xi) + k(2r)[\delta_2 r(\xi) - \delta_1 r(\xi)],$$

where $\dot{}$ denotes the time derivative d/dt and

$$\delta_1\ell(\xi) = \delta_2 r(\xi - 2r) - \delta_1 r(\xi)$$
$$\delta_2\ell(\xi) = \delta_2 r(\xi) - \delta_1 r(\xi + 2r)$$

are the changes in the widths of the excited areas. Then, we have

$$\tau\begin{bmatrix} \dot{X}_1(\xi) \\ \dot{X}_2(\xi) \end{bmatrix} = A \begin{bmatrix} X_1(\xi) \\ X_2(\xi) \end{bmatrix}$$

$$A = \begin{bmatrix} w(m - n) + k(m - n\Delta) & w(m - n)\Delta^{-1} + k(m - n\Delta^{-1}) \\ w(m - n)\Delta + k(m - n\Delta) & -w(m - n) + k(m - n\Delta^{-1}) \end{bmatrix},$$

where $w = w(2r)$, $k = k(2r)$,

$$m = (\alpha - w(0))/D, \quad n = - w(2r)/D,$$

$$D = [\alpha - w(0)]^2 - [w(2r)]^2,$$

$$\Delta X(\xi) = X(\xi + r), \quad \Delta^{-1}X(\xi) = X(\xi - r).$$

In order to show the stability of the above differential-difference equations, we put

$$X_i(\xi, t) = \sum_{N=-\infty}^{\infty} x_i(N, t) \exp\{- N\nu\xi\}, \quad i = 1,2$$

Then, we have separate equations for each of the Fourier components $x_1(N, t)$ and $x_2(N, t)$. We do not write down its explicit form. By examining the eigenvalues of the coefficient matrices of the equations, we have the following theorem.

Theorem. When $k(2r) < 0$, the homogeneous equilibrium solution is stable, and when $k(2r) > 0$, it is unstable.

The theorem shows that the continuous homogeneous representation $\bar{U}(\xi, \eta)$ becomes unstable when $k(2r) > 0$. This condition implies that the continuous representation becomes unstable if the correlation of the two signals $\underline{x}(\eta)$ and $\underline{x}(\eta + 2r)$ is strong enough. In this case, the homogeneous solution breaks down and a discretized mosaic group structure emerges. This explains the reason that a nerve field can stably maintain the columnar microstructure during self-organizing processes.

5. Formation of Microstructure in Computer Simulated Experiments

We show some results of computer simulation, where the parameters are set to satisfy $k(2r) > 0$. We start from a continuous homogeneous representation shown in Fig. 6a. Since the homogeneous solution is unstable, a microstructure emerges as a final result in Fig. 6b.

We next consider the case where the signals are two-dimensionally distributed and are paremetrized by η and θ as

$$\underline{x}(\eta, \theta) = [\cos\eta, \sin\eta, \cos\theta, \sin\theta].$$

It is difficult to map these signals in a one-dimensional field. However, assume that the signals

$$[\cos\eta, \sin\eta, 0, 0]$$

are first presented to the one-dimensional field in such a way that the mosaic group structure shown in Fig. 7a appears. After formation of such a structure, the signals $\underline{x}(\eta, \theta)$ are presented to the field at a later stage of development. It is expected that a topological map appears in each group in the field. Fig. 7b shows a computer simulated experiment, in which such a phenomenon is observed.

References

Amari, S. [1977]: Dynamics of Pattern Formation in Lateral-Inhibition Type Neural Fields. Biological Cybernetics, Vol.27, pp.77-87

Amari, S. [1980]: Topolographic Organization of Nerve Fields. Bull. of Math. Biology, Vol.42, pp.339-364

Amari, S. [1983]: Field theory of self-organizing neural nets, IEEE Trans. Systems, Man and Cybernetics, Vol.SMC-13, No.9 & 10, pp.741-748

Kohohen, T. [1984]: Self-Organization and Associative Memory, Springer

Takeuchi, A. and Amari, S. [1979]: Formation of topographic maps and columnar microstructures, Biological Cybernetics, Vol.35, pp.63-72

Willshaw, D.J. and Malsburg, C. von der [1976]: How patterned neural connections can be set up by self-organization, Proc. Roy. Soc. Lond., B-194, pp.431-445

Fig. 6

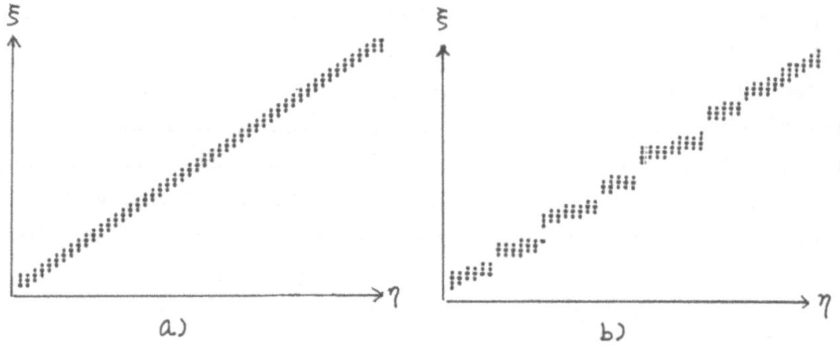

Fig. 7

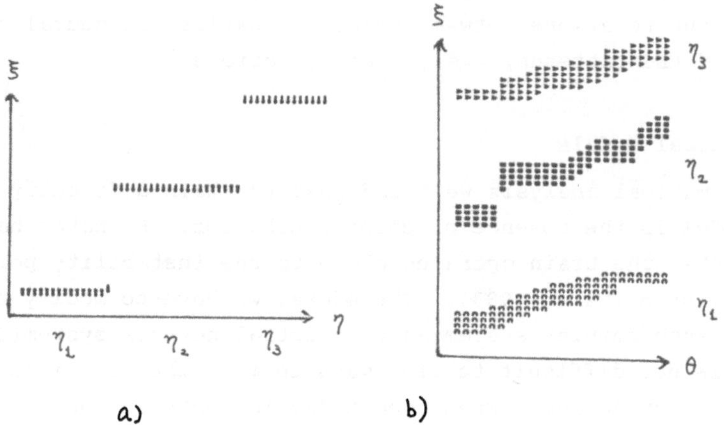

The Dynamics of a Glia-Modulated Neural Network and its Relation to Brain Functions

Xu Jinghua and Li Wei

Shanghai Institute of Biochemistry
Chinese Academy of Sciences, Shanghai, China

1. Introduction

The complex patterns of electroencephalograms(EEGs) have made it very difficult for physiologists and pathologists to interpret the relationships to brain functions. A possible way to approach this problem is to study the dynamic behaviors of a mathematical model of large scale neuron network based on the fundamental electrical properties of neurons and their interrelationships, as done by J.D. Cowan and his co-workers (Cowan 1974; Ermentrout and Cowan 1979, 1980). They proposed two coupled non-linear integro-differential equations, comprising two types of neurons (excitatory and inhibitory), and analysed them by bifurcation theory. Because of the statistical nature of their treatment, all complexities of neurobiological details, such as the plasticities of neurons and synapses, may be taken into consideration, but their results still are very difficult to compare with real EEG patterns, especially those of conscious human beings, which display highly irregular, quasi-random forms. These kinds of behaviors may also be found in general dynamic systems and are called chaotic. This paper is an investigation of the problems concerning the relations between chaotic behaviors of neural network dynamics and EEG patterns, i.e., brain functions.

2. Mathematical Models

By mathematical analysis we found that the main difficulty of Cowan's model is the absence of chaotic behavior. H. Haken has suggested that the brain operates close to the instability points of a dynamical system (Haken 1983). Therefore, we have to modify the model. In such complex system as the central nervous system(CNS), it certainly is not difficult to find ways to make the equations higher order or non-autonomous. From recent developments in neurobiology, however, we think that the most natural way is to take the role of neuroglial cells into consideration. The reasons are:

1. Glial cells outnumber neurons in many parts of the central nervous system of human beings, and their population can change under different physiological and pathological conditions.
2. The distribution of glia-neuron contacts varies greatly among brain regions as well as different types and portions of neurons. The contacts exist mostly the synaptic clefts (Wolff and Guldner 1978).
3. Glial cells are joined to one another by so-called gap-junctions, which have low resistance to electric current. Connecting through gap-junctions, they form local networks referred to as functional "syncytium" (Somjen 1983).
4. Neuronal activities alter the extracellular concentration of potassium. Glial cells influence this change in two ways: by acting as potassium buffers, redistributing the ions across their membranes and by restricting their free diffusion. Both mechanisms keep the potassium concentration within certain narrow limits. Within these limits glial cells may facilitate extra-synaptic interactions among those nerve processes sharing a common microenvironment. Such interactions would be greater under increased activities of the excitable membranes. Each role may follow a different time-course (Watson 1974).
5. Neuronal discharge evokes slow depolarization of glia, which is mediated by potassium ions released from neurons. During slow depolarization, the glia take up potassium ions by means of some active or passive processes to maintain the constancy of the neuronal environments, and glia play an essential role in the ion homeostasis of the brain (Wolff and Guldner 1978).
6. Glial cells exhibit characteristic slow depolarizations, which are closely parallel to the changes in potassium concentration. During epileptogenesis, potassium concentrations may reach levels that are 3 or 4 times greater than normal resting values. It has been proposed that the accumulation of potassium ions in the brain extracellular spaces increases the excitability to the point that sustained epileptic forms of discharge result (Prince et al. 1978).
7. The glia form a neurotransmitter pool. Total glutamate most likely is a major excitatory transmitter in the CNS. In nerve endings glutamate is derived directly from glucose and the glutamine in glia also may be used as a precursor for glutamate (Hamberger et al. 1978). Glia serve as dams limiting free diffu-

sion of transmitters (Somjen 1983).

8. Glia have specialized metabolic systems, for instance, the activity of Na-K-ATPase is markedly higher in freshly isolated astrocytes than that in neurons (Atterwill et al. 1984); a Ca independent S-100 protein modulated protein kinase is found in CNS (Qi 1985) and in astrocytes (Chen 1985).
9. Glia are excitable cells (Abbott 1985).
10. Glia can react to more complex patterns of stimulations. Rats kept in an enriched environment are better at solving problems; they have more neuroglia and higher rates of division (Watson 1974).

Taking all this experimental evidence into consideration, we have presented a glia-modulated neural network model to study the dynamics and its relation to brain functions (Xu and Li 1985).

The simplified models in the following have been proposed for the statistical study of a population of N neurons, which are divided into two types (excitatory and inhibitory). The i-th cell is connected to the j-th cell with K_{ij}, which is a spatial function representing the mean number of synapses. In our treatment, we take a statistical average over space. Each cell sums up its input currents linearly to generate a net change in membrane potential, which in a non-linear fashion propagates as current pulses. Such current pulses then stimulate other cells to start the process again. By the above cited experimental evidence and by the Nernst formula (Schmidt et al. 1979), we established the mathematical relation among the intercellular potassium of neurons and glia, the extracellular potassium, the neuron membrane potential and the effect of glia:

$$dE/dt = -E + (1 - E) \times J_e(C_1E - C_2I + G + P)$$

$$dI/dt = -I + (1 - I) \times J_i(C_3E - C_4I + Q)$$

$$dG/dt = (C_1 dE/dt - C_2 dI/dt) \times [\exp C_5(C_1E - C_2I + G + P)] \quad (1)$$

where $E(t)$ and $I(t)$ are the populations of excitatory and inhibitory neurons firing per unit time at the instant t; C_1, C_2, C_3, and C_4 represent the interconnections among neurons giving the mean numbers of synapses and C_5 is a control parameter; P and Q are the external stimuli to the net. The sigmoid functions J_e and J_i are taken as

$$J(z) = [1 + \exp(-\nu(z - \theta))]^{-1} - [1 + \exp(\nu\theta)]^{-1} \quad (2)$$

where z corresponds to the neuron membrane potential and θ the net threshold.

Eqs.(1) represent what happens in an epileptic seizure. Under normal functional states of the brain, we may have

$$dE/dt = -E + (1 - E) \times J_e(C_1E - C_2I + G + P)$$
$$dI/dt = -I + (1 - I) \times J_i(C_3E - C_4I + Q)$$
$$dG/dt = aE - bI + F(t) \quad (3)$$

To make the problem simpler, we assume that G has the form:

$$G = AM \times \cos(PDt) + BA$$

and substitute it into Eqs.(3).

3. The Numerical Results

Without external stimuli, $P = Q = 0$, the stationary solutions to the model systems are similar to those of the system involving only the first two equations in Eqs.(1) or (3). Under certain conditions on the parameters ($C_1 = 16$, $C_2 = 12$, $C_3 = 15$, $C_4 = 3$, $\nu_e = 1.3$, $\theta_e = 4$, $\nu_i = 2$, $\theta_i = 3.7$) there is a critical point G_0 of G such that the steady state will be unstable when G is below G_0 and stable otherwise. This point determines the initial value of G in numerical integration on Eqs.(1) and (3). The numerical results show the beautiful fit with the experimental results. Fig.(1) shows the behaviors of neurons and glia in Eqs.(1), which are synchronous and found also in real epileptic EEG patterns.

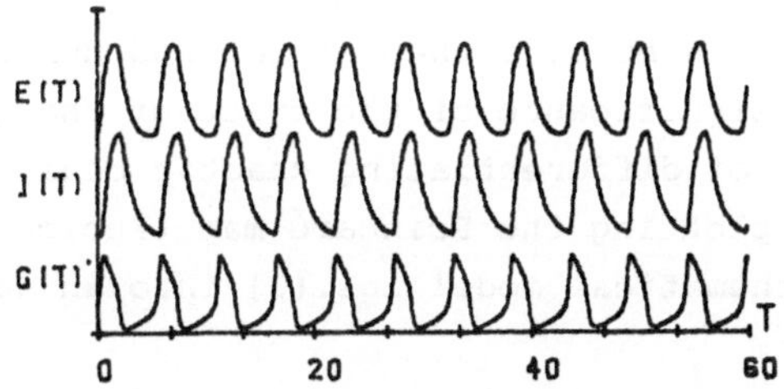

Fig. (1)
The synchronous activity in neuron-glia network of Eqs.(1). The control parameter $C_5 = 0.05$

Being an intrinsic non-linear oscillator, Eqs.(3) display a wide variety of periodic, multi-periodic, quasiperiodic and chaotic nature as a series of control parameters (AM, PD, BA) pass critical values. Figs.(2-3) show the chaotic behavior in phase space and time course. Comparing these results with common EEGs of a conscious human being, we also can see the qualitative fit of the model calculations with experimental recordings.

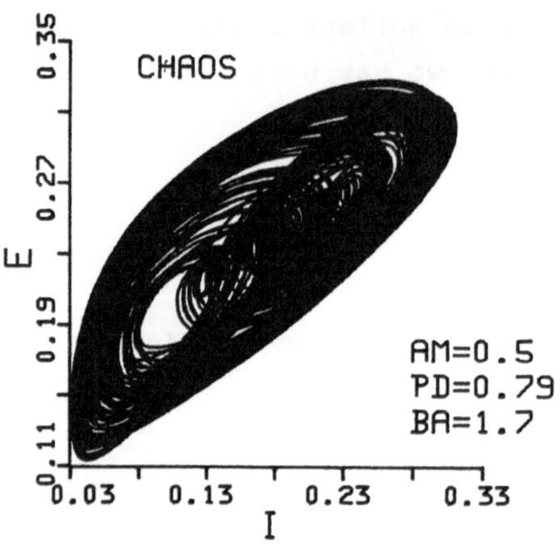

Fig. (2)
Phase plane diagram showing the evolution of the trajectories of chaotic state.

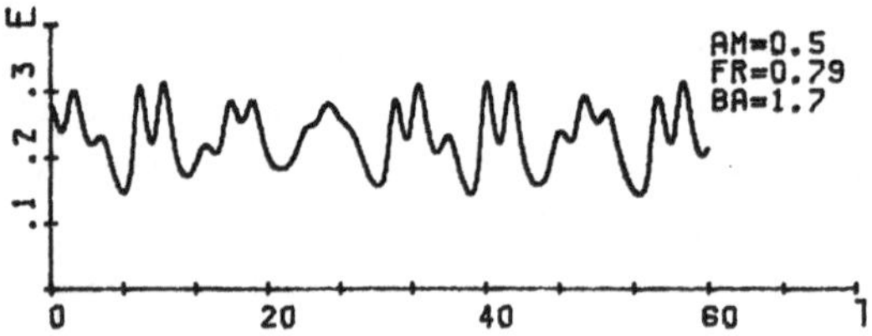

Fig. (3)
The time-course of excitatory neurons displaying chaotic behaviors.

4. Poincaré Maps and Fractals

To make the point clear, we present more theoretical and numerical work to explain the biological significance of the onset of chaos. There is another convenient way of differentiating chaotic behavior from quasiperiodic behavior by plotting the Poincaré map (Kubicek and Marek 1979). We change the mathematical model Eqs.(3) into an autonomous form:

$$dY_1/dt = -Y_1 + (1 - Y_1) \times J_e(C_1Y_1 - C_2Y_2 + Y_3)$$
$$dY_2/dt = -Y_2 + (1 - Y_2) \times J_i(C_3Y_1 - C_4Y_2)$$
$$dY_3/dt = -PD \times Y_4$$
$$dY_4/dt = PD \times Y_3 \qquad (4)$$

Analogous calculations yield a time sequence, which we have drawn in the phase plane of two variables (E, I), i.e., (Y_1, Y_2). A

Poincaré section is made at $Y_3 = AM$, $Y_4 = 0$, i.e., at the transient $t = 2\pi/PD \times k$ (k is an integer). Fig.(4) shows the irregular point routes of chaos, while Fig.(5) shows almost a closed circle representing the quasiperiodic behavior of the same system. The remarkable difference can be seen easily.

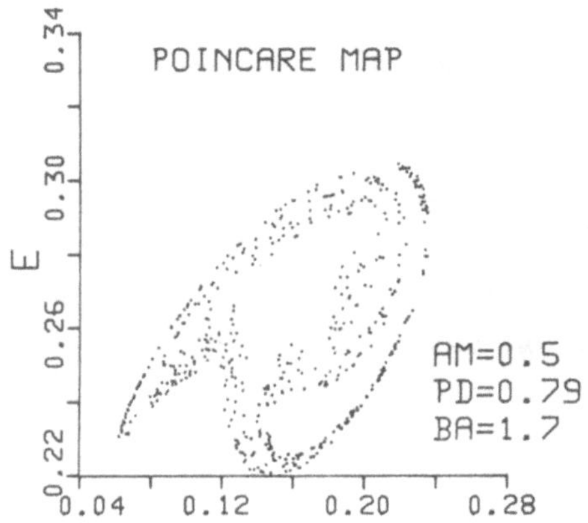

Fig.(4) The Poincare Map of Chaos

The values of the parameters $(C_1, C_2, C_3, C_4, \nu_e, \theta_e, \nu_i, \theta_i)$ are (16,12,15,3,1.3,4,2,3.7) respectively.

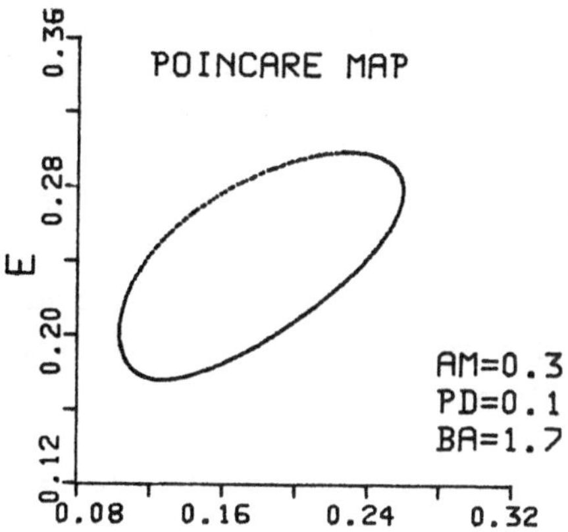

Fig.(5) The Poincare Map of a quasiperiodic solution.

The fractal dimension, which can give a useful criterion for characterizing the chaotic flows and strange attractors of dissipative dynamical systems, has been derived in numerous ways. Kaplan and Yorke(1979) suggested that, if in the nonmonotonous increasing sequence of Lyapounov exponents, from the first to the j-th are non-negative, whereas the $(j+1)$th becomes negative, then we have

$$D_0^1 = j + 1/|\lambda_{j+1} + 1| \times \sum_{i=1}^{j} \lambda_i \quad . \tag{5}$$

In another definition, Mori and Fujisaka(1980) showed that the fractal dimension of chaotic flow may be given as

$$D_0^2 = m^0 + m^+ \times (1 + |\lambda^+/\lambda^-|) \tag{6}$$

where m^0 and m^+ are the numbers of zero and positive Lyapounov exponents (LE) λ_i, and $\lambda^\pm$ are the mean values of positive and negative LE respectively. From numerical experiments for chaos as shown by LE (0.05, 0.00, 0.00, -0.14), we have a sequence of characteristic LE.

Then the order-0 dimensions D_0^1 and D_0^2 of the original system Eqs.(3) are about 2.4.

In yet a different manner, D_0, Grassberger(1983a, 1983b) defined the correlation dimension, i.e., the order-2 dimension D_2 as

$$D_2 = \lim_{\varepsilon \to 0} \mathrm{Log}C(\varepsilon)/\mathrm{Log}(\varepsilon) \tag{7}$$

where the correlation integral $C(\varepsilon)$ was defined as a time series x_1, $x_2, \ldots\ldots, x_k$ as

$$C(\varepsilon) = \lim_{M \to \infty} M^2 \times (\text{number of pairs } (x_i, x_j) \text{ with } |x_i - x_j| < \varepsilon). \tag{8}$$

M is the number of measurements. The distance $|x_i - x_j|$ is Euclidean. The actual time sequence $(x_1, x_2, \ldots\ldots, x_M)$ was embedded in a space of dimension n

$$x_i = (x_i, \ldots\ldots, x_{i+n-1}) \ . \tag{9}$$

There are K vectors $(K = M - n - 1)$ formed serially from the data. Then the discrete form of $C_n(\varepsilon)$ is

$$C_n(\varepsilon) = (1/M^2) \times \sum_{i=1}^{k} \sum_{j=i+1}^{k} \beta(\varepsilon - |x_i - x_j|) \ , \tag{10}$$

where

$$\beta(z) = \begin{cases} 0, & z \leq 0 \\ 1, & z > 0 \end{cases} \ . \tag{11}$$

If M and n are large enough, ε sufficiently small, it can be shown that

$$C_n(\varepsilon) \propto \varepsilon^{D_2} \tag{12}$$

where D_2 is the correlation dimension.

Considering the recording sequence $E(t)$, in Fig.(3) we plot $\mathrm{Log}C_n(\varepsilon)$ vs. $\mathrm{Log}(\varepsilon)$ for $n = 40, 41, \ldots, 44$. The time series consisted of $M = 4000$ points. We see a group of straight lines with the slopes of 2.0 ± 0.1, which is the correlation dimension as shown in Fig.(6). More attention is paid to comparison with P.E. Rapp's work (1985a, 1985b) on EEG experiments. By his results, the order-2 dimension is about 2.4 for a resting subject with eyes closed and 3.0 for a working subject who is order to count number backwards from 100 to 1. We reproduce these results qualitatively fairly well with our model.

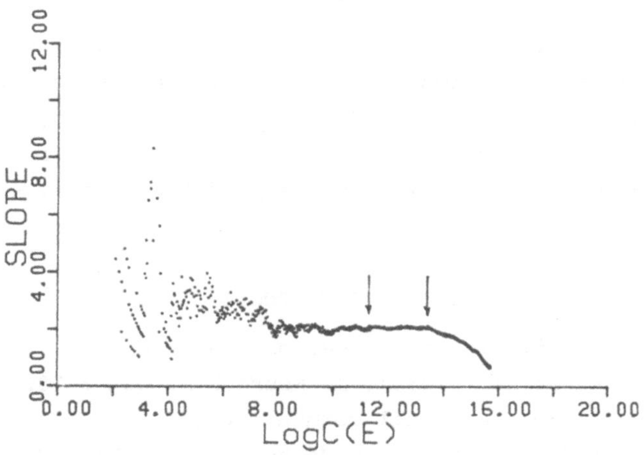

Fig.(6) Estimating attractor dimension. The plateau between arrows showing the dimension is about 2.0 ± 0.1.

5. Discussion

The role of glia long has been overlooked. Once we take it into consideration, the whole picture is changed, especially the neuron-glia network dynamics displaying chaotic behaviors. It is shown (Shaw 1980) that a simple system of equations displaying chaotic behavior is capable of acting as an information source, bringing into the macroscopic variables information that is not implicit in the initial conditions. Lyapounov exponents can in some simple cases measure the average amount of information flowing into (with negative real part) and out of (with positive real part) the system.

Numerical experiments have shown that in our dynamic models of the brain, the activities of glia cells modulate the neuron network to produce chaotic behaviors, which is just a process going from the normal to the strange attractor regime. From our point of view, the neuron network serves as a transmission apparatus by producing volleys of action potentials, just as in the case of radio communication the electronic circuit produce electromagnetic oscillations. All of these waves carry no information of their own. The information comes from frequency or amplitude modulations by some other sources outside the neuron network. The glia act as the internal information sources, while the outside stimulations serve as the external sources.

Since the glial population may change under different physiological and pathological conditions, our model also gives more room for further studies on problems such as memory, learning, etc. However, for more definite conclusion about the role of glia one has to wait for the progress of both experimental and theoretical works to be done in the future.

REFERENCES

Abbott, N.J. (1985). Trends in Neuroscience **8**, 191.

Atterwill, C.K. et al. (1984). J. Neurochem. **43**, 8-18.

Chen, Li-jun (1985). personal communication.

Cowan, J.D. (1974). Lectures on Mathematics in Life Science **6**, 121-133.

Ermentrout, G.B. and Cowan, J.D. (1979). J. Math. Biol. **7**, 205-280.

Ermentrout, G.B. and Cowan, J.D. (1980). SIAM J. Appl. Math. **38**, 1-21.

Grassberger, P. (1983a). Phys. Lett. **97A**, 227-230.

Grassberger, P. (1983b). Phys. Rev. A. **28**, 2591-2593.

Haken, H. (1983). Synergetics of Brain. 3-25. ed. E. Basar et al. Springer-Verlag.

Hamberger, A. et al. (1978). ibid. 163-172.

Kaplan, J. and Yorke, J. (1979). Lecture Notes in Math. **730**, 204.

Kubicek, M. and Marek, M. (1979). Computational Methods in Bifurcation Theory and Dissipative Structures. N.Y. Springer c1983.

Mori, H. and Fujisaka, H. (1980). Prog. Theor. Phys. **63**, 1044-1047.

Prince, D.A. et al. (1978). Dynamic Properties of Glia Cells. 281-303. ed. E. Schoffeniels et al. Pergamon Press.

Qi, Defang (1985). J. Neurochem. **43**, 256-260.

Rapp, P.E. et al. (1985a). Phys. Lett. **110A**, 335.

Rapp. P.E. et al. (1985b). In: "Non-linear Oscillations in Chemistry and Biology" ed. H.G. Othmer. Springer-Verlarg

Schmidt, R.F. et al. (1979). Grunriss der Neurophysiologie. Springer-Verlag.

Shaw, R. (1980). Z. Naturforsch. **36A**, 80-112.

Somjen, G. (1983). Neurophysiology: The Essential. Baltimore, Williams and Wilkins.

Watson, W.E. (1974). Physiol. Rev. **54**, 245-271.

Wolff, J.R. and Guldner, F.H. (1978). Dynamic Properties of Glia Cells. 115-118. ed. E. Schoffeniels et al. Pergamon Press.

Xu Jinghua and Li Wei (1985). Commun. in Theor. Phys. (Beijing, China). in press.

Self-Organization in Nervous Systems: Some Illustrations

P. Érdi and G. Barna

Central Research Institute for Physics
of the Hungarian Academy of Sciences
H-1525 Budapest, P.O.B. 49. HUNGARY

1. Introduction

The term "self-organization" is one of the most popular concepts of modern science. According to the synergetic approach, self-organizing systems are systems which can acquire macroscopic spatial, temporal, or spatio-temporal structures by means of internal processes (Haken 1985). Since dynamic behaviour of hierarchically arranged structures might be interpreted in terms of "self-organization", the nervous system can be considered as a prototype of self-organizing systems both conceptually (Szentágothai 1984, 1985) and mathematically (Amari 1983).

Two examples are given here (see also: Barna and Érdi 1986; Érdi and Barna 1986). First, an illustration is presented for self-organizing phenomena at the synaptic level in connection with interactions of certain neurochemical oscillators. Rhythmic behaviour is characteristic for biological systems. The stability of most biological rhythmic phenomena can be understood by the concept of limit cycle. The independence of the amplitudes and frequencies from the initial conditions might be associated with regular temporal patterns. The mutual interaction among different biological oscillators, however, might imply different phenomena as entrainment (phase locking), complex oscillations, quasiperiodicity, even chaos. The regular periodic operation of a synaptic level rhythmic generator requires a fine-tuned neurochemical control system. Even mild impairment of metabolism leading to acetylcholine synthesis might imply "abnormal" dynamic synaptic activity to be associated with neurological disorders. Similarly, chaotic behaviour associated with schizophrenics was found in a model of the central dopaminergic neuronal systems(King et al., 1984). In general terms, many neural disorders are characterized by changing the "normal" temporal patterns to "abnormal" behaviour. This abnormal dynamics might be interpreted as a result of disease occurring in a physiological control system operating within a range of control parameters (Guevara et al., 1983).

Second, a model for the formation and plasticity of ocular dominance columns is discussed. The algorithm to be presented is a direct extension of our model for the formation and plastic behaviour of retinotectal connections (Érdi and Barna 1984, 1985; Érdi and Szentágothai 1985) to a situation where axons coming from two presynaptic sheets compete for forming synapses with neurons of layer IV of visual cortex, establishing partially overlapping bands. A selection-based model for the ontogeny and plasticity of ocular dominance columns was given by von der Malsburg (1979). While in his model existence of a chemical marker molecule has been assumed, we adopt the hypothesis of the role of patterns of spontaneous activity for the segregation of the ocular dominance bands. Furthermore, von der Malsburg adopts a pure deterministic model, while according to our approach the formation of ordered neural structures can be interpreted with the concept of noise-induced transition. In accordance with the spirit of the theory (Horsthemke and Lefever 1984), fluctuations may operate as "organizing forces" during structure formation. The undoubtfully vague notion of organization might get a more precise meaning in this context.

The critical remark of Pellionisz and Llinás (1985) on the history of the notion of "self-organization" can be accepted. Still, our intention is to illustrate its conceptual power. "Self-organization" is of course not qualified as the only theoretical concept to catch the essential aspects of the nervous system. In any case, it seems to be the most appropriate framework to treat dynamic phenomena, while other approaches (e.g. the tensor network theory of the metaorganization, Pellionisz and Llinás 1985) might be explored as the functional geometrical theory of the nervous system.

2. Periodic perturbation of the transmitter-recycling system

2.1. Neurochemical background

The motivation for this work is the very fact that at least three different neurochemical and neurobiophysiological oscillatory phenomena appear at different hierarchical levels of cholinergic synaptic transmission (Érdi and Tóth 1981, Érdi 1983). A complete theory of the hierarchical regulation mechanism of dynamic synaptic activity would require the study of cooperation and competition among the three oscillators. As a first step, the model for the transmitter-recycling hypothesis adopted to explain the "integrated" synaptic activity has been supplemented by an independent, harmonic oscillator.

We restrict ourselves to study of the skeleton model of slow oscillation due to the "integrated synaptic activity". According to the basic assumptions of our four compartmental model, the state of the system is characterized by cytoplasmic ACh concentration (X), ACh concentration at the postsynaptic membrane surface (Z), Ch concentration near postsynaptic cell (W) and Ch concentration near presynaptic cell (Y).

The subprocesses of the model:

1.) transmitter release, cleft processes transmitter-receptor interaction;
2.) ACh hydrolysis (this is the most rapid subprocess);
3.) metabolic products (mostly choline) diffuse to the vicinity of the presynaptic cell;
4.) (re)uptake of Ch;
5.) autocatalytic synthesis of ACh.

The hypoxia and hypoglycemia due to disturbance of the metabolism leads to reduced ACh synthesis and neurological disorders (Gibson and Blass 1976a,b).

Many neurological diseases can be associated with metabolic disturbance of precursor of neurotransmitter synthesis (Wurtman et al., 1981).

According to the idealized neurochemical picture the rhythmic integrated synaptic activity is perturbed by another oscillation due to the presynaptic metabolism, and "complex oscillatory" phenomena might occur in consequence of the competition and cooperation of the distinct frequencies.

2.2. The model

A four-compartmental formal chemical model

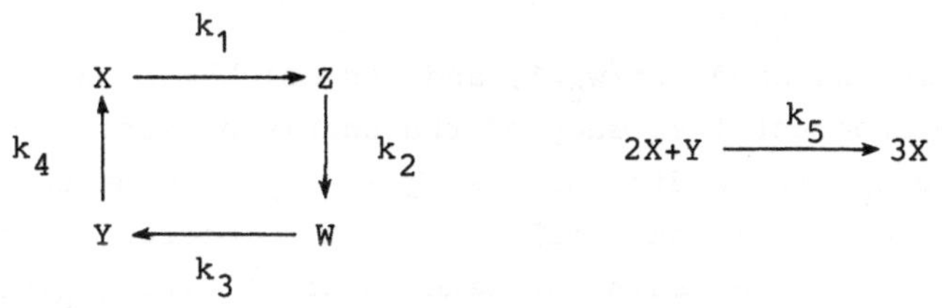

can be associated with the five subprocesses of integrated synaptic

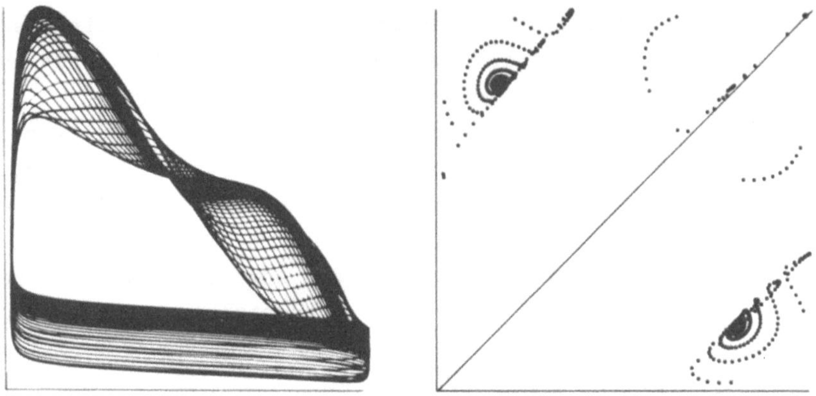

Figure 1. Phase plane and Lorenz plot of the perturbed variable for $w^*/w_0=2$.

activity mentioned in Sect.2.1. The massconserving model reaction having a single nonlinearity can exhibit sustained oscillation. The "limit shell" character of the oscillation has been visualized (Tóth 1985). Utilizing the neurochemical assumption according to which the choline uptake (however, not exclusively) controls the process, an independent choline oscillation is assumed. Therefore the model is:

$$\dot{x}(t)=-k_1x(t)+k_4y(t)+k_5x^2(t)y(t)$$
$$\dot{z}(t)=k_1x(t)-k_2z(t)$$
$$\dot{w}(t)=k_2z(t)-k_3w(t)$$
$$\dot{y}(t)=k_3w(t)-k_4y(t)-k_5x^2(t)y(t)+a\cdot\cos(wt+\phi)$$

To give a rough estimate for the first order rate constants, the $k_3 \approx k_4 \ll k_1 < k_2$ relations were adopted. (Simulation experiments to be documented here were done with parameter values $k_1=1$, $k_2=10$, $k_3=k_4=0.01$, $k_5=0.1$, and the value of the total mass = 110.)

2.3. Results and conclusions

i.) Phase locking

The main entrainment band ($w^*/w_0=1$) and its neighbourhood were examined (w_0 is the inherent frequency of the undriven system). In a certain range around w_0, its width increasing with a, the perturbing signal ($a/a_0=0.01$, where a_0 is the amplitude of the inherent oscillation), the width of the main entrainment band is about 0.02 ($w_0=0.23$). Other entrainment bands also have been studied (e.g. Fig.1.)

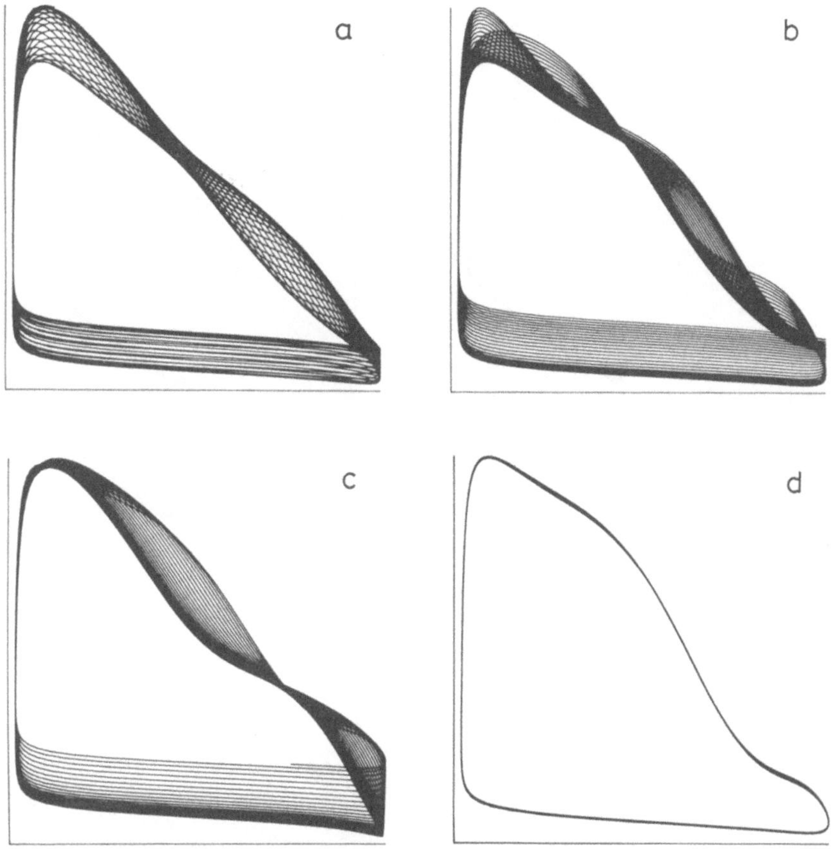

Figure 2. Phase plane diagram in different time regime a.)approaching to the phantom attractor b.-c.)intermediate regime d.)near the true attractor

ii.) Transients: they can be interesting

In principle it is well known that infinite computational length would be necessary to obtain "complicated", asymptotic behaviour. The study of attractors does not make it superfluous to follow the complicated transient dynamics during intermediate time regimes (for the role of transient chaos, see Kantz and Grassberger 1985). In the vicinity of $w^*/w_0=2$ for $a=0.5$, numerical calculations demonstrated the existence of a phantom attractor, which seems to show 22T periodicity. It can be interpreted as a quasistable temporal structure. Though the "phantom structure" seemed to be stabilized, a very slight spontaneous modification of the trajectory implied "destabilization" leading to "simple" limit cycles (Fig.2-3).

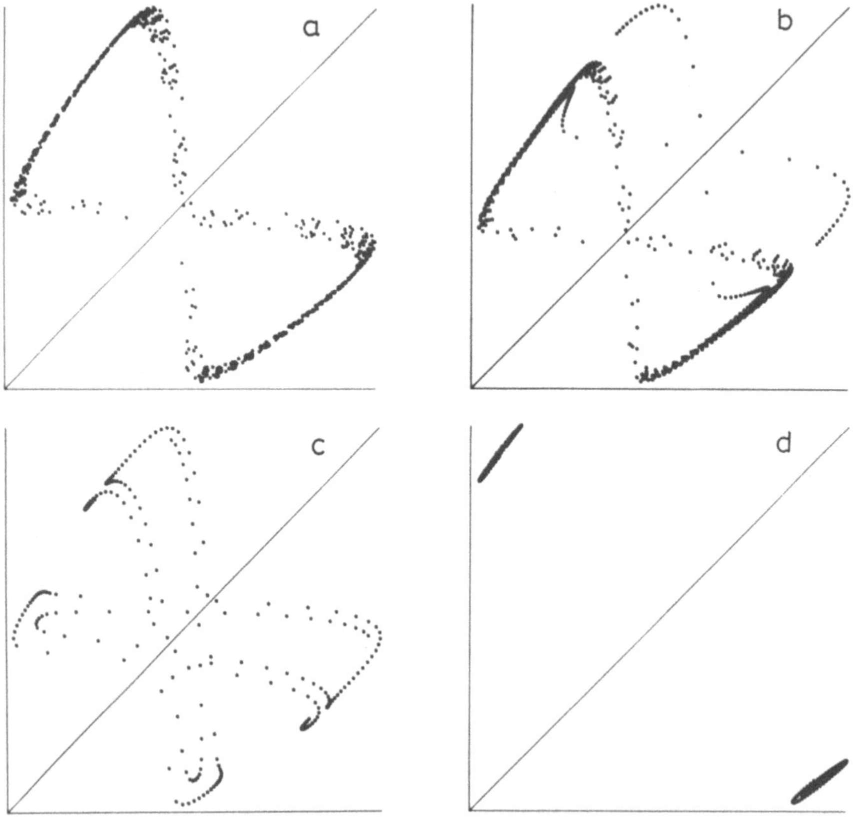

Figure 3. Stroboscopic representation of perturbed variable a.)quasi-stable "butterfly" shape b.)decay of the quasi-stable shape and start to tend a new shape c.)tending to a new attractor d.)arriving to a fix point of the Poincaré plot.

From a formal point of view it was clearly demonstrated that slight modification of the parameters of perturbation might imply transition to different kinds of complex periodic phenomena. It is well-known that periodically perturbed oscillatory systems might give rise to "complex phenomena, such as phase locking, quasiperiodicity and chaos, and these facts have been illustrated as well in neurophysiological experiments and models (see reference in Guevara et al., 1983). However, neurochemistry adopts a much more static approach than the Hodgkin-Huxley equation based neurophysiology of single nerve cells. What we wanted to verify is that the regular periodic operation of synaptic level rhythmic generators requires a fine-tuned neurochemical control system. Mild impairment of transport processes might imply "abnormal" dynamic synaptic activity. From a formal point of view the model - being a four-dimensional autonomous system (four variables minus one constraint plus perturbation) - could be a candidate to exhibit hyper-toroid and hyperchaotic (Rössler 1983) phenomena. The detailed analytic examination of the model would be very important and feasible (Rössler and Hudson 1985). Speculations based on calculations partially presented here seem to be beyond the experimental realm.

3. Formation and plasticity of ocular dominance columns

3.1. Neurobiological background

One of the greatest challenges for neurobiology is the problem of the mechanism of ontogenetic formation of ordered neural structures. It is more or less accepted that the establishment of neural networks during embryonic and postnatal development does not result exclusively from genetic mechanisms, but interactions with the environment contribute to the formation of adult connectivity. The environment could operate either through direct, instructive, or indirect, selective, mechanisms (Changeux 1984).

Ocular dominance columns or stripes are characteristic examples of ordered neural structures. It is well known, that the term "ordered" is here in connection with the fact that projections between the eye and brain are arranged retinotopically. This means that neighbouring presynaptic cells project through their fibers onto neighbouring cells of the postsynaptic sheet. Retinotopic map of the cortex might be considered point-to-patch rather than point-to-point representations (e.g. Gilbert 1985).

In normal cats and monkeys the afferents from the lateral geniculate nucleus (LGN) laminae corresponding to the the two eyes innervete common target structures, e.g. layer IV of visual cortex and form partially overlapping bands. This alternating termination of LGN afferents showing about 300-400 μm wide periodicity is thought to be the anatomic substrate for the ocular domains (Hubel and Wiesel 1972).

It is likely that ocularity domains are incompletely formed in monkeys and absent in kittens immediately after birth. While the normal development leads to near-periodic spatial patterns, visual deprivation during the critical period results in severe symmetry-breaking of the width of ocularity domains (Wiesel 1982). The plastic behaviour of ocular dominance columns is still under investigation (see e.g. Tieman 1984, Mower et al., 1985).

Activity patterns heavily influence the formation and plastic behaviour of ocularity domains. It was demonstrated (Fraser 1985) that treatment with tetrodotoxin (TTX) of one eye of a kitten from 2 to 6 weeks of age prevent the normal development of ocular dominance columns. Since ocular dominance columns develop at least to some extent before birth in monkeys if activity does play a role in columnar segregation, spontaneous activity must be sufficient (Stryker 1982).

3.2. The model

The formation of mapping between two one-dimensional presynaptic and one postsynaptic array of cells is investigated (k_L and k_R are the length of the two presynaptic chains, ℓ is the length of the post-synaptic chains; the subscripts refer to the terms left and right), m^L, m^R and $\underline{a}$ denote the two presynaptic and one postsynaptic activity vectors. The activity of the presynaptic chains is determined by unstructured stimuli, only one randomly selected element in both presynaptic activity vector has non-zero value. (The physiological relevance can be improved by taking into consideration correlations among presynaptic activities. Amari, personal remark).

The activity of the postsynaptic chain increases due to the transfer of presynaptic information and decreases by a first-order decay process. The evolution equation is:

$$a_j := a_j + k_2\left(\sum_{i=1}^{m_L} m_i^L\, S_{ij}^L + \sum_{i=1}^{m_R} m_i^R\, S_{ij}^R\right) - k_3 a_j \qquad \forall\ j \qquad (1)$$

where S_{ij}^L and S_{ij}^R are the elements of the matrices of synaptic strengths. Synaptic strength can be modified by three different factors (the superindex means that it refers both to L and R):

$$S_{ij}^* := S_{ij}^* + k_4 m_i a_j - k_5 (m_i - S_{ij}^* a_j) \qquad \forall\ i,j \qquad (2)$$

This equation might be associated with Hebb's rule supplemented with a selective decreasing term (Hirai 1980). The second factor is the modification due to the effects of immediate neighbours (the motivation of this step comes from the work of Kohonen 1982). For a general element (being not on the boundary):

$$S_{ij}^* := S_{ij}^* + k_a (S_{i-1,j}^* + S_{i+1,j}^*)/2 \qquad \forall\ i,j \qquad (3)$$

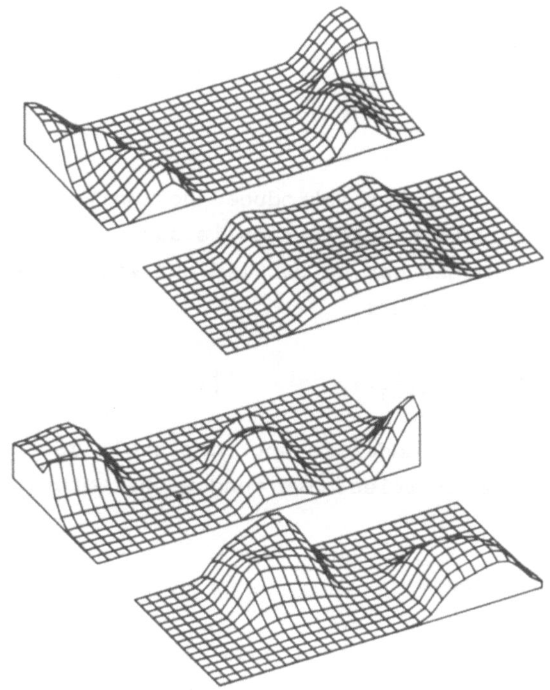

<u>Figure 4.</u> Locally and globally ordered structure with pure deterministic and superimposed random learning rules respectively.

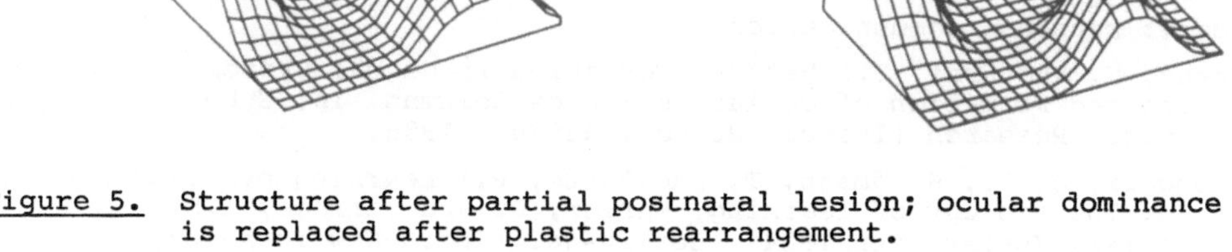

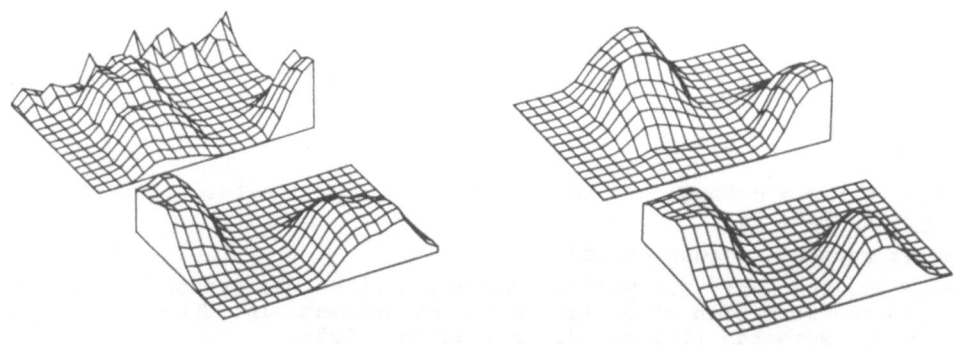

<u>Figure 5.</u> Structure after partial postnatal lesion; ocular dominance is replaced after plastic rearrangement.

Similarly, the effect of the other two elements is taken into consideration, the precise formulation will not be given here. The third factor is the result of a normalization procedure.

$$S^*_{ij} := \frac{S^*_{ij}}{\left(\frac{\ell}{k_R + k_L}\right)^{\frac{1}{2}} \sum_{n=1}^{k_*} S^*_{nj}} \qquad \forall\, i,j \qquad (4)$$

Normalization procedure is introduced not only for rows as in (4), but also for columns, specification is again neglected.

Simulation experiments can be carried out with a deterministic and stochastic "learning rule":

$$S^*_{ij} := \max(0, S^*_{ij} + k_6 \xi^*_{ij}), \qquad \xi^*_{ij} \in N(0,1) \qquad \forall\, i,j \qquad (5)$$

(Even in the case $k_6=0$ the model is not completely deterministic, since the presynaptic activities have random character).

3.3 Results and conclusions

Deterministic simulation experiments (i.e. $k_6=0$) resulted mainly in locally ordered structures. The inclusion of external noise can drastically modify the macroscopic structure: globally ordered near-periodic structure appears during the simulation of ontogeny (Fig.4). The model is also capable of describing plastic properties. Partial postsynaptic lesion and rearrangements of synaptic connection are illustrated (Fig.5).

References

Amari, S.: Field theory of self-organizing neural nets. IEEE SMC-13(741-748)1983.

Amari: personal communication.

Barna, G. and Érdi, P.: Pattern Formation in Neural Systems II. Noise-induced Formation of Ocular Dominance Columns In: Cybernetics and System Research (Trappl, R. ed.) Reidel, 1986.

Changeux, J.-P., Heidmann, T. and Patte, P.: Learning by selection. In: The biology of learning, Marler, P. and Terrace, H.S., eds. Dahlem, Konferenzen 1984. Berlin-Heidelberg-New York-Tokyo: Springer, pp.115-133.

Érdi, P.: Hierarchical thermodynamic approach to the brain. Int. J. Neurosci. 20(193-216)1983.

Érdi, P. and Barna, G.: Self-organizing mechanism for the formation of ordered neural mappings. Biol. Cybernetics 51(93-101)1984.

Érdi, P. and Barna, G.: Self-organization of neural networks: noise-induced transition. Phys. Lett. 107A(287-290)1985.

Érdi, P. and Barna, G.: Pattern Formation in Neural Systems. I. Autorhythmicity, Entrainment, Quasiperiodicity and Chaos in Neurochemical Systems. In: Cybernetics and System Research (Trappl, R. ed.) Reidel, 1986.

Érdi, P. and Szentágothai, J.: Neural connectivities: between determinism and randomness. In: Dynamics of macrosystems. Eds. Aubin, J.-P., Saari, D. and Sigmund, K. Lect. Notes in Economics and Mathematical Systems. Vol.257, Springer, 1985. pp.21-29.

Érdi, P. and Tóth, J.: Oscillatory phenomena at the synapse. Adv. Physiol. Sci. 34(113-121)1981.

Fraser, S.E.: Cell interactions involved in neuronal patterning: an experimental and theoretical approach. In: Molecular basis of neural development. Edelman, G.M.: Gall, W.E. and Cowan, W.M., eds. Neuroscience Res. Found. 1985. pp.481-507.

Gibson, G.E., and Blass, J.: Inhibition of acetylcholine synthesis and of carbohydrate utilization by maple-syrupurine disease metabolites. J. Neurochem. 26(1073-1078)1976a.

Gibson, G.E. and Blass, J.: Impaired synthesis of acetylcholine in the brain accompanying mild hypoxia and hypoglycemia. J. Neurochem. 27(37-42)1976b.

Gilbert: Horizontal integration in the neocortex. TINS8(160-165)1985.

Guevara, M.R., Glass, L., Mackey, C. and Shrier, A.: Chaos in Neurobiology. IEEE Trans. Systems, Man and Cybernetics SMC-13(790-797) 1983.

Haken, H.: Application of the maximum information entropy principle to self-organizing systems. Z. Phys. B. 61(335-338)1985.

Hirai, Y.: A new hypothesis for synaptic modification: an interactive process between postsynaptic competition and presynaptic regulation. Biol. Cybernetics 36(41-50)1980.

Horsthemke, W. and Lefever, R.: Noise-induced transition. Theory and applications in physics, chemistry and biology, Springer: Berlin-Heidelberg-Tokyo, 1984.

Hubel, D.H. and Wiesel, T.N.: Laminar and columnar distribution of geniculo-cortical fibers in the macaque monkey. J. Comp. Neurol. 146(421-450)1972.

Kantz, P. and Grassberger, P.: Repellers, semi-attractors and long-lived chaotic transients. Physica 17D(75-86)1985.

King, R. Barchas, J.D. and Huberman, B.A.: Chaotic behaviour in dopamine neurodynamics. Proc. Natl. Acad. Sci. USA 81(1244-1247) 1984.

Kohonen, T.: Analysis of a simple self-organizing process. Biol. Cybernetics 44(135-140)1982.

Malsburg, Ch. von der: Development of ocularity domains and growth behaviour of axon terminals. Biol. Cybernetics 32(49-62)1979.

Mower, G.D., Caplan, C.J., Christen, W.G. and Duffy, F.H.: Dark rearing prolongs physiological but not anatomical plasticity of cat visual cortex. J. Comp. Neurol. 235(448-466)1985.

Pellionisz, A. and Llinás: Tensor network theory of the metaorganization of functional geometries in the CNS. Neuroscience 16(245-273) 1985.

Rössler, O.E.: The Chaotic hierarchy. Z. Naturforsch. 38a(788-801) 1983.

Rössler, O.E. and Hudson, J.L.: A piecewise-linear hierarchy. (Int. Symp. on Math. Biol. Nov. 1985, Kyoto, Japan: Abstract)

Stryker, M.P.: Role of visual afferents activity in the development of ocular dominance columns. In: Rakic, P. and Goldman-Rakic, P.S. Development and modifiability of the cerebral cortex. Neurosci. Res. Progr. Bull. 20(1982).

Szentágothai, J.: Downward causation. Ann. Rev. Neurosci. 7(1-11)1984.

Szentágothai, J.: Theorien zur Organisation und Funktion des Gehirns. Naturwissenschaften 72(203-309)1985.

Tieman, S.B.: Effects of monocular deprivation on geniculocortical synapses in the cat. J. Comp. Neurol. 222(166-176)1984.

Tóth, J.: A mass action kinetic model of neurochemical transmission. In: Dynamic phenomena in neurochemistry and neurophysics: theoretical aspects (Érdi, P. ed.), Budapest, 1985. pp.52-55.

Wiesel, T.N.: Postnatal development of the visual cortex and the influence of environment. Nature 299(583-591)1982.

Wurtman, R.J., Hefti, F. and Melamed, E.: Precursor control of neurotransmitter synthesis. •Pharmacol. Rev. 32(315-355)1981.

TOWARD MOLECULAR SENSORY PHYSIOLOGY: MATHEMATICAL MODELS

Lee A. Segel
Department of Applied Mathematics
Weizmann Institute of Science
Rehovot 76100, Israel

Introduction

A major endeavor in modern biology is the effort to unravel the molecular mechanisms of sensory signalling. Here I shall briefly review some of the work I have been doing in this field together with several collaborators. The research to be described has centered in two areas, models for the control of neurotransmitter release and a new molecular model for adaptation at the receptor level.

Control of neurotransmitter release

The work on neurotransmitter release was done in collaboration with Prof. Hanna Parnas of the Department of Neurobiology, Hebrew University of Jerusalem. Our strategy was to seek out striking qualitative experimental findings and to try to construct a model of release that would explain a number of such findings. With a few parameters at hand one can often easily explain one or two major qualitative results, but confidence grows in a theory that can explain several such results.

Our goal was to arrive at theories at the molecular level, but as will be seen, more phenomenological approaches were called for until both theory and experiment had progressed further than the state they were in when we started.

Nerves usually communicate with other nerves or muscles via the release (probably from vesicles) of a chemical neurotransmitter that diffuses across a narrow intracellular gap and affects the target cell. The release of neurotransmitter is triggered by the arrival of an electrical impulse. Among other things, this impulse alters the permeability of the signalling cell's membrane to calcium, which flows in from the extracellular medium and somehow plays a major role in inducing transmitter release. For background neurophysiology see Kuffler, Nichols, and Martin (1984).

It has been generally agreed for some time that intracellular calcium controls release, but it is difficult to manipulate intracellular conditions. Thus the total release L induced by a single impulse has been measured as a function of extracellular calcium C_e. The result is an S-shaped increasing curve, beginning at the origin, which saturates (i.e. approaches a horizontal asymptote) when C_e gets large. Plots of $\log L$ vs $\log C_e$ appear to have horizontal asymptotes both at low and high values of $\log C_e$. Various features of the curves depend on the magnitude of the stimulating impulse, i.e. on the depolarization of the cell (Cooke, Okamoto, and Quastel, 1973).

Suppose a control impulse is given and a release L_1 is measured. If a second impulse is given a few milliseconds later, the second release L_2 on the average is larger than the first. The quotient $F = L_2/L_1$ is called the *facilitation*. When the time interval between the two impulses is relatively short, F is found to be a decreasing function of C_e; when the time interval is relatively long, F increases with C_e (Rahamimoff, 1968).

If intracellular calcium controls release, then models must contain an equation for the total calcium entry E induced by the impulse. Another equation must represent evidence that an elevation in intracellular calcium is countered by homeostatic mechanisms: an excess in the intracellular calcium concentration C above the normal resting level C_r is removed via uptake to cell organelles and/or resecretion. A third equation must express the dependence of the total release L on C.

Parnas and Segel (1980, 1981) postulated the following simple phenomenological equations and showed that they could explain several major experimental findings such as those that we have described.

Entry:

$$E = \frac{\varepsilon C_e^r}{K_\varepsilon + C_e^r}. \tag{1}$$

Release:

$$L(C) = \frac{\lambda C^n}{K_\lambda + C^n}, \qquad C = C_r + E(C_e). \tag{2}$$

Removal:

$$\frac{dC}{dt} = -\frac{\mu C^s}{K_\mu + C^s} \text{ when } C > C_r; \qquad \frac{dC}{dt} = 0 \text{ when } C = C_r. \tag{3}$$

Here the constants r, n and s measure the degree of cooperativity; K_ε, K_λ, and K_μ are measures of saturation; ε, λ, and μ represent the maxima of the three processes.

One example of the use of this approach is the following general formula (Parnas and Segel, 1981), merely an application of the chain-rule, for the observed slope in a log-log plot of release $\mathcal{L}$ as a function of C_e:

$$\frac{d\log \mathcal{L}(C_e)}{d\log C_e} = \left.\frac{d\log L(C)}{d\log C}\right|_{C=C_r+E(C_e)} \cdot \frac{d\log E(C_e)}{d\log C_e} \cdot \frac{E(C_e)}{C_r + E(C_e)}. \tag{4}$$

The standard rationale for examining the log-log slope is this. If the experimental results can be approximated by

$$\mathcal{L}(C_e) \approx \frac{\mathcal{L}_{\max} C_e^{\ell}}{K + C_e^{\ell}} \tag{5}$$

then measurements of the maximum slope (which occurs at low values of C_e) in a log-log plot provide an estimate of ℓ. Formula (4) shows how the "phenomenological cooperativity" ℓ is related to the cooperativity of the constituent processes that yield the overall release. In spite of the complications that stem from the detailed modelling, the theory can be used to design experiments to pinpoint the effects of drugs that alter release (Parnas and Segel, 1982).

If a preliminary train of impulses is applied to the preparation, measurements of the decay of facilitation F as a function of the time t between a control and a test impulse show as many as four different time scales. A number of investigators concluded that the measurements of $F(t)$ reflected the superposition of several linear processes. There were two main reasons for these conclusions: the decay curves could be well approximated by exponentials with different time constants, and these "exponents" could be individually modified by various treatments – for example the addition of various agents or changes in the extracellular calcium concentration C_e. Parnas and Segel (1983) showed that nonetheless linear models could not fit the facilitation data while their nonlinear model could describe changes with C_e of straight line segments that fit data in semilog plots of $F(t)$. A general moral was inferred by Parnas and Segel: "Even if data seems to be representable by several independently alterable exponentials one must be cautious in drawing inferences concerning the number, linearity, or independence of the underlying processes."

It is conventionally assumed that neurotransmitter release is terminated by the rapid return of the intracellular calcium concentration, owing to the removal processes, to values not far from the resting level. It was demonstrated by Parnas and Segel (1984) that such an assumption leads to the conclusion that total release is an unsaturated function of intracellular calcium and that the duration of release is strongly affected by the level of extracellular calcium. These results are contrary to experiment. Parnas and Segel (1984) were thus led to assert that the change of another quantity, not intracellular calcium, must be the primary of cause of release termination.

A major reason for the good progress made in the studies described so far was close cooperation between the theorists and experimentalists. Hanna Parnas was the central figure, participating not only in the theoretical work but also directly cooperating in the design of experiments carried out by Prof. Itzhak Parnas of the Department of Neurobiology, Hebrew University of Jerusalem, and Prof. Joseph Dudel of the Physiology Institute, Technical University, Munich. The majority of the papers by the Parnas' and Dudel appeared in *Pflügers Archiv.* The papers cited here contain references to other works in the series.

The model (1), (2), (3) was based on experiments in the literature. Parnas, Dudel, and Parnas (1982) and Parnas, Parnas, and Dudel (1982) performed experiments especially designed to test the model. They found good qualitative agreement with the theory. Semi-quantitative agreement was obtained when the removal equation was generalized to include the sum of two cooperative saturating processes, one of which is inhibited by high extracellular calcium and needs high extracellular sodium – and is thus probably a sodium-calcium exchange process. Another important experimental conclusion was an independent demonstration, reinforcing earlier conclusions of Llinas, Steinberg, and Walton (1981), that depolarization not only effects calcium entry, as had been quite well established by earlier investigators, but also directly influences release (Dudel, Parnas, and Parnas, 1983). Such experiments led Dudel, Parnas, and Parnas (1983) to the hypothesis that depolarization was the fast changing "other" quantity that plays a major role in the control of release. Parnas, Dudel, and Parnas (1986) postulated that there is a depolarization-regulated molecule or complex S that can bind intracellular calcium C, and that several complexes of S and C are required to initiate release.

The models finally have reached the molecular level. This is appropriate because it has now been shown that such models can be challenged by adapting an experimental technique pioneered by Katz and Miledi (1965) to study the time course of release. This technique was employed in a paper by Parnas, Parnas, and Segel (1986) that combines theoretical and experimental findings. Comparison of theoretical results with measurements of the early rise of the time course of release was shown capable of rejecting the first of the following two major kinetic schemes to account for the release L:

$$T \underset{k_{-1}(D)}{\overset{k_1(D)}{\rightleftarrows}} S, \qquad S + nC \underset{k_{-2}}{\overset{k_2}{\rightleftarrows}} (SC^n), \qquad (SC^n) + V \xrightarrow{k_3} L. \tag{6}$$

$$T \underset{k_{-1}(D)}{\overset{k_1(D)}{\rightleftarrows}} S, \qquad S + C \underset{k_{-2}}{\overset{k_2}{\rightleftarrows}} (SC), \qquad n(SC) + V \xrightarrow{k_3} L. \tag{7}$$

Here T is the passive form of S. At high (low) depolarization D, k_1 is much larger (smaller) than k_{-1}.

More than five years of cooperative work between experimentalists and theoreticians has led to the postulation of scheme (7) as the core molecular mechanism of controlling neurotransmitter release. A great deal remains to be done, and it is likely that the model will be modified as further experimental results are obtained. Certainly the model must eventually be supplemented, since the actual biophysical mechanism for vesicle release remains a mystery. Probabilistic models must be constructed to fit data that concern individual release events. [A start has been made by Lustig, Parnas, and Segel (1986).] Nonetheless, research is now at the molecular level, and it appears that depolarization has joined calcium as a major factor regulating neurotransmitter release. Moreover, although further work is again required, the paper of Parnas, Parnas, and Segel (1986) provides a new and powerful experimental tool for measuring the release cooperativity parameter n.

Adaptation in cAMP signalling

Cyclic AMP (cAMP) is a chemoattractant for the cellular slime mold amoebae of species *Dictyostelium discoideum*. A few hours after developmental programs are initiated by starvation, the *D. discoideum* amoebae aggregate because of centripetal chemotactic steps controlled by centrifugal waves of cAMP, relayed by excitable cells, and probably initially triggered by autonomous periodic cAMP secretors. Some time ago, Albert Goldbeter (Service de Chimie Physique, Free University, Brussells) and I constructed a molecular-level theory that can explain many aspects of cAMP secretory behavior in *D. discoideum*, notably the shift from essentially inert, to excitable, to autonomous periodic secretion. Some later experimental findings required modification of the theory. For example Goldbeter pointed out that the addition of a further term in one of the equations could remedy a defect caused by "inappropriate" (for the original model) timing of a decrease in extracellular phosphodiesterase, an enzyme that degrades cAMP. An exposition of the theory can be found in Segel (1984).

Certain results of Devreotes and Steck (1979) could not be explained by the Goldbeter-Segel theory. These results concerned response of the cells to experimentally imposed step increases in extracellular cAMP concentration. Excitable cells reacted to higher extracellular cAMP levels by elevating their cAMP secretion rate, but later (as predicted by the Goldbeter-Segel theory) adapted to the higher level of excitation and ceased secreting. What the theory did not predict was observations that the adaptation would be delayed if an additional increase in extracellular cAMP was imposed a few minutes after the original increase was established.

Goldbeter and Martiel (1982) found one way to meet the challenge of the new experiments by invoking receptor desensitization, some evidence for which had earlier been reported. (Also see Martiel and Goldbeter, 1984). Here I wish to discuss another approach in which I have been involved together with Peter Devreotes and Barry Knox of the Department of Biological Chemistry, Johns Hopkins School of Medicine.

We wish to explain the observed seemingly exact adaptation of a sensory system to an elevation in the concentration L of a stimulating molecule. We assume that the mechanism for adaptation resides entirely in the receptor molecule that binds L. (We employ the same letter to designate the molecule and its concentration.)

Suppose that in the absence of ligand L the receptor spontaneously shifts back and forth between two conformations, R and D. Suppose further that the liganded forms $RL \equiv X$ and $DL \equiv Y$ constitute yet other conformations. Suppose finally that the *activity* $\mathcal{A}$ of the aggregate of receptor molecules, a measure of their contribution to physiological response, is given by the linear combination

$$\mathcal{A} = a_1 R + a_2 X + a_3 Y + a_4 D. \tag{8}$$

At any given fixed ligand concentration L, the receptor forms R, X, Y, and D will attain steady state concentrations that depend on L. It is a central and surprising finding of our analysis that the coefficients in (8) can be chosen so that the steady state value of $\mathcal{A}$ is independent of L. Hence whatever function the response $\mathcal{R}$ is of the activity $\mathcal{A}$, the response will exactly adapt to its basal value, the response when $L = 0$. Moreover the system provides an adequate transient response before adapting. All this is true whether the principle of microscopic reversibility directly applies to the cyclic interconversion of R, D, X, and Y, or whether interconversion is via covalent modification and hence application of microscopic reversibility is to subcycles of the main process.

To obtain a molecular implementation of our ideas, suppose that response is a function f of the concentration M of some intracellular effector molecule that binds to each of the four receptor forms with dissociation constants a_i, $i = 1, 2, 3, 4$. If binding is sufficiently fast we may use the equilibrium relationships $RM = a_1 \cdot M$, etc. for the complexes RM, XM, YM, and DM. If M_T is the total amount of effector then

$$M_T = M + RM + XM + YM + DM, \qquad \text{i.e. } M_T = M + \mathcal{A} \cdot M, \qquad \mathcal{R} = f[M_t/(1+\mathcal{A})].$$

(If bound effector drives response then $\mathcal{R} = f(M_T - M) = f[M_T \mathcal{A}/(1+\mathcal{A})]$.) Full theoretical details can be found in Segel, Goldbeter, Devreotes, and Knox (1985), (1986).

Comparison of the model with experiments on *D. discoideum* is discussed by Knox, Devreotes, Goldbeter, and Segel (1986). Recent observations have shown that indeed the receptor exists in two major different configurations, which we identify with R and $RL \equiv X$ on the one hand and with D and $DL \equiv Y$ on the other. There is quite good evidence that the conformational change occurs via covalent modification (Devreotes and Sherring, 1985; Klein, Fontana, Knox, Thiebert, and Devreotes, 1985). Knox et al. (1985) show how to obtain the kinetic constants of the covalent modification and demodification from measurements of the time variation of these processes. When these constants are obtained, there are in essence no free parameters left in the model, which well predicts the elevation of the cAMP secretion rate and its adaptation. The model also predicts deadaptation when the stimulus is removed, but so far no measurements have been made of a decrease of the rate of cAMP secretion below its low basal value.

Chemotaxis in bacteria also involves adaptation in response (tumble frequency) following changes in extracellular concentrations of a stimulant (chemoattractant). In this case Knox et al. (1985) also show encouraging agreement with experimental results taken from the literature. More theoretical and experimental work of course remains to be done. To mention one example, analysis is in progress to test whether the observed transitions in secretion behavior occur when the present model for receptor behavior is embedded in the earlier models for cAMP synthesis and secretion.

Conclusions

The two examples presented here seem to permit the drawing of the following general conclusions. (i) Mathematical modelling can play an important role in sorting out the complex interactions that occur among the molecules involved in sensory physiology. (ii) As usual, progress is greatly enhanced by close cooperation between theoreticians and experimentalists. (iii) Years of work seem required to make significant improvements in our understanding of sensory phenomena. It is thus important to choose problems that are worthy of a major effort, and then to make that effort.

It is also worth making a general comment concerning agreement with experiment. The issue is familiar to many, yet misunderstandings frequently arise.

We have sought semi-quantitative agreement with a range of experiments. It would not make our theory more convincing if we could adjust parameters to obtain quantitative agreement. The reason is that our assumptions [such as the form of the equations (1)–(3)] are only intended to capture the principal features of the most important processes. One should not expect that models embodying assumptions of such a nature will capture the details of a class of phenomena.

Acknowledgements

Much of the author's research reported here was supported by grants from the U.S. Army Research Office, Durham, particularly during summer work at the Department of Mathematical Sciences, Rensselaer Polytechnic Institute. Thanks are due to H. Parnas for suggesting several improvements in the manuscript.

References

Cooke, J.D., Okamoto, K., and Quastel, D.M.J. (1973). The role of calcium in the depolarization-secretion coupling at the motor nerve terminal. *J. Physiol.*, 228:459–497.

Devreotes, P.N., and Steck, T.L. (1979). Cyclic 3′, 5′ AMP relay in *Dictyostelium discoideum*. Requirements for initiation and termination. *J. Cell Biol.*, 80:300–9.

Devreotes, P.N., and Sherring, J. (1985). Kinetics and concentration dependence of reversible cAMP-induced modification of the surface cAMP receptor in *Dictyostelium*. *J. Biol. Chem.*, 260:6378–84.

Dudel, J., Parnas, I., and Parnas, H. (1983). Neurotransmitter release and its facilitation in crayfish muscle. VI. Release determined by both, intracellular calcium concentration and depolarization of the nerve terminal. *Pflügers Arch.*, 399:1-10.

Goldbeter, A., and Martiel, J.L. (1982). A critical discussion of plausible models for relay and oscillation of cyclic AMP in *Dictyostelium* cells. In: *Lecture Notes in Biomathematics, Vol. 49: Rhythms in Biology and Other Fields of Application*, edited by M. Cosnard, J. Demongeot, and A. Le Breton, pp. 173–88. Berlin: Springer-Verlag.

Katz, B., and Miledi, R. (1965). The measurements of synaptic delay, and the time course of acetylcholine release at the neuromuscular junction. *Proc. Roy. Soc. (London)*, B161:483–95.

Klein, P., Fontana, F., Knox, B., Thiebert, A., and Devreotes, P. (1985). *Cold Spring Harbor Symp. Quant. Biol.*, in press.

Knox, B., Devreotes, P., Goldbeter, A., and Segel, L. (1986). A molecular mechanism for sensory adaptation based on ligand-induced receptor modification. *Proc. Nat. Acad. Sci. (USA)*, in press.

Kuffler, S.W., Nicholls, J.G., and Martin, R.A. (1984). *From Neuron to Brain*, 2nd ed. Sunderland, MA: Sinauer Associates.

Llinas, R., Steinberg, I.Z., and Walton, K. (1981). Relationship between presynaptic calcium current and postsynaptic potential in squid giant synapse. *Biophys. J.*, 33:323–51.

Lustig, C., Parnas, H., and Segel, L.A. (1986). On the quantal hypothesis of neurotransmitter release: An explanation for the calcium dependence of the binomial parameters. *J. Theor. Biol.*, in press.

Martiel, J.L., and Goldbeter, A. (1984). Oscillations et relais des signaux d'AMP cyclique chez *Dictyostelium discoideum*: analyse d'un modèle fondé sur la modification du récepteur pour l'AMP cyclique. *Compte Rendu Acad. Sc. Paris*, 298:549–52.

Parnas, H., Dudel, J., and Parnas, I. (1982). Neurotransmitter release and its facilitation in crayfish. 1. Saturation kinetics of release, and of entry and removal of calcium. *Pflügers Arch.*, 393:1–14.

Parnas, H., Dudel, J., and Parnas, I. (1986). Neurotransmitter release and its facilitation in crayfish. VII. Depolarization, not Ca inflow, determines the time course of phasic release. *Pflügers Arch.*, in press.

Parnas, H., Parnas, I., and Segel, L. (1986). A new method for determining cooperativity in neurotransmitter release. *J. Theor. Biol.*, in press.

Parnas, I., Parnas, H., and Dudel, J. (1982). Neurotransmitter release and its facilitation in crayfish. II. Duration of facilitation and removal processes of calcium from the terminal. *Pflügers Arch.*, 393:232–6.

Parnas, H., and Segel, L.A. (1980). A theoretical explanation for some effects of calcium on the facilitation of neurotransmitter release. *J. Theor. Biol.*, 84:3–29.

Parnas, H., and Segel, L.A. (1981). A theoretical study of calcium entry in nerve terminals, with application to neurotransmitter release. *J. Theor. Biol.*, 91:125–69.

Parnas, H., and Segel, L.A. (1982). Ways to discern the presynaptic effect of drugs on neurotransmitter release. *J. Theor. Biol.*, 94:923–41.

Parnas, H., and Segel, L.A. (1983). A case study of linear versus non-linear modelling. *J. Theor. Biol.*, 103:549–80.

Parnas, H., and Segel. L.A. (1984). Exhaustion of calcium does not terminate evoked neurotransmitter release. *J. Theor. Biol.*, 107:345–65.

Rahamimoff, R. (1968). A dual effect of calcium ions on neuromuscular facilitation. *J. Physiol.*, 203:121–33.

Segel, L.A. (1984). *Modelling Dynamic Phenomena in Molecular and Cellular Biology*. Cambridge: Cambridge University Press.

Segel, L.A., Goldbeter, A., Devreotes, P.N., and Knox, B.E. (1985). A model for sensory response and exact adaptation mediated by receptor modification. In: *Sensing and Reponse in Microorganisms*, edited by M. Eisenbach, and M. Balaban, pp. 175–83. Amsterdam: Elsevier.

Segel, L.A., Goldbeter, A., Devreotes, P.N., and Knox, B.E. (1986). A mechanism for exact sensory adaptation based on receptor modification. *J. Theor. Biol.*, in press.

OUTLINE OF SOME RECENT RESULTS ON THE FIRST-PASSAGE-TIME PROBLEM IN BIOLOGICAL MODELING

Luigi M. Ricciardi
Dipartimento di Matematica e Applicazioni
Università degli Studi
Via Mezzocannone 8, Napoli, Italy

1. Introduction.

There are problems in population biology and in neurobiology that require the evaluation of first-passage-time probability density functions (p.d.f.) for diffusion or for Gaussian processes. For instance, the fixation of a gene may be viewed as the process leading the frequency of that gene to attain the value one for the first time; the extinction time of a population can be interpreted as the time when first the population size shrinks to some preassigned small level; the firing of a neuron can at times be modeled as the first crossing through a critical threshold value by the random process depicting the time course of the membrane potential. The literature on this subject is so vast that it cannot and need not be referenced here. It may suffice to mention that an explicit reference to these problems can be found in Ricciardi (1977) and in the literature quoted therein. Instead, we would like to stress that the determination of first-passage-time densities and of its features is in general a formidable task. It is fair to claim that so far only computational methods have been employed, which require the use of large scale computers, are extremely time consuming and hence not quite suitable to carry out the systematic analyses imposed by the nature of the biologically interesting problems that usually involve numerous parameters to be specified by best fitting procedures (Favella et al. 1982) Ricciardi and Sato, 1983a; Ricciardi et al., 1983).

In this talk we shall leave aside Gaussian processes which have been the object of specific investigations (see, for instance, Ricciardi and Sato, 1983b and 1985) and limit our considerations to first-passage-time problems for one dimensional diffusion processes. Our aim is to point out that recently some noteworthy progress has been made both along analytical and computational directions that, in our view, has disclosed novel and exciting perspectives in some of the biological modelling. For the sake of conciseness, we shall refer to first-passage-time problems for the Ornstein-Uhlenbeck (O.U.) process which plays an important role in theoretical neurobiology (Capocelli and Ricciardi, 1971; Ricciardi, 1977; Ricciardi and Sacerdote, 1979; Cerbone et al., 1981). However, generalizations and extension to other situations are possible.

2. Asymptotic Behavior of the Firing p.d.f.

The interest of the O.U. process in neurobiology was pointed out long ago by numerous authors. In particular, Capocelli and Ricciardi (1971) and Ricciardi (1976) provided a somewhat rigorous proof that, under certain assumptions, the neuron's membrane potential can be modelled by a one dimensional diffusion process $\{X(t);\ t \geq 0\}$ with constant infinitesimal variance and affine drift. Denoting by $S(t)$ the neuron's threshold, the firing time T is the r.v. defined by

$$T = \inf\{t:\ X(t) > S(t)\ \mid X(0) = x_0\}\ , \tag{1}$$

where $x_0 < S(0)$ denotes the resting potential. The firing p.d.f. $g[S(t), t \mid x_0]$ is then the first-passage-time p.d.f. of $X(t)$ through $S(t)$:

$$g\,[S(t), t \mid x_0\,] = \frac{\partial}{\partial t}\ P\{T < t\}\ . \tag{2}$$

Even in the simplest instance, i.e. when $S(t)=S$ is an arbitrary constant, a closed form expression for g is lacking. It is instead possible to write down the moment generating function of g, namely its Laplace transform $g_\lambda(S|x_0)$. This is given by

$$g_\lambda\,(S|x_0\,) = \exp\left(\frac{x_0{}^2 - S^2}{2\sigma^2\theta}\right) \frac{D_{-\lambda\theta}\left(-\frac{x_0}{\sigma}\sqrt{\frac{2}{\theta}}\right)}{D_{-\lambda\theta}\left(-\frac{S}{\sigma}\sqrt{\frac{2}{\theta}}\right)} \tag{3}$$

where $D_\nu(x)$ denotes the Parabolic Cylinder Function, σ^2 denotes the infinitesimal variance of $X(t)$ and $-x/\theta$ $(\theta>0)$ its drift. Expression (3) has been used by Ricciardi and Sacerdote (1979) and by Cerbone et al. (1981) to obtain some information on the statistical features of the firing time T . However, even this simpler task is very cumbersome, stressing and expensive due to the very slow convergence of the series appearing when the derivatives of (3) with respect to λ are performed and then evaluated at $\lambda = 0$.

An alternative procedure to evaluate g is to make use of some Volterra integral equations of first and second kind (Favella et al., 1982; Ricciardi et al. 1984) in which g is the unknown function. However, these are weakly singular equations as their kernels contain a singularity at t=0 of the kind $t^{-1/2}$, which so far has forced one to resort to cumbersome numerical solution methods.

While referring to Section 3 for a sketch of a much more promising solution method, we wish here to point out a quite unexpected feature of the function $g(S,t \mid x_0)$ which was discovered by looking into some computational results. This is, in my own experience, one of those rather rare instances in which the validity of some theorems has been clearly suggested by a careful analysis of numerical evaluations.

More precisely, looking at some numerical tables of Cerbone et al. (1981) it was discovered that for positive boundaries of the order of a couple of units or more, whatever the initial state x_0 the evaluated variance V of the first-passage-time and the square of its mean value t_1 are equal to an excellent degree of approximation. Furthermore, the skewness Σ turns out to be very close to 2 and the goodness of these approximations increases with the magnitude of the threshold S. This is clearly shown by Table 1 in which a normalized O.U. process is considered (Cerbone et al., 1981). All this suggests that the function (3) can be approximated by an exponential p.d.f. for a wide range of threshold and resting potential values. It turns out that this is indeed the case, as paradigmatically shown by Table 2 in which for S=4 and $x_0 = 0$ the first-passage-time p.d.f. $\tilde{g}$, evaluated by the algorithm of Favella et al. (1982), is listed for some values of t. Table 2 also lists the values of an exponential density $g^* = \alpha^{-1} \exp(t/\alpha)$ where α is taken as the numerically evaluated average first-passage-time. The agreement between the values of $\tilde{g}$ and g^* is truly impressive.

A mathematical justification of these results can be found in Nobile et al. (1985a) while its extension to a wider class of diffusion processes has been provided in Nobile et al. (1985b). Here, we only wish to emphasize the great potential usefulness for the applications of the fact that the impossible task of calculating the inverse Laplace transform of (3) is fully bypassed since the function g can be well approximated by a specified exponential density even for small threshold values.

3. A New Computation Method.

We now turn to the problem of determining the first-passage-time p.d.f. for an arbitrary time homogeneous one dimensional diffusion process $\{X(t);\ t \geq t_0,\ t_0 \in R\}$ defined over an interval $I \equiv (r_1, r_2)$, with r_i (i=1,2) natural boundaries and such that $P\{X(t_0) = x_0\} = 1$, $x \in (r_1, r_2)$. Let for all $\tau < t$ and $x, y \in [r_1, r_2]$

$$F(x,t \mid y,\tau) = P\{X(t) \leq x \mid X(\tau) = y\} \tag{4a}$$

$$f(x,t|y,\tau) = \frac{\partial}{\partial x} F(x,t|y,\tau) \tag{4b}$$

denote the transition distribution and the transition p.d.f., respectively. In the sequel for the sake of conciseness we shall assume that the boundary $S(t)$ in (1) is such that $x_0 < S(t_0)$. It is then possible to prove the following theorem (Buonocore et al., in preparation):

Theorem 1. Let $S(t)$ and $k(t)$ be continuous functions in $[t_0, +\infty)$. Setting for all $y \in I$ and $\tau < t$

$$\psi[S(t),t|y,\tau] = \frac{d}{dt} F[S(t),t|y,\tau] + k(t)\, f[S(t),t|y,\tau] \tag{5}$$

one has:

$$g[S(t),t|x_0,t_0] = -2\psi[S(t),t|x_0,t_0]$$

$$+ 2\int_{t_0}^{t} d\tau \;\; g[S(\tau),\tau|x_0,t_0]\, \psi[S(t),t|S(\tau),\tau] \tag{6}$$

For the sake of conciseness we refer to Buonocore et al. (loc.cit.) for the proof. Note that eq. (6) is a Volterra second kind equation whose kernel ψ defined by (5) contains the arbitrary function $k(t)$. The question then naturally arises: can one specify $k(t)$ in such a way that (i) the kernel ψ is identically zero (hence obtaining the closed form solution $g[S(t), t|x_0, t_0] = -2\psi[S(t), t|x_0, t_0]$) or ii) the kernel ψ is continuous as τ approaches t (so that eq. (6) admits of a simple numerical solution)? The answer is yes. In particular, it is possible to show that all known closed form first-passage-time densities for Wiener and O.U. processes are immediately recovered by imposing the vanishing of the function $\psi[S(t),t|S(\tau),\tau]$ in (6). Furthermore, for the O.U. process one can prove (Buonocore et al., loc. cit.) that the following theorem holds.

Theorem 2. Let $S(t)$ be a $C^2[t_0, +\infty)$-class function and let $X(t)$ be the O.U. process. Then,

$$\lim_{\tau \uparrow t} \psi[S(t),t|S(\tau),\tau] = 0 \qquad \text{iff } k(t) = -\tfrac{1}{2}[S(t)/\theta - S'(t)] \tag{7}$$

where $S'(t)$ is the derivative of $S(t)$.

Hence, choosing the originally arbitrary function k(t) as specified by (7), the integral equation (6) ceases to be singular at $\tau = t$ so that simpler and straightforward solution methods can be used to solve it. In general, whenever the limit in (7) is zero, using the integration step h an approximation g_1 to g can for instance be obtained by use of a composite trapezium rule:

$$g_1[S(t_0+h),t_0+h|x_0,t_0] = -2\psi[S(t_0+h),t_0+h|x_0,t_0]$$

$$g_1[S(t_0+kh),t_0+kh|x_0,t_0] = -2\psi[S(t_0+kh),t_0+kh|x_0,t_0] \tag{8}$$

$$+ 2h\sum_{j=1}^{k-1} g_1[S(t_0+jh),t_0+jh|x_0,t_0]\,\psi[S(t_0+kh),t_0+kh|S(t_0+jh),t_0+jh]$$

$$(k=2,3,\ldots).$$

The convergence of this procedure is finally secured by the following theorem (Buonocore et al., loc. cit.):

Theorem 3. Let $T = t_0 + Nh$ with $N \in N_0$ and let

$$\Delta_{kh} = g[S(t_0+kh),t_0+kh|x_0,t_0] - g_1[S(t_0+kh),t_0+kh|x_0,t_0] \quad (k=1,2,\ldots,N).$$

Then:

$$\lim_{h\downarrow 0} |\Delta_{kh}| = 0 \qquad (k = 1,2,\ldots,N;\ kh \text{ fixed}).$$

In conclusion, we should like to stress that the above sketched procedure not only cuts computation times in a very relevant manner but, above all, it possesses a very remarkable feature: an extraordinary simplicity which allows the solution algorithm to be implemented on personal computers. Hence, problems of biological modeling of the type mentioned in the foregoing can now be re-considered with novel encouraging perspectives.

Acknowledgments

This work has been performed under CNR-JSPS Scientific Cooperation Programme, Contract no. 84.00227.01, CNR Contract No. 85.00002.01 and under MPI support.

TABLE 1

	S	t_1	V	t_1^2	$\sum$
$x_o = 3$	4	$0.193146\text{x}10^{4}$	$0.406163\text{x}10^{7}$	$0.373053\text{x}10^{7}$	2.00533
	5	$0.140654\text{x}10^{6}$	$0.198076\text{x}10^{11}$	$0.197835\text{x}10^{11}$	2.00000
	6	$0.282674\text{x}10^{8}$	$0.799052\text{x}10^{15}$	$0.799045\text{x}10^{15}$	2.00000
	7	$0.159800\text{x}10^{11}$	$0.255359\text{x}10^{21}$	$0.255360\text{x}10^{21}$	2.00000
$x_o = 0$	1	$0.209341\text{x}10^{1}$	$0.584193\text{x}10^{1}$	$0.425820\text{x}10^{1}$	2.30002
	2	$0.104284\text{x}10^{2}$	$0.105275\text{x}10^{3}$	$0.108751\text{x}10^{3}$	2.01654
	3	$0.869316\text{x}10^{2}$	$0.742438\text{x}10^{4}$	$0.750358\text{x}10^{4}$	2.00033
	4	$0.201839\text{x}10^{4}$	$0.406906\text{x}10^{7}$	$0.407389\text{x}10^{7}$	2.00000
	5	$0.140741\text{x}10^{6}$	$0.198076\text{x}10^{11}$	$0.198080\text{x}10^{11}$	2.00000
	6	$0.282675\text{x}10^{8}$	$0.799052\text{x}10^{15}$	$0.799051\text{x}10^{15}$	2.00000
	7	$0.159800\text{x}10^{11}$	$0.255359\text{x}10^{21}$	$0.255360\text{x}10^{21}$	2.00000
$x_o = -2$	-1	$0.523296\text{x}10^{0}$	$0.215579\text{x}10^{0}$	$0.273819\text{x}10^{0}$	3.95988
	0	$0.142520\text{x}10^{1}$	$0.106657\text{x}10^{1}$	$0.203119\text{x}10^{1}$	5.30883
	1	$0.351861\text{x}10^{1}$	$0.690850\text{x}10^{1}$	$0.123806\text{x}10^{2}$	2.11054
	2	$0.118536\text{x}10^{2}$	$0.106341\text{x}10^{3}$	$0.140507\text{x}10^{3}$	1.99161
	3	$0.883568\text{x}10^{2}$	$0.742544\text{x}10^{4}$	$0.780685\text{x}10^{4}$	1.99991
	4	$0.201982\text{x}10^{4}$	$0.406906\text{x}10^{7}$	$0.407967\text{x}10^{7}$	2.00000
	5	$0.140742\text{x}10^{6}$	$0.198076\text{x}10^{11}$	$0.198083\text{x}10^{11}$	2.00000
	6	$0.282675\text{x}10^{8}$	$0.799052\text{x}10^{15}$	$0.799051\text{x}10^{15}$	2.00000
	7	$0.159800\text{x}10^{11}$	$0.255359\text{x}10^{21}$	$0.255360\text{x}10^{21}$	2.00000

TABLE 2

t	$\tilde{g}$	g^*
6	$0.493983\text{x}10^{-3}$	$0.493973\text{x}10^{-3}$
15	$0.491770\text{x}10^{-3}$	$0.491776\text{x}10^{-3}$
30	$0.488129\text{x}10^{-3}$	$0.488134\text{x}10^{-3}$
90	$0.473834\text{x}10^{-3}$	$0.473837\text{x}10^{-3}$
180	$0.453173\text{x}10^{-3}$	$0.453173\text{x}10^{-3}$
270	$0.433413\text{x}10^{-3}$	$0.433410\text{x}10^{-3}$
360	$0.414515\text{x}10^{-3}$	$0.414509\text{x}10^{-3}$
450	$0.396440\text{x}10^{-3}$	$0.396432\text{x}10^{-3}$

References

Capocelli R.M. and Ricciardi L.M. (1971) Diffusion approximation and first passage time problem for a model neuron. Kybernetik 8, 214-223.

Cerbone G., Ricciardi L.M. and Sacerdote L. (1981) Mean, variance and skewness of the first-passage time for the Ornstein-Uhlenbeck process. Cybernetics and Systems 12, 395-429.

Favella L.F. and De Griffi R.M. (1981) On a weakly singular Volterra integral equation. Calcolo, Vol. XVIII, 153-195.

Favella L., Reineri M.T., Ricciardi L.M. and Sacerdote L. (1982) First passage-time problems and some related computational methods. Cybernetics and Systems 13, 95-128.

Nobile A.G., Ricciardi L.M. and Sacerdote L. (1985a). Exponential trends of Ornstein Uhlenbeck first-passage-time densities. J. Appl. Prob. 22, 346-359.

Nobile A.G., Ricciardi L.M. and Sacerdote L. (1985b). Exponential trends of first-passage-time densities for a class of diffusion processes with steady-state distribution. J. Appl. Prob. 22,611-618.

Ricciardi L.M. (1976) Diffusion approximation for a multi-input model neuron. Biol. Cybernetics 24, 237-240.

Ricciardi L.M. (1977) Diffusion Processes and Related Topics in Biology. Lecture Notes in Biomathematics, Vol.14 Springer-Verlag, Berlin.

Ricciardi L.M. and Sacerdote L. (1979) The Ornstein-Uhlenbeck process as a model for neuronal activity. Biol. Cybernetics 35, 1-9.

Ricciardi L.M., Sacerdote L. and Sato S. (1983) Diffusion approximation and first passage time problem for a model neuron.II. Outline of a computation method. Math. Biosciences 64, 29-44.

Ricciardi L.M. and Sato S. (1983) A note on the evaluation of first-passage-time probability densities. J. Appl. Prob. 20, 197-201.

Ricciardi L.M. and Sato S. (1983) A note on first passage time problems for Gaussian processes and varying boundaries. IEEE Trans. Information Theory IT-29, 454-457.

Ricciardi L.M. and Sato S. (1985) On the evaluation of first-passage-time densities for Gaussian processes (preprint).

Active Rotator Model for Large Populations of Oscillatory and Excitable Elements

Yoshiki KURAMOTO, †Shigeru SHINOMOTO and Hidetsugu SAKAGUCHI
Department of Physics, Kyoto University, Kyoto 606, Japan
†Research Institute for Fundamental Physics, Kyoto University, Kyoto 606, Japan

§1. Introduction

The purpose of the present paper is to demonstrate the utility of what we call an active rotator model. We will study some dynamical processes for which the interaction of infinitely many degrees of freedom is essential. The processes with which we are concerned take place in a spatially distributed system that is more or less homogeneous in its constitution; the system consists of many similar units, and each unit is assumed to be functionally active in some sense. More specifically, it is assumed that each element (which will ordinarily be called a cell) represents either a limit-cycle oscillator or an excitable functional unit. Since, as is well known, oscillation and excitability are dynamical modes commonly shared by cellular membrane and biochemical reactions, the present subject is relevant to many physiological processes in living organisms. Examples of what are considered to be populations of such active elements include the population of pacemaker cells in the sino-atrial node of a mammal's heart and neural nets in the cerebral cortex. Our system need not be restricted to discrete populations of well-defined units; even continuous or almost continuous tissues such as the smooth muscle of the small intestine or other digestive organs may also comprise possible examples.

As a mathematical idealization of our system, we introduce a lattice (Kuramoto 1984a) with a cell on each lattice point. Generally, the cells are supposed to mutually interact, and many interesting behaviors are anticipated. First of all, wave activity in various forms will arise, including synchronizing waves sent out from pacemaker centers, and rotating waves of action potential such as are observed in heart muscle. (Allessie et al 1977) As a second possibility, we may have very complicated spatio-temporal structures or turbulent behavior caused by intrinsic stochasticity. (Kuramoto 1981, 1984a,b) Another interesting class of phenomena are statistical in nature. There the system is supposed to involve randomness; for instance, the natural frequencies of cell oscillators may differ from cell to cell. Depending on the degree of randomness involved, the system may exhibit highly organized collective behavior or, conversely, such organized motion may disintegrate into fragmented cellular activities. (Kuramoto 1984a,b; Kuramoto 1975) This kind of order-disorder behavior may have much in common with phase transitions in equilibrium statistical mechanics. In the present paper, some general features of such cooperative transitions will be considered, but waves or turbulence will not, the latter having already been studied extensively in the past —— especially by means of reaction-diffusion models.

§2. Model

There exists a method to describe each cell very simply; cell state is assumed to be described by a single phase variable ϕ. Equivalently, the cell state can be represented by a point of angle ϕ on a unit circle, a vector or a complex number $\exp(i\phi)$. This is known as a ring device and was employed by Winfree (1967,1980), Kuramoto (1975), Ermentrout (1985), Ermentrout and Kopell (1984), Ermentrout and Rinzel (1981) and others with considerable success. The present theory also employs this kind of rotator model but does so somewhat more extensively than was done in previous models. In the absence of interaction, each cell is assumed to obey the equation

$$\frac{d\phi}{dt} = f(\phi) \quad , \qquad f(\phi + 2\pi) = f(\phi) \; . \tag{1}$$

Physiological variables such as membrane potential are expected to be periodic functions of ϕ, denoted by $X(\phi)$; if the variables are collective, they will be given as a sum of such quantities $X(\phi)$ over the population, e.g. $\sum_i X(\phi_i)$. The cell-cell interaction is now introduced as

$$\frac{d\phi_i}{dt} = f_i(\phi_i) + \sum_j \Gamma_{ij}(\phi_i,\phi_j) \quad , \tag{2}$$

where Γ_{ij} is periodic both in ϕ_i and ϕ_j. If necessary, one may add to the right-hand side external effects such as periodic forcing or random noises. In order to proceed, we need explicit forms for f_i and Γ_{ij}. Here we propose a pair of model equations which are considered to represent two typical situations. They are

$$\text{(I)} \quad \frac{d\phi_i}{dt} = \omega_i + \sum_j k_{ij} \sin(\phi_j - \phi_i + \alpha) \quad , \qquad |\alpha| < \frac{\pi}{2} \quad , \tag{3a}$$

and

$$\text{(II)} \quad \frac{d\phi_i}{dt} = \omega_i - b \sin\phi_i + \sum_j k_{ij}\, \delta(\phi_j + \frac{\pi}{2}) \; , \qquad b \gtrsim \omega_i > 0 \quad . \tag{3b}$$

In model I, (Kuramoto 1984a,b; Sakaguchi and Kuramoto 1986) f_i is given by a constant ω_i, so that the cell is represented by a smooth oscillator with frequency ω_i. Such smoothness is reflected in the coupling form which is assumed to depend only on the phase difference between the interacting pair. Only the fundamental harmonic is retained, and it generally involves a phase constant α that one is free to restrict to the range indicated. If $K_{ij} > 0$, the interacting pair will tend to minimize their phase difference, whereas $K_{ij} < 0$, will favor out-of-phase behavior. For this reason, we will call the case of positive K_{ij} "attractive coupling" and that of negative K_{ij} "repulsive coupling".

Let us look at model II, where f_i involves the additional term $b \cdot \sin\phi_i$. Since b is assumed to be slightly greater than ω_i, the cell on our circular state space has a stable-unstable pair of equilibria, which are relatively close to each other. This gives the simplest possible excitable system. It normally stays at equilibrium; but once it is excited beyond the unstable equilibrium, it no longer can retrace its way back and must make a long journey around the circle. One interesting coupling form between such excitable cells is the delta-function assumed above. Note that our delta-function does not depend on ϕ_i, which makes our model appropriate for neural nets. The ϕ_i in Eq.(3b) is then considered as representing a post-synaptic cell, and ϕ_j a pre-synaptic cell; the i-th cell receives a signal at the moment that the j-th cell makes a spike discharge. The net effect of the stimulus on the i-th cell is an increase in ω_i if K_{ij} is positive, and, if this effect is strong enough, the i-th cell will also be excited beyond threshold and fire. This is essentially what occurs in neural systems when coupling is excitatory. If, in contrast, $K_{ij} < 0$ the coupling is inhibitory. Models I and II will be used to study transition phenomena in the same spirit as the employment of the Ising and other simplified models in the statistical mechanical theory of phase transition. Most of the following discussion will concern model I, while model II will be considered only briefly.

§3. First type cooperative transition

We begin with model I, which is defined on a lattice; its spatial extent is assumed to be essentially infinite. Cells interact with their nearest neighbors with strength K, which is assumed to be positive. The natural frequencies ω_i change from cell to cell, and they take random values according to a Gaussian distribution of variance W^2. Qualitatively, we expect that if the frequency difference between a given nearest-neighbor pair is small enough or, equivalently, their mutual coupling is strong enough, then they will exhibit perfect mutual entrainment. In general, a number of neighboring cells that are relatively close in intrinsic frequency will form a cluster having a common frequency, and the average size of such synchronized clusters will become larger as the coupling strength K increases. At this point, one may raise the interesting question whether there exists some critical K value K_c such that, above K_c, the maximum size of clusters attains O(N). Such a large cluster would imply that a finite fraction of the oscillators find themselves in identical frequency states, thereby forming a condensate. Let the number of such condensed oscillators be N_s and the total cell number N. Then the ratio

$$r = \lim_{N\to\infty} N_s/N \qquad (4)$$

will be useful as an order parameter. The answer to our question concernig the possibility of transition between vanishing and nonvanishing r should of course depend on dimensionality.

We now show some results from our computer simulations. In space of dimension 1, there is no such phase transition. Figure 1 shows some frequency patterns for different coupling strengths. (Sakaguchi and Kuramoto, in prep.) Each frequency plateau implies perfect local entrainment. The average cluster size clearly increases with K, but it by no means goes to O(N) while K remains finite. In space of dimension 2, the system behaves quite differently. (Shinomoto and Kuramoto, in prep.) In Fig. 2 the dark areas represent locally synchronized regions which are separated from each other by white boundaries, so that frequency changes discontinuously across the

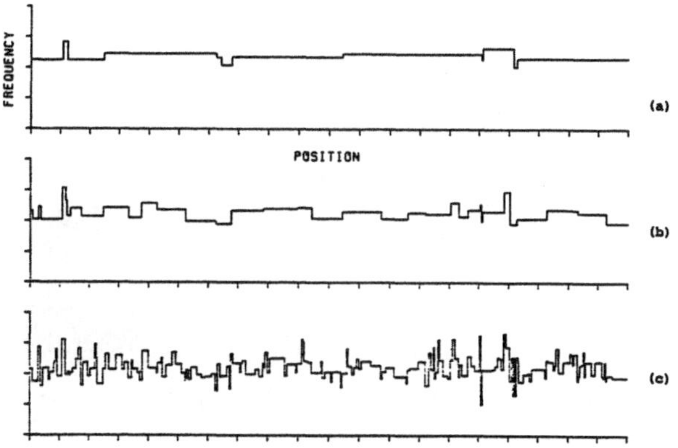

Fig. 1 Frequency patterns in an oscillator chain represented by model I with $\alpha=\pi/6$. K=3.0(a), 2.0(b) and 1.0(c).

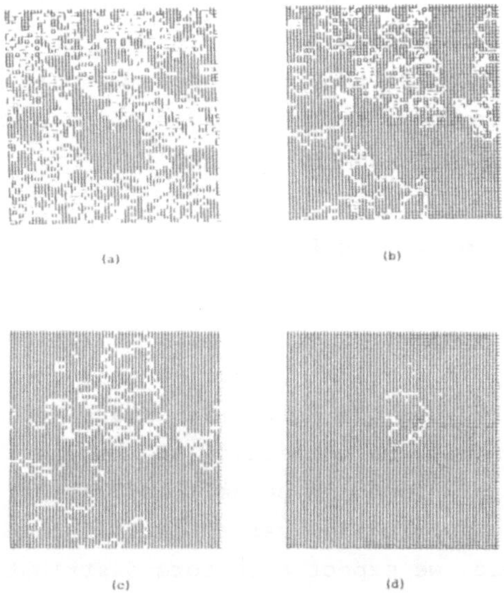

Fig. 2 Distribution of synchronized clusters in a two-dimensional oscillator lattice represented by model I with $\alpha=0$. Nearest-neighbor coupling strengths (a)k=1.0, (b)1.2, (c)1.4 and (d)1.6.

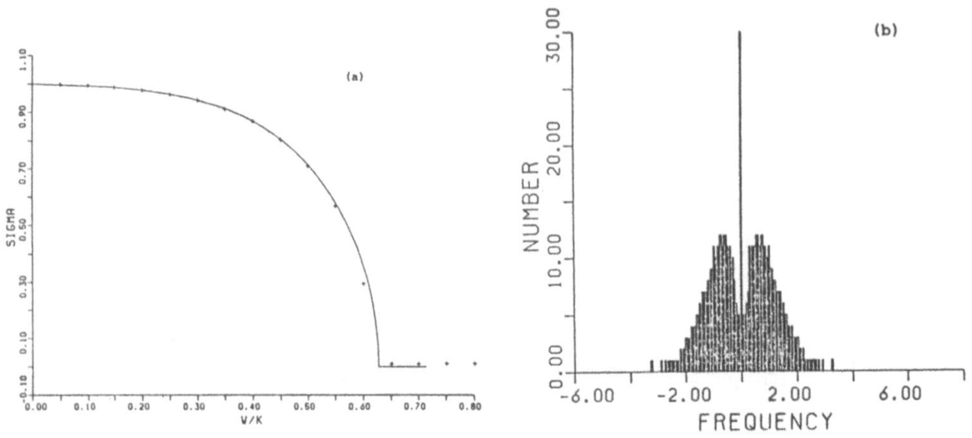

Fig. 3(a) Order parameter σ versus W/K for system described by Eq.(6) with α=0.

(b) Frequency distribution in the ordered state (W/K=0.6). The system considered is the same as in (a).

boundary. Much more elaborate computer simulations have failed to yield clear evidence for phase transition in 2-dimensional systems. In the 3-dimensional case, there is some numerical evidence that a critical value K_c exists. (Shinomoto and Kuramoto, in prep.) It now seems useful to introduce another order parameter, denoted by $\sigma\exp(i\theta)$, which is a complex number defined by

$$\sigma \exp(i\theta) = \lim_{N\to\infty} \frac{1}{N} \sum_j \exp[i\phi_j] \quad . \tag{5}$$

Its intuitive meaning will be clear from the following. Suppose σ is nonvanishing. This is equivalent to saying that the oscillators are distributed non-uniformly on the circular state space. This non-uniform distribution would be expected to rotate (rigidly), which implies the existence of some in-phase oscillation throughout the system. Macroscopic physiological variables will then show a rhythmic behavior. If σ is zero, we expect a uniform distribution without such collective rhythmicity. In two-dimensional space, our computer simulations imply that such collective oscillations are absent. (Ermentrout 1985; Ermentrout and Kopell 1984; Ermentrout and Rinzel 1981)

We would expect collective oscillations to arise in higher dimensions. In connection with this, the following extreme case would be of considerable interest. This is an oscillator aggregate in which each oscillator couples to all the others with equal strength of $O(N^{-1})$. (Kuramoto 1975, 1984a,b; Winfree 1967) Let this coupling strength be K/N. Fortunately, some analytic solutions are then available because a certain form of the mean field approximation holds exactly. We can see this by rewriting Eq.(3a) with $K_{ij} = K/N$ in the form

$$\frac{d\phi_i}{dt} = \omega_i - K\sigma \sin(\phi_i - \alpha + \theta) \quad . \tag{6}$$

Under the assumption that σ tends to a constant as $t \to \infty$ and θ to $\tilde{\omega}t$ with $\tilde{\omega}$ still to be determined, Eq.(6) can be solved for each ϕ_i in this long-time limit. The solutions thus obtained are then substituted into (5), yielding a self-consistent equation for σ and $\tilde{\omega}$. As an illustration, Fig. 3a shows an analytically calculated curve of σ as a function of K, and this is compared to our computer simulation of the same uniform-coupling model. (Sakaguchi and Kuramoto 1986) The agreement is excellent, as was true for various other parameter values. The onset of collective oscillation is clearly seen. For the present special model, the appearance of the two order parameters σ and r is simultaneous. Figure 3b, which is obtained from our computer simulation (Sakaguchi and Kuramoto 1986), shows a frequency distribution in the ordered state. The central delta peak implies that a macroscopically large number of oscillators form a condensate of an identical frequency. The broad background corresponds to those oscillators which failed to synchronize with the collective oscillation.

§4. Second type cooperative transition

In this section, we briefly discuss the behavior of model II. To make clear the point of discussion, the mean field model will again be employed, with k_{ij} equated to K/N and the coupling assumed to be excitatory. It is theoretically unclear as to whether model II shows mutual synchronization, but our intuition is strongly skeptical as to its possibility. Rather, it seems natural to expect the cells to be completely random with respect to their mutual phase relationship, and we should, therefore, now be looking for a macroscopic solution corresponding to such a situation. The simplest such macroscopic state will be that in which the quantity σ defined by

$$\sigma = \lim_{N\to\infty} \frac{1}{N} \sum_i \delta(\phi_i + \frac{\pi}{2}) \tag{7}$$

takes a constant value. The equation for each ϕ_i then reduces to an equation for a free cell with ω_i simply replaced by $\omega_i + K\sigma$, i.e.

$$\frac{d\phi_i}{dt} = \omega_i + K\sigma - b \sin\phi_i \quad . \tag{8}$$

The above is solved in the form

$$\phi_i(t) = \begin{cases} \tilde{\omega}_i t + p(\tilde{\omega}_i t) & ((\omega_i+K\sigma)/b \geq 1) \ , \quad (9a) \\ \sin^{-1}[(\omega_i+K\sigma)/b] & ((\omega_i+K\sigma)/b < 1) \ , \quad (9b) \end{cases}$$

where $\tilde{\omega}_i = \sqrt{(\omega_i + K\sigma)^2 - b^2}$, and $p(x)$ is some 2π-periodic function of x. If the sum $\omega_i + K\sigma$ exceeds b, the cell is repeatedly excited, and its frequency or firing rate is given by $\tilde{\omega}_i$. Obviously, only firing cells such as these can contribute to the order parameter σ. We now substitute the solution in (9a) into Eq. (7). The above-mentioned assumption that the phases ϕ_i are totally uncorrelated with each other leads to the self-consistent equation

$$\sigma = F(\sigma) \quad ,$$
$$F(\sigma) = \frac{1}{2\pi}\int_{b-K\sigma}^{\infty} d\omega \; g(\omega) \frac{\tilde{\omega}(\sigma)}{\omega+K\sigma+b} \quad , \tag{10}$$

where $g(\omega)$ denotes the distribution function of ω_i. The above expression for F may be comprehended by noticing that the factor $\tilde{\omega}$ represents the frequency with which a cell repeatedly visits the phase point at $-\pi/2$ and that the factor $(\omega+K\sigma+b)^{-1}$ comes to equal the inverse rate of change of the phase when passing through this particular phase value.

The detailed form of $F(\sigma)$ depends on $g(\omega)$, but it typically takes a sigmoidal form, as is depicted schematically in Fig. 4. If the

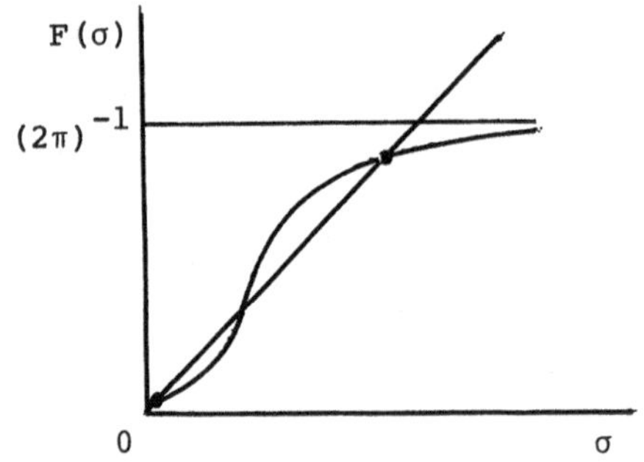

Fig. 4 Graphical solution of the self-consistent equation (11) (schematic).

excitatory coupling K is strong enough, the uprise of this curve occurs close to the origin, which gives rise to three intersections with the 45 degree line, i.e. the system shows bistability. Note that σ measures the level of activity, so that the two stable macroscopic states thus obtained differ greatly in activity. Such an argument concerning the origin of bistability may seem to be a rather trivial repetition of what is already well known in statistical neurodynamics. (See for example, Wilson and Cowan, 1972) However, we simply intended in the present paper to demonstrate the usefulness of phase models, and, in fact, they turned out at least to show essentially different types of cooperative transitions; i.e. one due to mutual entrainment and another due to mutual adjustment of activity level and not the adjustment of their mutual phase relationship. It is hoped that the phase model or rotator model as proposed here will find many other applications.

Acknowledgment

The present work is supported by a Grant-in-Aid for Fusion Research from the Ministry of Education, Culture and Science of Japan.

References

M.A. Allessie, F.I.M. Bonke and F.J. G. Shopman, Circ. Res. 41 1977, 9.

G.B. Ermentrout and J. Rinzel, J. Math. Biol. 11 1981 269.

G.B. Ermentrout and N. Kopell, SIAM J. Math. Anal. 15 1984 215.

G.B. Ermentrout, J. Math. Biol. 22 1985, 1.

Y. Kuramoto, Lecture Notes Phys. Vol. 39 Springer-Verlag, 1975 p.420.

Y. Kuramoto, Prog. Theor. Phys. Suppl. 64 1978 346.

Y. Kuramoto, Physica 106A 1981, 128.

Y. Kuramoto, Prog. Theor. Phys. Suppl. 79 1984a, 223.

Y. Kuramoto, 1984b, Chemical Oscillations, Waves, and Turbulence, Springer-Verlag, Heidelberg.

H. Sakaguchi and Y. Kuramoto, Prog. Theor. Phys. 76 1986, 576

H. Sakaguchi and Y. Kuramoto, in preparation.

S. Shinomoto and Y. Kuramoto, in preparation.

H.R. Wilson and J.D. Cowan, Biophys. J. 12 1972, 1.

A.T. Winfree, Theor. Biol. 16 1967, 15.

A.T. Winfree, The Geometry of Biological Time Springer-Verlag, 1980.

Y. Yamaguchi, K. Kometani and H. Shimizu, J. Stat. Phys. 26 1981, 719.

Y. Yamaguchi and H. Shimizu, Physica 11D 1984, 212.

THE HARNESSING AND STABILITY OF STRIATED MUSCLE

B.J. GANNON[1] and M.E. JONES[1,2],
Department of Anatomy and Histology[1],
The Flinders University of South Australia, and
Department of Anaesthetics[2],
Repatriation General Hospital,
Daw Park, South Australia.

1. PROBLEMS

We discuss three problems related to the contraction of striated muscle. These problems involve the matching of the force to the load, the longitudinal stability of a muscle fibre, and the harnessing of the force developed by the contractile elements.

1a. Matching the force to the load

For many muscles in the human body, the mechanical disadvantage at which the muscle acts is greatest when the muscle is longest. An example is the biceps brachii muscle, the insertion of which into the radial tuberosity is nearly in line with the axis of rotation of the elbow joint when the elbow is extended. As the elbow flexes and the biceps contracts, the tuberosity is rotated anterior to the joint, and the biceps gains better leverage. It would seem, therefore, that the muscle needs to generate most force when it is extended and has least leverage.

This requirement needs to be reconciled with the theory of interdigitating or sliding myofilaments (Huxley and Niedergerke, 1954, Huxley and Hanson, 1954), whereby muscle contraction is thought to involve the successive making and breaking of cross-connections between myosin and actin filaments. In such an arrangement maximum force developed depends on the number of cross-bridges and should, as Huxley and Niedergerke pointed out, be greatest where overlap of actin and myosin is greatest. The well known experimental finding of a length-tension relationship for whole muscle in which actively developed tension is maximal when the muscle is maximally extended in the body, would seem to indicate that developed tension correlates poorly with the extent of actin-myosin overlap.

1b. **Longitudinal stability**

When a muscle contracts, a very large number of sarcomeres would appear to be pulling one against another. If we consider just two sarcomeres in series and developing tension isometrically, it is clear that a stability problem exists. If the sarcomeres are at the same length and are developing the same tension, then they will be in equilibrium. It is, however, an unstable equilibrium if we accept that developed tension increases as a function of the number of contributing cross-bridges. In isometric contraction, should one sarcomere contract incrementally more than the other, the second must lengthen and a disequilibrium of force will develop. The more contracted sarcomere, having the greater overlap and number of cross-bridges will develop proportionally more tension and will contract further at the expense of the other. Equilibrium is only re-established when the shorter sarcomere is maximally contracted or the longer one is prevented from damaging itself by the intervention of some poorly understood passive mechanism preventing overextension.

In fact, this must be a very naive view, because a contracting muscle does not simultaneously exhibit maximally contracted and maximally extended sarcomeres; in a contracting muscle there is near-uniform contraction of the individual sarcomeres. The mechanism whereby this uniformity is maintained and the apparent longitudinal instability avoided, is not clear.

1c. The harnessing of contraction

Most muscles, when they contract, exert forces on bones via intervening tendons composed largely of collagen. The collagen from tendons is known to merge with that of the periosteum and to be continuous with Sharpey's fibres which pass into the bone matrix itself. Collagen, however, is extracellular; actin and myosin, and the remainder of the contractile apparatus of which they are the most obvious part, are all intracellular. The force of contraction associated with the progressive interdigitation of actin and myosin must be transmitted from its intracellular origin to the extracellular collagen. This is supposed to take place at the ends of the myofibres in a longitudinal direction, at the 'myotendinous junction'. However, electron microscopic studies have failed to demonstrate any convincing structural component at the myotendmous junction passing from actin and myosin within the cell to collagen outside, (Nakao, 1976: Mackay et al., 1969). Indeed, the transfer of force at the myotendinous junction has been attributed, in the absence of any obvious structural elements, to 'viscous cohesion' (Warwick and Williams, 1973).

In summary, three factors suggesting that our understanding of muscle contraction is unsatisfactory are

(i) the lack of a convincing structural element harnessing the contraction of actin and myosin,

(ii) the theoretical longitudinal instability of striated muscle, and

(iii) discrepancy between the length-tension relationship which exists macroscopically and that which would be predicted from the traditional interpretation of the sliding filament hypothesis.

These lead us to question one premise of the current model.

2. DO MUSCLES PULL OR DO THEY PUSH?

The evidence in favour of the sliding filament hypothesis is such that we will not question it. Depolarization leads to calcium influx which in turn leads to the interdigitation of actin and myosin, during which process the chemical energy of ATP is converted partly into mechanical energy which overcomes resistance to interdigitation. Given all that, and accepting that some of the molecular details have yet to be explained to everybody's satisfaction, the fundamental process is that chemical energy is expended while the muscle changes shape against resistance. Since the total volume of a muscle fibre is constant during contraction, a decrease in fibre length is associated with an increase in its diameter. The harnessing of the energy of contraction could potentially be from pulling at the ends of the fibre, or by pushing out at its sides. It is the latter possibility that has received very little attention in the past, and which we pursue here. Two preliminary comments in relation to the conjecture are in order. The first is that Mullins and Guntherot (1965) proposed a similar mechanism for smooth muscle cells. We believe, because of considerations of stability, that the lateral harnessing of a muscle's change in shape is more applicable to striated muscles (vide infra). The second comment is that, whenever there are two or more strategies available to nature, all will eventually be found to have been adopted somewhere.

2a. The mechanical advantage of lateral thrust

In a grossly simplified view (Fig. 1) consider A,D to be the tendinous origin and insertion of a muscle with collagen passing between these points along ABD and ACD both paths being of length 2l, the angles CAD etc. being ϕ = arctan x/y. (length AD = 2y; length BC = 2x).

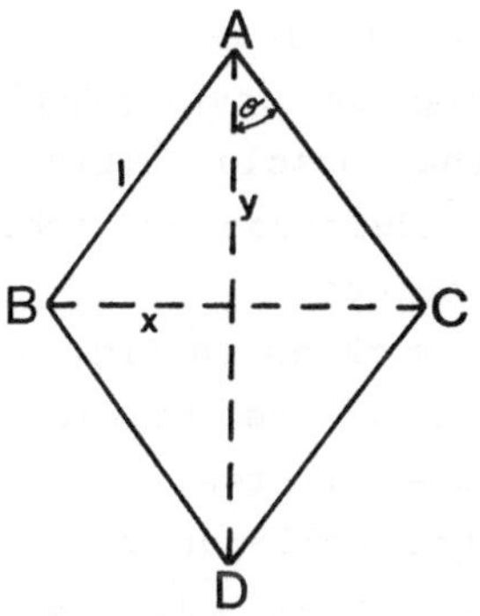

Fig. 1 Mechanical advantage resulting from lateral thrust.

If an element spanning BC expands outwards keeping lengths ABD and ACD constant, thus shortening AD, we have

$$x^2 + y^2 = \text{constant} = l^2$$

whence

$$\frac{dy}{dx} = -\frac{x}{y} \qquad \ldots \text{ eq.(1)}$$

i.e. the mechanical advantage obtained when approximating A,D by pushing out between B and C is cot ϕ.

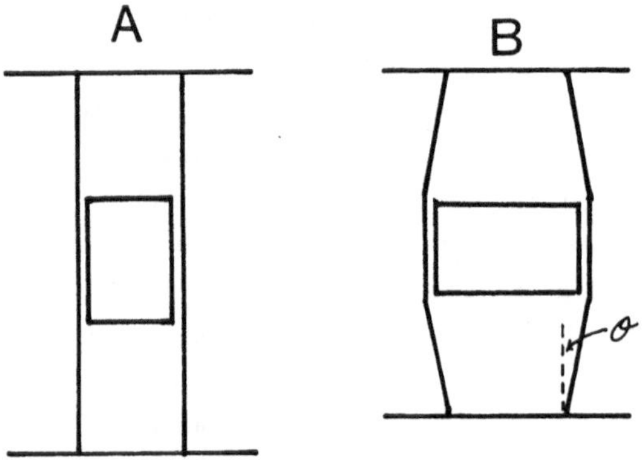

Fig. 2 A sarcomere in its relaxed state (A) is contained between parallel collagen fibres. As it expands laterally (B) the mechanical advantage decreases as ϕ increases.

If we now translate this to the marginaly more realistic scheme of Fig. 2 in which a sarcomere pushes out against extracellular collagen passing along the parallel sides of a muscle fibre, the initial contraction of the sarcomere, when actin and myosin overlap is least, occurs under conditions of infinite mechanical advantage. This mechanical advantage progressively decreases as the angle ϕ , of the enveloping collagen with the long axis of the fibre increases. By harnessing contraction using an enveloping sheath of longitudinal collagen against which the muscle fibre expands laterally, an initially high mechanical advantage is achieved and progressively decreases as contraction proceeds.

By using a collagen network as in Fig. 3, the initial mechanical advantage can be reduced from the infinite value consequent upon parallel longitudinal collagen fibres. The smaller the angle ϕ , and the more nearly parallel the collagen fibres, the greater would be the mechanical advantage. If we imagine such a network about a cylindrical muscle fibre, we can compare the forces resulting from longitudinal and lateral harnessing as a function of the angle ϕ .

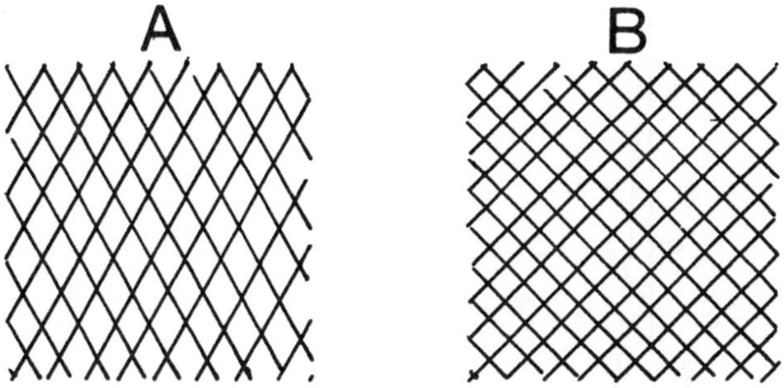

Fig. 3. Using a collagen mesh around the muscle fibre, the mechanical advantage depends on the obliquity of the fibres, and is higher in A than in B.

When a cylinder contracts longitudinally at constant volume we have $r^2 l$ = constant where r = radius and l = length.

$$\text{i.e. } dr/dl = -r/2l \qquad \ldots \text{ eq.(2)}$$

On the other hand, for an enveloping collagen network with a shape as depicted in Fig. 1, the longitudinal and circumferential span being in the proportion y:x we can recast eq.(1) in the form

$$\frac{dy/y}{dx/x} = -\frac{x^2}{y^2} = -\tan^2\phi$$

$$\text{i.e.} \quad dy/y = -\tan^2\phi \,.\, dx/x \qquad \ldots \text{ eq.(3)}$$

We can then compare the change in muscle fibre length, and the change in the length of a surrounding network of collagen, resulting from a small change in radius. Let $dr/r = \eta$

Then $dl/l = -2\eta$ (Fractional change in muscle fibre length - eq.2)
$dy/y = -\eta\tan^2\phi$ (Fractional change in collagen net length - eq.3)

In general, the mechanical advantage resulting from expanding against the net rather than pulling at the ends of the myofibrils is (Fig 4):

$$\frac{dl/l}{dy/y} = 2\cot^2\phi$$

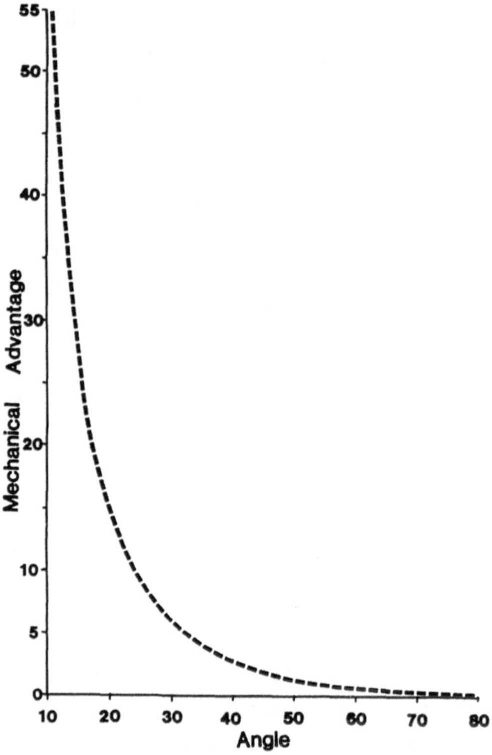

Fig. 4 Mechanical advantage, defined as fractional decrease in sarcomere length divided by fractional decrease in collagen net length, plotted as a function of the angle ø.

Different values of ø result in the muscle working at different mechanical advantage; the smaller ø the greater the mechanical advantage. As the muscle contracts, ø increases; the maximally extended muscle, the sarcomeres of which have minimum actin-myosin overlap and which therefore produce least force, is working at the lowest values of ø and hence at the maximum mechanical advantage. As contraction proceeds, ø increases and the mechanical advantage decreases at the same time as the sarcomeres become capable of producing greater force. The collagen seen in the basal lamina is not as regular a mesh as we have considered here. In fact, as a muscle fibre shortens longitudinally and expands laterally, it will apply tension successively to increasingly oblique populations of collagen fibres in the basal lamina. This may well be more important in reducing the overall mechanical advantage than the fact that any one collagen fibre comes to lie more obliquely during longitudinal contraction.

2b. **Longitudinal stability**

If the sarcomeres are not tethered at the ends, but are harnessed by pushing out at the sides, the system is longitudinally stable. Consider three contractile elements pushing outwards against a collagen net and let the middle element be in some sense weak, and less able to develop tension. Above and below, the contractile elements become shorter and broader, so that the weaker central element now finds itself surrounded by a tense, concave collagen sheath tending to expand it laterally and induce it to shorten. The interaction between elements is in fact co-operative; an element which does not contract is brought passively to a configuration in which any subsequent contraction by it would be at an infinite mechanical advantage. For a sheath in which the collagen passes directly from origin to insertion, the configuration around a passive length of sarcomeres would involve the sheath having zero curvature. Were the sheath a network of intersecting fibres as in Fig. 3, then in 3 dimensions the sheath would assume a shape around the muscle fibre such that the individual collagen fibres pass in a straight line over the surface of the non-contracting sarcomeres. This 'ruled surface' is a hyperboloid of one sheet, or a hyperboloid of revolution in the special case of the muscle fibre being is circular in cross section (Fig. 5).

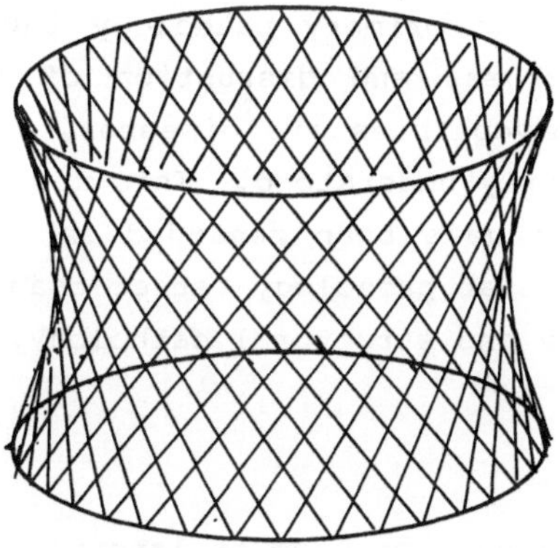

Fig. 5 A regular network held out above and below by contracting sarcomeres, and enclosing inactive sarcomeres between, will adopt a shape in which collagen fibres follow a straight path over the inactive sarcomeres.

2c. Harnessing

For a longitudinal pull at the ends of the sarcomeres to be the mechanism of harnessing their contraction it was necessary to postulate some structural element, or the presence of a poorly defined 'viscous cohesion', to couple intracellular proteins to extracellular collagen. If two elements are to push against each other, however, no such coupling is required, other than the presence of a collagen net around each muscle fibre to harness the lateral push. Such a net is in fact known to exist, although its function has been unclear. The basal lamina surrounding muscle fibres contains a mesh of type IV collagen (see, for instance the photograph by Heuser in Fawcett, D.W. (1981) and also Schmalbruch, 1974).

3. IMPLICATIONS

3a. Myopathy and cardiomyopathy

If the harnessing and stability of striated muscle is a function of its ability to expand laterally against a type IV collagen network in the basal lamina then primary pathology involving that collagen will manifest as weakness and disruption of the striated muscle. Myopathies and cardiomyopathies are diseases, the aetiology of which is, in many cases, poorly understood. This may be because changes in intramuscular collagen have been regarded as secondary consequences of primarily intracellular pathology rather than as the primary cause of the pathophysiology and structural pathology.

3b. The dual problem of lateral stability

The conjecture of lateral harnessing resolves the problem of longitudinal stability, but it introduces the dual problem of lateral stability. If we consider three contractile elements side by side with force harnessed laterally (Fig. 6a), then if the central element is weaker the adjoining elements contract more causing the central one to remain longer, have less overlap, and remain weaker. (Fig. 6b).

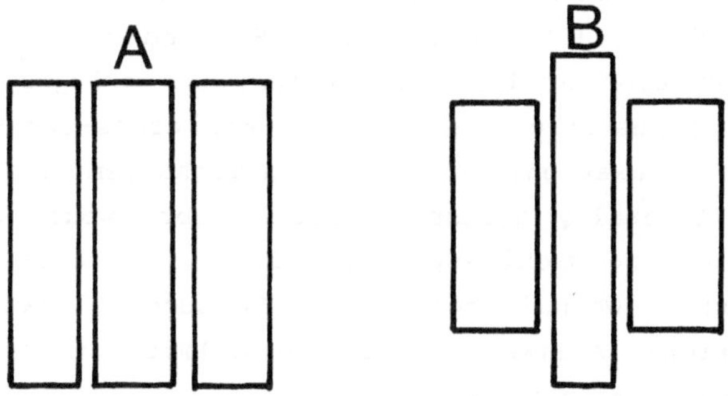

Fig. 6 Lateral harnessing leads to lateral instability: the more extended a muscle fibre the less it resists squeeze from its neighbours.

The problem is the same regardless of whether we interpret the elements as single fibres or groups of fibres. This problem is, however, essentially different from the longitudinal stability problem in that the elements there were intracellular sarcomeres which could not be individually harnessed. The lateral stability problem is resolved by noting that each muscle fibre is surrounded by collagen, and hence a muscle fibre which contracts more is thereby constrained to work at a lesser mechanical advantage. The resolution of this problem at the level of the single muscle fibre, the small group of muscle fibres and the large group of muscle fibres may correspond to the well documented hierarchy of intramuscular collagen (endomysium, perimysium, epimysium) surrounding progressively larger populations of cells. This collagen hierarchy, however, cannot solve the lateral stability problem that must exist within each muscle fibre where contractile elements can push against each other. Since collagen is extracellular, a different strategy must be adopted at the intracellular level to induce lateral stability. One answer could be mechanically to join elements in phase laterally across a muscle fibre so that the force of contraction between any two actin and myosin elements assists the contraction of its neighbours. This, of course, is precisely what the Z lines would do, and may be why striated muscles need the internal ordering which we see as striations. This solves the problem for individual bundles of myofilaments, but muscle fibres contain many parallel bundles of myofilaments each separated from their fellows. Nevertheless, extraction experiments have demonstrated a system of lateral connections between the Z lines of adjacent myofilbrils (Street, 1983; Wang and Ramirez-Mitchell, 1983; Lazarides, 1982 and references therein). A postulated function of these interfibrillar bridges has been the maintenance of transverse stability of the cross-striations during contraction and relaxation. Indeed, such mechanisms to keep contraction in phase laterally across a muscle fibre would seem to be a necessary part of the machinery of any cell which transfers force by expanding laterally when it contracts. We suggest that the presence of cross-striations is likely to be the hallmark of any muscle fibre which has adopted that strategy.

REFERENCES

FAWCETT, D.W., 1981, in: 'THE CELL'. Fig. 29 pp 59-61. W.B. Sanders.

HUXLEY, A.F., and NIEDERGERKE, R., 1954, Interference microscopy of living muscle fibres. NATURE, 173: 971-973.

HUXLEY, H.E., and HANSON, J., 1954, Changes in the cross-striations of muscle during contraction and stretch and their structural interpretation. NATURE, 173: 974-976.

LAZARIDES, E., 1982, Intermediate filaments: A chemically heterogeneous, developmentally regulated class of proteins. Am. Rev. Biochem., 51: 219-50.

MACKAY, B., HARROP, R.J., and MUIR, A.R., 1969, The fine structure of the muscle-tendon junction in the rat. ACTA. ANAT., 1973: 588-604.

MULLINS, G.L., and GUNTHEROTH, W.G., 1965, A collagen network hypothesis for force transference in smooth muscle. NATURE, 206: 592-594.

NAKAU, T., 1976, Some observations on the fine structure of the myotendinous junction in myotomal muscles of the tadpole tail. CELL TISSUE RES., 166: 241-254.

SCHMALBRUCH, H., 1974, The sarcolemma of skeletal muscle fibres as demonstrated by a replica technique. CELL TISSUE RES., 150: 377-387.

STREET, S.F., 1983, Lateral transmission of tension in frog myofibers: A myofibrillar network and transverse cytoskeletal connectors are possible transmitters. J. CELL. PHYS. 114: 346-364.

WANG, K. and RAMIREZ-MITCHELL, R., 1983, A network of transverse and longitudinal intermediate filaments is associated with sarcomeres of adult vertebrate skeletal muscle. J. CELL BIOL., 96: 562-570.

WARWICK, R., and WILLIAMS, P.L., 1973, in: GRAY'S ANATOMY, 35th Edition, p.483. Longman, Edinburgh.

Lecture Notes in Biomathematics

Vol. 1: P. Waltman, Deterministic Threshold Models in the Theory of Epidemics. V, 101 pages. 1974.

Vol. 2: Mathematical Problems in Biology, Victoria Conference 1973. Edited by P. van den Driessche. VI, 280 pages. 1974.

Vol. 3: D. Ludwig, Stochastic Population Theories. VI, 108 pages. 1974.

Vol. 4: Physics and Mathematics of the Nervous System. Edited by M. Conrad, W. Güttinger, and M. Dal Cin. XI, 584 pages. 1974.

Vol. 5: Mathematical Analysis of Decision Problems in Ecology. Proceedings 1973. Edited by A. Charnes and W.R. Lynn. VIII, 421 pages. 1975.

Vol. 6: H. T. Banks, Modeling and Control in the Biomedical Sciences. V. 114 pages. 1975.

Vol. 7: M.C. Mackey, Ion Transport through Biological Membranes, An Integrated Theoretical Approach. IX, 240 pages. 1975.

Vol. 8: C. DeLisi, Antigen Antibody Interactions. IV, 142 pages. 1976.

Vol. 9: N. Dubin, A Stochastic Model for Immunological Feedback in Carcinogenesis: Analysis and Approximations. XIII, 163 pages. 1976.

Vol. 10: J.J. Tyson, The Belousov-Zhabotinskii Reaktion. IX, 128 pages. 1976.

Vol. 11: Mathematical Models in Medicine. Workshop 1976. Edited by J. Berger, W. Bühler, R. Repges, and P. Tautu. XII, 281 pages. 1976.

Vol. 12: A. V. Holden, Models of the Stochastic Activity of Neurones. VII, 368 pages. 1976.

Vol. 13: Mathematical Models in Biological Discovery. Edited by D.L. Solomon and C. Walter. VI, 240 pages. 1977.

Vol. 14: L.M. Ricciardi, Diffusion Processes and Related Topics in Biology. VI, 200 pages. 1977.

Vol. 15: Th. Nagylaki, Selection in One- and Two-Locus Systems. VIII, 208 pages. 1977.

Vol. 16: G. Sampath, S.K. Srinivasan, Stochastic Models for Spike Trains of Single Neurons. VIII, 188 pages. 1977.

Vol. 17: T. Maruyama, Stochastic Problems in Population Genetics. VIII, 245 pages. 1977.

Vol. 18: Mathematics and the Life Sciences. Proceedings 1975. Edited by D.E. Matthews. VII, 385 pages. 1977.

Vol. 19: Measuring Selection in Natural Populations. Edited by F.B. Christiansen and T.M. Fenchel. XXXI, 564 pages. 1977.

Vol. 20: J.M. Cushing, Integrodifferential Equations and Delay Models in Population Dynamics. VI, 196 pages. 1977.

Vol. 21: Theoretical Approaches to Complex Systems. Proceedings 1977. Edited by R. Heim and G. Palm. VI, 244 pages. 1978.

Vol. 22: F.M. Scudo and J.R. Ziegler, The Golden Age of Theoretical Ecology: 1923–1940. XII, 490 pages. 1978.

Vol. 23: Geometrical Probability and Biological Structures: Buffon's 200th Anniversary. Proceedings 1977. Edited by R.E. Miles and J. Serra. XII, 338 pages. 1978.

Vol. 24: F.L. Bookstein, The Measurement of Biological Shape and Shape Change. VIII, 191 pages. 1978.

Vol. 25: P. Yodzis, Competition for Space and the Structure of Ecological Communities. VI, 191 pages. 1978.

Vol. 26: M.B Katz, Questions of Uniqueness and Resolution in Reconstruction from Projections. IX, 175 pages. 1978.

Vol. 27: N. MacDonald, Time Lags in Biological Models. VII, 112 pages. 1978.

Vol. 28: P.C. Fife, Mathematical Aspects of Reacting and Diffusing Systems. IV, 185 pages. 1979.

Vol. 29: Kinetic Logic – A Boolean Approach to the Analysis of Complex Regulatory Systems. Proceedings, 1977. Edited by R. Thomas. XIII, 507 pages. 1979.

Vol. 30: M. Eisen, Mathematical Models in Cell Biology and Cancer Chemotherapy. IX, 431 pages. 1979.

Vol. 31: E. Akin, The Geometry of Population Genetics. IV, 205 pages. 1979.

Vol. 32: Systems Theory in Immunology. Proceedings, 1978. Edited by G. Bruni et al. XI, 273 pages. 1979.

Vol. 33: Mathematical Modelling in Biology and Ecology. Proceedings, 1979. Edited by W.M. Getz. VIII, 355 pages. 1980.

Vol. 34: R. Collins, T.J. van der Werff, Mathematical Models of the Dynamics of the Human Eye. VII, 99 pages. 1980.

Vol. 35: U. an der Heiden, Analysis of Neural Networks. X, 159 pages. 1980.

Vol. 36: A. Wörz-Busekros, Algebras in Genetics. VI, 237 pages. 1980.

Vol. 37: T. Ohta, Evolution and Variation of Multigene Families. VIII, 131 pages. 1980.

Vol. 38: Biological Growth and Spread: Mathematical Theories and Applications. Proceedings, 1979. Edited by W. Jäger, H. Rost and P. Tautu. XI, 511 pages. 1980.

Vol. 39: Vito Volterra Symposium on Mathematical Models in Biology. Proceedings, 1979. Edited by C. Barigozzi. VI, 417 pages. 1980.

Vol. 40: Renewable Resource Management. Proceedings, 1980. Edited by T. Vincent and J. Skowronski. XII, 236 pages. 1981.

Vol. 41: Modèles Mathématiques en Biologie. Proceedings, 1978. Edited by C. Chevalet and A. Micali. XIV, 219 pages. 1981.

Vol. 42: G.W. Swan, Optimization of Human Cancer Radiotherapy. VIII, 282 pages. 1981.

Vol. 43: Mathematical Modeling on the Hearing Process. Proceedings, 1980. Edited by M.H. Holmes and L.A. Rubenfeld. V, 104 pages. 1981.

Vol. 44: Recognition of Pattern and Form. Proceedings, 1979. Edited by D.G. Albrecht. III, 226 pages. 1982.

Vol. 45: Competition and Cooperation in Neutral Nets. Proceedings, 1982. Edited by S. Amari and M.A. Arbib. XIV, 441 pages. 1982.

Vol. 46: E. Walter, Identifiability of State Space Models with applications to transformation systems. VIII, 202 pages. 1982.

Vol. 47: E. Frehland, Stochastic Transport Processes in Discrete Biological Systems. VIII, 169 pages. 1982.

Vol. 48: Tracer Kinetics and Physiologic Modeling. Proceedings, 1983. Edited by R.M. Lambrecht and A. Rescigno. VIII, 509 pages. 1983.

Vol. 49: Rhythms in Biology and Other Fields of Application. Proceedings, 1981. Edited by M. Cosnard, J. Demongeot and A. Le Breton. VII, 400 pages. 1983.

Vol. 50: D.H. Anderson, Compartmental Modeling and Tracer Kinetics. VII, 302 pages. 1983.

Vol. 51: Oscillations in Mathematical Biology. Proceedings, 1982. Edited by J.P.E. Hodgson. VI, 196 pages. 1983.

Vol. 52: Population Biology. Proceedings, 1982. Edited by H.I. Freedman and C. Strobeck. XVII, 440 pages. 1983.

Vol. 53: Evolutionary Dynamics of Genetic Diversity. Proceedings, 1983. Edited by G.S. Mani. VII, 312 pages. 1984.

Vol. 54: Mathematical Ecology. Proceedings, 1982. Edited by S.A. Levin and T.G. Hallam. XII, 513 pages. 1984.

Vol. 55: Modelling of Patterns in Space and Time. Proceedings, 1983. Edited by W. Jäger and J.D. Murray. VIII, 405 pages. 1984.

Vol. 56: H.W. Hethcote, J.A. Yorke, Gonorrhea Transmission Dynamics and Control. IX, 105 pages. 1984.

Vol. 57: Mathematics in Biology and Medicine. Proceedings, 1983. Edited by V. Capasso, E. Grosso and S.L. Paveri-Fontana. XVIII, 524 pages. 1985.

ctd. on inside back cover

Lecture Notes in Biomathematics

Vol. 58: C.A. Macken, A.S. Perelson, Branching Processes Applied to Cell Surface Aggregation Phenomena. VIII, 122 pages. 1985.

Vol. 59: I. Nåsell, Hybrid Models of Tropical Infections. VI, 206 pages. 1985.

Vol. 60: Population Genetics in Forestry. Proceedings, 1984. Edited by H.-R. Gregorius. VI, 287 pages 1985.

Vol. 61: Resource Management. Proceedings, 1984. Edited by M. Mangel. V, 138 pages 1985.

Vol. 62: B. J. West, An Essay on the Importance of Being Nonlinear. VIII, 204 pages. 1985.

Vol. 63: Carme Torras i Genís, Temporal-Pattern Learning in Neural Models. VII, 227 pages, 1985.

Vol. 64: Peripheral Auditory Mechanisms. Proceedings, 1985. Edited by J.B. Allen, J.L. Hall, A. Hubbard, S. Neely and A. Tubis. VII, 400 pages. 1986.

Vol. 65: Immunology and Epidemiology. Proceedings, 1985. Edited by G.W. Hoffmann and T. Hraba. VIII, 242 pages. 1986.

Vol. 66: Nonlinear Oscillations in Biology and Chemistry. Proceedings, 1985. Edited by H.G. Othmer. VI, 289 pages. 1986.

Vol. 67: P.H. Todd, Intrinsic Geometry of Biological Surface Growth. IV, 128 pages. 1986.

Vol. 68: The Dynamics of Physiologically Structured Populations. Edited by J.A.J. Metz and O. Diekmann. XII, 511 pages. 1986.

Vol. 69: S.H. Strogatz, The Mathematical Structure of the Human Sleep-Wake Cycle. VIII, 239 pages. 1986.

Vol. 70: Stochastic Methods in Biology. Proceedings, 1985. Edited by M. Kimura, G. Kallianpur and T. Hida. VI, 229 pages. 1987.

Vol. 71: Mathematical Topics in Population Biology, Morphogenesis and Neurosciences. Proceedings, 1985. Edited by E. Teramoto and M. Yamaguti. IX, 348 pages. 1987.